Statistical Mechanics

Statistical Mechanics

Fundamentals and Model Solutions

Second Edition

Teunis C. Dorlas

CRC Press
Taylor & Francis Group
Boca Raton London New York

CRC Press is an imprint of the
Taylor & Francis Group, an **informa** business

Second edition published [2021]
by CRC Press
6000 Broken Sound Parkway NW, Suite 300, Boca Raton, FL 33487-2742

and by CRC Press
2 Park Square, Milton Park, Abingdon, Oxon, OX14 4RN

© 2021 Taylor & Francis Group, LLC

[First edition published by CRC Press 1999]

CRC Press is an imprint of Taylor & Francis Group, LLC

Library of Congress Cataloging-in-Publication Data

Names: Dorlas, T. C. (Teunis Christiaan), 1955- author.
Title: Statistical mechanics : fundamentals and model solutions / Teunis C. Dorlas, School of Theoretical Physics Dublin Institute for Advanced Studies.
Description: Second edition. | Boca Raton : CRC Press, 2021. | Includes bibliographical references and index.
Identifiers: LCCN 2020055180 | ISBN 9780367471750 (paperback) | ISBN 9780367478810 (hardback) | ISBN 9781003037170 (ebook)
Subjects: LCSH: Statistical mechanics.
Classification: LCC QC174.8 .D68 2021 | DDC 530.13--dc23
LC record available at https://lccn.loc.gov/2020055180

ISBN: 9780367478810 (hbk)
ISBN: 9780367471750 (pbk)
ISBN: 9781003037170 (ebk)

Typeset in Computer Modern font
by KnowledgeWorks Global Ltd.

Access the support material at: www.routledgetextbooks.com/textbooks/instructor_downloads/

In memory of my parents

Contents

Preface

Statistical mechanics is the physical theory of the thermal properties of macroscopic systems. It relates these properties to the mechanics of the constituent microscopic particles. Owing to the enormous number of particles, their average behaviour is well-defined and this is what determines their thermal properties. This theory was originally formulated by Ludwig Boltzmann and Josiah Willard Gibbs in the latter part of the 19th century. However, since these early beginnings the theory has been extended and developed considerably. In particular **equilibrium statistical mechanics**, which is the subject of this book, has been put on a much firmer mathematical basis. Nevertheless, the subject is normally still taught in a rather traditional fashion.

This book is an introduction to the subject with more emphasis on the mathematical structure than is usual. In particular it is emphasized that it is essential that the number of particles in the system is large. This justifies taking the thermodynamic limit where the number of particles and the volume of the system tend to infinity in such a way that the particle density remains constant. Also, in particular in Part I about thermodynamics, the **convexity** of many of the thermodynamic functions is shown to play a central role. This important property is often ignored in standard texts, which makes the definition of thermodynamic potentials somewhat obscure.

The greater reliance on mathematical concepts naturally means that the student is assumed to be familiar with some basic undergraduate mathematics. Specifically, the student is assumed to be familiar with the basic concepts of real analysis (continuous functions, open and closed sets, lim inf and lim sup) and of probability theory (random variables and distribution functions). (There is, however, an appendix about probability theory at the end of the book.) More non-standard theory is developed in the text: there is a chapter in Part I devoted to some properties of convex functions (as well as an appendix with more advanced theory) and a chapter in Part II which is an introduction to the theory of large deviations. However, the proofs of theorems in these chapters are not essential, and by retaining the dimensional factors in most formulas, I expect that this book will still be suitable for physics students.

I taught parts of this book to final-year undergraduate mathematics students at the University of Wales Swansea. Essential to the course are chapters 1-10, 18 (if the students are unfamiliar with quantum mechanics), 19-22 and 24. These can be supplemented by some of the remaining chapters. Chapters 11-17 illustrate some applications of thermodynamics. Chapters 25 and 26 are more mathematical, giving a firmer basis for the formalism of chapter 24 in the case of classical spin systems. This gives a flavour of the mathematical justification of the theory. Chapter 23 is about quantum gases and is a prerequisite for chapters 31-35. The latter chapters treat some more advanced topics and are more difficult than the rest of the text. In particular chapters 34 and 35 require more extensive knowledge of quantum mechanics.

In this second edition, I have added two more chapters. Chapter 36 is about the Mayer expansion for classical gases. It is treated in many standard texts, but the simple proof of absolute convergence is usually omitted. Chapter 37 is more advanced and concerns polymer models and the cluster expansion. It relies on chapter 36 as well as on chapter 25. There is extensive mathematical literature about this subject which is not so easily penetrated. It therefore seemed useful to include an introduction to this broad subject, which is also used extensively in constructive quantum field theory.

I am very indebted to Professor John Lewis for introducing me to the theory of large deviations and pointing out its importance for statistical mechanics, as well as for encouraging me to write this book. Unfortunately, since the publication of the first edition, Professor Lewis has passed away. It is a great privilege to have been one of his many post-docs and he is sorely missed, especially for his encouraging style, optimism and clarity of thought. I am also grateful for the many years of very enjoyable research collaboration with Professor Joe Pulé and I thank Professor Aubrey Truman for allowing me to teach a course on statistical mechanics to final-year undergraduates in Swansea.

Teunis C Dorlas
Dublin Institute for Advanced Studies
March 2020

Introduction

This book is about statistical mechanics, an important area of physics with many applications. We shall concentrate mainly on the basic principles of the subject, but several applications are mentioned as we discuss a variety of models for macroscopic physical systems. A rough definition of the subject could be:

> **Statistical mechanics** *is the theory of the physical behaviour of macroscopic systems starting from a knowledge of the microscopic forces between the constituent particles.*

As all every-day bodies are *macroscopic*, i.e. they consist of a very large number of particles (of the order of Avogadro's number $N_0 = 6.0 \cdot 10^{23}$), this definition may seem to encompass virtually all of physics. Indeed, statistical mechanics has a very broad range of applications, but in practice it concerns only systems where the influence of the *temperature* is of importance. In this book we shall only be concerned with systems in *thermodynamic equilibrium*. This is the state that a macroscopic system ends up in when it is left undisturbed for a sufficiently long time. (The time it takes for a system to return to equilibrium after a disturbance is called the **relaxation time**.) The theory of the relations between the various macroscopic observables like temperature, volume, pressure, magnetization, polarization, etc. of a system is called **thermodynamics**. We shall see that in principle all thermodynamic quantities can be derived from one single function. Statistical mechanics is an endeavour to relate the macroscopic observables of a system to the motion of the individual particles and to derive the thermodynamic relations from a knowledge of the forces between these particles. The behaviour of specific macroscopic systems can then be understood in more detail by studying a model for that system.

> A ***model*** *of a physical system is a caricature of the system obtained by extracting only the essential features of the phenomenon to be studied so that it becomes manageable for mathematical investigation.*

A particularly important phenomenon in macroscopic systems is the occurrence of **phase transitions**. Phase transitions occur when two different thermodynamic states can coexist at a given temperature. This means that at that temperature certain other thermodynamic variables can have two or more values when the system is in equilibrium. Thermodynamic variables that distinguish between phases are called **order parameters**.

Examples of phase transitions are the liquid-gas transition and the solid-liquid transition of stable materials, where the order parameter is the density. But there are others like the Curie transition in ferromagnets, where the order parameter is the magnetization, and the order-disorder transition in alloys, where the order parameter is the relative concentration.

EXAMPLE 1: *The ideal gas.*

As an example of a thermodynamic system let us consider the ideal or perfect gas. In general, the thermodynamic behaviour of a low-density gas at sufficiently high temperatures is given by the **ideal gas law**:

$$p V = n R_0 T. \tag{1}$$

Here p is the pressure, V is the volume and T is the temperature of the gas. R_0 is a universal constant called the **gas constant**: $R_0 \sim 8.3$ J/K (Joules per Kelvin), and n is the number of moles of the gas. A **mole** is the number of hydrogen atoms in a gram. (This is called **Avogadro's number** and denoted N_0.) The temperature here is to be measured in Kelvins, i.e. degrees Celsius above the **absolute zero** of temperature: 0 K $= -273\,°$C. The ideal gas law is a combination of three separate laws: Boyle's law, which says that, at constant temperature, the pressure of a gas is inversely proportional to the volume; Charles' law, which states that the volume of a gas at constant pressure is linear in the temperature; and Avogadro's law, which states that at a given temperature and pressure, equal volumes of different gases contain equal numbers of molecules. We shall discuss all this in more detail later. It turns out that not all thermodynamic quantities can be derived from this single law. One needs another law which can be formulated as:

$$U = C_V T. \tag{2}$$

U is the **internal energy** of the gas, a quantity we shall introduce later. In general it depends on both the volume and the temperature, but for an ideal gas it is independent of the volume. The constant C_V is called the **heat capacity at constant volume**. It is in fact independent of the volume for an ideal gas.

In a $p - V$ diagram the graphs of (1) for constant temperature are hyperbolas. As the temperature of a real gas is lowered it no longer obeys the ideal gas law: the graph develops a bump and eventually a straight portion. This is illustrated in Figure 1.

The flat part corresponds to the *coexistence of two phases*: the liquid phase and the vapour phase. (Gas below the **critical temperature** is usually called a **vapour**.) The details of this picture are explained in Part I, chapter 9.

Thermodynamics cannot derive the above picture nor the laws (1) or (2). It only deals with the macroscopic behaviour of general systems. Hence, *given*

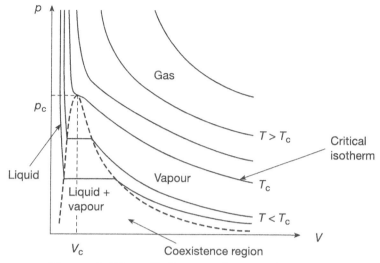

Figure 1. The p–V diagram of a typical material.

this picture or these laws it can predict the behaviour of the system under various conditions. We shall discuss examples like the car engine and refrigerators. To derive the laws (1) and (2) one needs to relate the quantities p, V, etc. to the constituent particles of the system, which is the subject of statistical mechanics.

This course consists of three parts:

Part I: Thermodynamics.
Part II.: Fundamentals of Statistical Mechanics.
Part III: Models in Statistical Mechanics.

In Part I we postulate the basic laws of thermodynamics and derive the main consequences. We also discuss a number of applications. Thermodynamics was developed in the 19th century by Sadi Carnot, Julius Robert Mayer, James Joule, Rudolf Clausius, Hermann Helmholtz, William Thomson (later Lord Kelvin), Josiah Willard Gibbs and others. It was inspired by the development of the steam engine by James Watt between 1765 and 1790. However, although the subject is already very old, it is still of great importance. In particular in chemistry and engineering it is indispensible. We shall discuss several interesting applications in the final chapters of Part I.

In Part II we derive the thermodynamic laws from the laws of motion for the microscopic particles. This theory is called statistical mechanics and was developed at the end of the 19th and the beginning of the 20th centuries by James Clerk Maxwell, Josiah Willard Gibbs, Ludwig Boltzmann and Max Planck. Originally this theory was based on classical mechanics, but when it was discovered that the laws of motion for microscopic particles are governed by *quantum mechanics* the theory had to be reformulated. It became clear that the introduction of quantum mechanics, although it corresponds with classical mechanics in the description of the *motion* of macroscopic bodies, does have profound consequences for the thermodynamic behaviour of macroscopic bodies. In particular it resolved serious difficulties with the so-called **equipartition theorem**. Instrumental in this development were Max Planck, Albert Einstein, Satendra Nath Bose, Enrico Fermi, Paul Adrian Maurice Dirac and Wolfgang Pauli. Fortunately, a rather rudimentary knowledge of quantum mechanics suffices to understand the fundamentals of statistical mechanics. We shall therefore not assume prior knowledge of quantum theory in the main part of this book. The few basic facts needed will be explained in chapter 18 of Part II. (A more complete overview of quantum mechanics is given in appendix B. This will not be needed for a proper understanding of Part II and the simpler models of Part III, but some of the more complicated models discussed in Part III do depend on a knowledge of quantum mechanics.) We shall assume a basic knowledge of Newtonian mechanics, however, including the concepts of velocity, acceleration, force, kinetic and potential energy and angular momentum.

In Part III we consider some simple model systems and apply the methods of statistical mechanics to derive their thermodynamic behaviour. In particular, we consider examples of systems that exhibit a phase transition.

The behaviour of a macroscopic system comes about as an *average* over the behaviour (i.e. motion) of all the individual particles. Taking averages is the subject of *probability theory* which is therefore of the essence for understanding statistical mechanics. We assume that the reader is familiar with mathematical probability theory, but an overview of the main facts is contained in appendix A.

EXAMPLE 2: *Free expansion of a gas.*
Consider a container partitioned into two halves, where the left half contains a gas while the right half has been evacuated by a pump.

A gas consists of individual molecules which move around in the container almost independently. When the partition is removed (the valve opened) the gas molecules will move into the right half of the box. The probability that they all move back into the left half at the same time is exceedingly small. Mathematically speaking, it is truly zero only in the limit $N \to \infty$, where N is the number of particles (molecules). This illustrates the necessity to study model systems in the **thermodynamic limit**, taking $N \to \infty$ and at the same time taking the volume $V \to \infty$ as well, but keeping the density of particles $\rho = N/V$ constant. There are good reasons for this procedure. For example, the fact that thermodynamic variables are either intensive or extensive (see chapter 4) can be seen to be a consequence of the large size of thermodynamic systems, and it is also crucial for the existence of phase transitions. Most treatments of statistical mechanics are rather cavalier about this limit. We shall pay more attention to it.

The last few sections of Part III are more advanced and contain an analysis of some more difficult models. The emphasis is on exact solutions of models. It

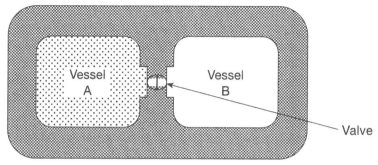

Figure 2. Free expansion of a gas.

must be stressed that most models in statistical mechanics cannot be solved exactly. One then has to take recourse to approximation techniques. These are not discussed in detail in this book.

Statistical mechanics is still a very active area of research and there are many open problems and poorly explained phenomena. For example, it has still proved too difficult to calculate the freezing point of water from first principles. The recently discovered high-temperature superconductors are also essentially still a mystery. In fact, the general phenomenon of phase transitions is by no means completely understood and is still actively being researched. Considerable progress has already been made, however. Important contributions in this respect were made by, among others, Lars Onsager, Joseph Edward Mayer, John G. Kirkwood, Tsung-Dao Lee, Chen Ning Yang, Lev D. Landau, Michael E. Fisher, Léon van Hove, David Ruelle, Yakov G. Sinai, Roland L. Dobrushin, Robert B. Griffiths, Jürg Fröhlich, Thomas Spencer, Barry Simon, Elliott H. Lieb and Rodney J. Baxter. To illustrate the wide applicability of statistical mechanics, here are some interesting phenomena that owe their explanation to it:

1. *Paramagnetism and ferromagnetism.* Paramagnetism is the normal, induced magnetization of a material when it is put into a magnetic field. The temperature dependence of the magnetization can be analysed using simple statistical mechanical models. Ferromagnetism is the spontaneous magnetization occurring in certain metals like iron and nickel. It is an example of a phase transition. We shall consider a simple model for this phenomenon, the famous **Ising model**, in Part III. For a more detailed description of magnetic phenomena, see for example the books on solid state theory by Kittel (1976) and Blakemore (1985), or the introductory text on magnetism by Brailsford (1966). (Note that these references do not give much explanation of the magnetic behaviour of materials. Indeed, magnetism is a complicated matter and one needs a good understanding of quantum mechanics to follow the complicated explanation of magnetic phenomena. (For references to more advanced treatments, see Mattis (1965), Morrish (1965) and Williams (1966).)

2. *Rubber, gels and liquid crystals.* All these materials consist of very long **macromolecules**. We discuss a very simple model for rubber in Part III and show why rubber shrinks when it is heated as opposed to most other materials like metals, which expand on heating. For more information on macromolecules or polymers in general, see for example Billmeyer (1984), Yang and Lovell (1991) and Gedde (1995). Gels are suspensions of macromolecules in a solvent. They have a high viscosity. It was discovered that they have a phase transition for high concentrations. A nice description of this phenomenon can be found in Tanaka (1981). Liquid crystals consist of long rigid molecules, which can be in various configurations corresponding to different phases. They are used in liquid-crystal displays. For an elementary discussion of liquid crystals, see the book by Guinier (1984), which is a simple introduction to the properties of materials of various kinds. Other monographs on liquid crystals are by Priestley *et al.* (1975) and by Chandrasekhar (1994).

3. *Metallic conduction.* Electrons in metals can be treated as a gas, which leads to predictions for the heat capacity and the thermal and electrical conductivity. We mention the simplest model of a metal, the free electron model, in chapter 31 but do not consider the heat capacity or the electrical and thermal conductivity. Most books on solid state theory discuss these matters in some detail. See for example Kittel (1976) and Blakemore (1985).

4. *Superconductivity.* Certain metals become perfect conductors at sufficiently low temperatures. This effect can be explained by statistical mechanics using an idealized model, which nevertheless yields quantitative information. This so-called BCS-model, due to John Bardeen, Leon Niels Cooper and J. Robert Schrieffer, will be analysed in chapter 35. For a description of the phenomenon and its applications, see also Kittel (1976, chapter 12).

5. *Black body radiation.* When a body is heated it radiates energy in the form of electromagnetic waves (light). The study of the relation between the frequency of the radiation and the temperature gave rise to the development of quantum mechanics. We consider the thermodynamics of black-body radiation in chapter 15.

6. *The physical constitution of stars.* The stability of a star is a result of a balance between the gravitational force and the pressure of the gas ball which the star is. This was first analysed by Sir Arthur Eddington and will be discussed in chapter 17. Massive stars collapse when their hydrogen fuel runs out. The gravitational force can then only be resisted by the pressure of a degenerate fermion gas of individual electrons, which is due to quantum effects. This was first analysed by Subrahmanyan Chandrasekhar. It is an application of the free fermion gas and will be discussed in detail in chapter 32. He found that if the mass is larger than

a certain limiting value then the electron pressure is also insufficient to counteract the gravitational force. Such stars become neutron stars or even black holes.

7. *Phonons in crystalline solids.* These are vibrations: high-frequency sound. They are important for the determination of the structure of these solids and can be modelled as a gas. This is discussed in most books on solid state physics (Kittel (1976), Blakemore (1985)). (See also chapters 13 and 33.)

Acknowledgments

The author and Taylor & Francis kindly acknowledge permission to reproduce the following figures.

Figure 12.1. Reproduced by permission from Atkins P W 1992 *The Elements of Physical Chemistry* (Oxford: Oxford University Press).

Figure 12.2 (left). Reproduced by permission from Poirier J-P 1991 *Introduction to the Physics of the Earth's Interior* (Cambridge: Cambridge University Press).

Figure 12.2 (right). Reproduced by permission from Saxena S K *et al.* 1993 Experimental evidence for a new iron phase and implications for Earth's core *Science* **260** 1312–1314, American Association for the Advancement of Science.

Figure 12.4. Reproduced from the hyperscript "Iron, Steel and Swords", $https://www.tf.uni-kiel.de/matwis/amat/iss/kap\_6/illustr/s6\_1\_2.html$ with permission from the creator Prof. Dr. H. Föll at the University of Kiel, Germany.

Figure 12.6 and Figure 34.3 Reproduced by permission from Atkins K R 1959 (paperback 2014) *Liquid Helium* (Cambridge: Cambridge University Press).

Figure 31.4. Reproduced by permission from Pippard, A B 1957 An experimental determination of the Fermi surface of copper *Phil. Transactions of Roy. Soc. London* **250** 325–357.

Figure 34.2. Swenson C A 1950 The liquid–solid transformation in helium near absolute zero *Phys. Rev.* **79** 626–631, American Physical Society.

Figure 35.1. Reproduced from Kamerlingh Onnes H 1911 *Commun. Phys. Lab. Leiden* vol 124c, with kind permission of the Kamerlingh Onnes Laboratory, Leiden.

I

Thermodynamics

Traditionally, the theory of thermodynamics is stated in terms of three basic laws: the zeroth law, the first law, and the second law. (There is a third law but it is not essential for the main thrust of thermodynamics. It will be mentioned briefly at the end of chapter 4.) We shall discuss these laws in turn and conclude that it follows from these laws that the thermodynamic behaviour of a system is completely determined when a certain quantity, called the **entropy**, is given as a function of a number of independent macroscopic variables. These independent variables characterize the **equilibrium state** of the system. The equilibrium states of the ideal gas, for example, are completely determined by the pressure p and the volume V.

> A **simple system** is a system for which two parameters suffice to determine its equilibrium state. In general, the **state space** of a thermodynamic system is the set of possible values of a minimal number of parameters determining the equilibrium state of the system. The state space of a simple system is two-dimensional.

The zeroth law is given in chapter 1, and the first law in chapter 2. In chapter 3, the concept of *differential form* is defined. This is a mathematical way of working with infinitesimal quantities used in the previous chapter. The important second law is introduced in chapter 4 together with the concept of **entropy**. A general application of this law is discussed in chapter 5. In chapter 6 it is shown that the entropy function has an important mathematical property, namely **concavity**. A number of important theorems about concave and convex functions are formulated and proved in chapter 7. In particular, the definition of the Legendre transform and theorems 7.2 and 7.3 are important for the proper understanding of the subsequent chapters, but the proofs could be omitted. Chapter 8 introduces some important thermodynamic potentials based on the Legendre transform, and chapter 9 discusses the thermodynamics of phase transitions. Chapter 10 is a very brief introduction to the thermodynamics of magnetism. These three chapters are short but essential. The final chapters discuss some applications of thermodynamics in more detail. These are of course not essential but give an idea of the wide range of applications and the continuing importance of the subject in physics and chemistry.

CHAPTER 1

The Zeroth Law

The zeroth law states:

> *There exists a function of state T, the* **temperature**, *such that two systems in thermal contact are in thermal equilibrium if and only if their temperatures are equal.*

Note that this law does not fix the temperature *scale*! On the *Celsius scale*, one defines 0 °C as the melting point of ice and 100 °C as the boiling point of water, both under a pressure of 1 atm. The temperature of a given system can then be determined by bringing it into thermal contact with a *thermometer* and measuring some property of the thermometer system using linear interpolation between 0 °C and 100 °C (and extrapolation beyond this range).

EXAMPLE 1.1: *Mercury vs. Resistance Thermometer.*
In a mercury thermometer for example, the temperature is determined by the length of a column of mercury as it expands due to heating. Thus:

$$\theta_m = \frac{\ell_\theta - \ell_0}{\ell_{100} - \ell_0} \times 100\,^\circ\text{C} \tag{1.1}$$

is the **empirical temperature** as it is measured by a mercury thermometer. In a resistance thermometer, the temperature is determined by the electrical resistance of a metal wire:

$$\theta_R = \frac{R_\theta - R_0}{R_{100} - R_0} \times 100\,^\circ\text{C}. \tag{1.2}$$

We now encounter the following problem: Suppose that the electrical resistance R varies nonlinearly with temperature as measured on the mercury scale:

$$R_\theta = R_0(1 + b\theta_m + c\theta_m^2). \tag{1.3}$$

Then

$$\theta_R = \frac{b\theta_m + c\theta_m^2}{b + 100c} \neq \theta_m. \tag{1.4}$$

Fortunately, it turns out that the nonlinearity in equation (1.3) is rather small over a not too large interval of temperatures. (i.e. $c \ll 1$.) Nevertheless, for an accurate definition of the temperature scale one needs a **standard thermometer** with which all other thermometers can then be gauged. For this we can use an inert gas. Indeed, using one of the above types of thermometer, it was discovered that most rare gases satisfy the **ideal gas law (1)** mentioned in the introduction:

$$\boxed{p\,V = n\,R_0\,T,}$$
(1.5)

where the constant R_0 is independent of the gas and where $T = \theta + 273$ if θ is measured in degrees Celsius.

This law was in fact arrived at in stages. First Robert Boyle discovered in 1660 that the pressure of a gas at constant temperature is inversely proportional to the volume. Then Jaques Charles and Joseph Gay-Lussac discovered (around 1800) that, at constant pressure, the volume of a gas is linear in the temperature: $V(\theta) = \text{constant} \cdot (\theta + \theta_0)$. Combining the two laws we get: $pV = K(\theta + \theta_0)$. It turned out that the constant K still depends on the gas, but the constant θ_0 is independent of the gas! It was William Thomson, later Lord Kelvin, who realized in 1848 that the point $\theta = -\theta_0 = -273\,°\text{C}$ is the **absolute zero** of temperature. It is therefore better to measure temperature from the absolute zero upwards. This is the Kelvin scale of temperature. Thus $0\,\text{K} = -273\,°\text{C}$. (To be precise, the zero of the Celsius scale is taken to be the triple point of water and a more accurate value for θ_0 is 273.15 K.) The dependence of the constant K on the gas was discovered by Amadeo Avogadro in 1811. He deduced from the atomic theory formulated by John Dalton in 1808 that, at the same temperature and pressure, equal volumes of different gases contain equal numbers of molecules. By extension this means that the constant K is proportional to the number of molecules: $K = k_B N$, where k_B is a universal constant called **Boltzmann's constant**. In practice it is very difficult to measure the number of molecules directly. Instead, one usually compares masses. Conventionally, one defines the **atomic mass unit** (amu) as one twelfth of the mass of a carbon atom (the ^{12}C isotope, to be precise). The lightest atom, the hydrogen atom, then has a mass of approximately 1 amu. Next one defines 1 *mole* (mol) as the number of molecules which have a combined mass in grams equal to the mass of a single molecule measured in atomic mass units. This number is **Avogadro's number**: $N_0 = 6.0225 \cdot 10^{23}$, that is 1 g equals N_0 amu. If M_0 is the molar mass and m_0 the mass of a molecule *now measured in grams*, then $M_0 = N_0\,m_0$. One mole of carbon atoms therefore has a mass of 12 g. One mole of hydrogen gas has a mass of approximately 2 g because a hydrogen molecule consists of 2 hydrogen atoms. Denoting the number of moles of a gas by n we have

$$N = n\,N_0 \text{ and } R_0 = N_0\,k_B = 8.3145 \text{ J K}^{-1}\,\text{mol}^{-1}.$$
(1.6)

The problem of nonlinearity of temperature scales can now be resolved by taking a gas as the *standard thermometer*. (The gas has to be sufficiently rarefied;

otherwise the ideal gas law breaks down.) This is called a **gasthermometer**. Figure 1.1 shows a schematic drawing of this kind of thermometer.

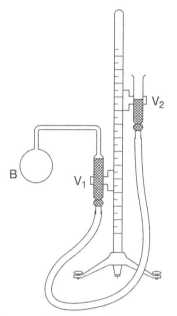

Figure 1.1 The gasthermometer.

The glass ball B containing an inert gas is dipped into a liquid whose temperature is to be determined. It is connected via a capillary tube to a glass vessel V_1 containing mercury. In turn, V_1 is connected to another vessel V_2 by a rubber tube. The height of V_2 can be adjusted so that the mercury level in V_1 remains at the tip of a needle point fixed inside the vessel. The volume of the gas in B (and the capillary) is thus kept constant and the pressure can be read off from the difference in mercury levels in V_2 and V_1.

The **absolute temperature** measured in Kelvin is then *by definition* given by

$$T = \frac{pV}{nR_0} \tag{1.7}$$

where p, V, and n are the pressure, volume and number of moles of a sufficiently dilute gas brought into thermal contact with the system of interest.

The gasthermometer is of considerable practical use as long as the temperatures do not become too low. For very low temperatures the ideal gas law breaks down and one has to match the temperature scale with other thermometers. Paramagnetic salts are often used as thermometers in this case.

CHAPTER 2

The First Law

The first law of thermodynamics is an extension of the law of conservation of energy to include thermal processes. It can be formulated as follows:

> *There exists a function of state U, the **internal energy**, such that if an amount of energy E is supplied to an otherwise isolated system, bringing it from an equilibrium state α to an equilibrium state β, then*
>
> $$E = U(\beta) - U(\alpha) \tag{2.1}$$
>
> *irrespective of the way in which this energy was supplied.*

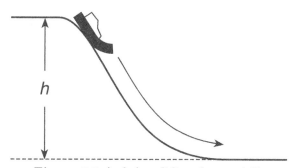

Figure 2.1 A sledge sliding down a hill.

EXAMPLE 2.1: *The sledge.*
Consider a sledge of mass M sliding down a hill of height h (figure 2.1). Assume that the speed of the sledge at the top of the hill is negligible. By conservation of energy, the potential energy $E_{pot} = M\,g\,h$ must have been transformed into kinetic energy: $E_{kin} = \frac{1}{2}M\,v^2$ at the bottom of the hill. The speed at the bottom of the hill must be $v = \sqrt{2gh}$. In fact the speed will be smaller due to friction. Where has the energy gone? The answer is, of course, that it was transformed into heat.

The first precise experiments about heat flow were done by James Joule in the 1840s. He did measurements on a thermally insulated container of gas to which he supplied energy in different ways: mechanical stirring, electrical heating, and compression. Figure 2.2 shows the three processes schematically. Joule found that the final temperature of the gas depends only on the amount of energy supplied; not on the way it was supplied, whether it was heat or work. Note that it was important that the container was well-insulated so that no heat could escape. In that case all the energy supplied must have been absorbed by the gas. In each of the different ways of heating the gas, it is possible to determine the amount of energy supplied, and it turns out that this determines the final state of the gas completely.

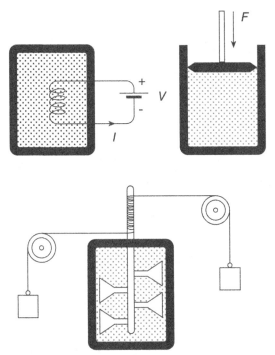

Figure 2.2 Joule's experiments.

In the case of electrical heating, the amount of heat supplied per unit time is given by

$$\dot{Q} = V I = I^2 R, \tag{2.2}$$

where V is the voltage over the heating element, I is the current flowing through it, and R is its electrical resistance (this relation is in fact due to Joule); in the case of mechanical stirring it is Mgh if h is the height over which the weight M is dropped.

The **mechanical work** performed in compressing the gas is given by

$$W = -\int_\alpha^\beta p \, dV, \tag{2.3}$$

where p is the pressure and the integral is performed along a path in state space from α to β. (Note that this depends on the path taken, but more about this later.) To prove this, consider a container with piston which is lowered by a small amount δz (figure 2.3).

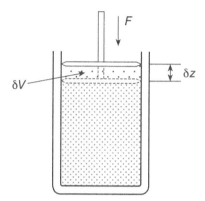

Figure 2.3 Derivation of the work differential.

Pressure *is defined as force per unit area*, so if A is the area of the piston, then the force applied to it is given by

$$F = p \, A. \tag{2.4}$$

The volume of the container changes by an amount $\delta V = -A \, \delta z$ (Note the minus sign: the volume decreases) and hence the work performed is given by

$$\delta W = F \, \delta z = -p \, \delta V. \tag{2.5}$$

Summing over all infinitesimal displacements, we obtain (2.3).

Because the final temperature depends only on the amount of energy supplied, Joule postulated the existence of an **internal energy** function $U(T)$. He found that it depends linearly on the temperature:

$$U(T) = C_V \, T. \tag{2.6}$$

This is the relation (2) mentioned in the introduction. Later experiments showed that the heat capacity C_V is in fact only weakly dependent on the type of gas: *for monatomic gases* it is given by

$$C_V = \frac{3}{2} n \, R_0 = 12.5 \, n \text{ J K}^{-1}, \tag{2.7}$$

and *for diatomic gases* by

$$C_V = \frac{5}{2} n R_0 = 20.7 \, n \text{ J K}^{-1}. \tag{2.8}$$

It turns out that (2.6) also holds for solids at sufficiently high temperatures. In that case

$$C_V = 3 \, n \, R_0 = 24.9 \, n \text{ J K}^{-1}. \tag{2.9}$$

This is called the **law of Dulong and Petit**. For lower temperatures this law does not hold and C_V decreases with decreasing temperature. For more general systems undergoing an adiabatic process the final state is also dependent only on the amount of energy supplied. We can therefore still postulate the existence of an internal energy function of state U. However, in general U depends also on the volume and any other thermodynamic parameters of the system.

If the system is not thermally isolated, heat can flow through the walls of the container and the total energy supplied is the sum of the heat Q supplied to the system and the work W performed on the system. Alternatively, as in example 2.1, no container is present and heat can escape freely. We then get, instead of equation (2.1),

$$\boxed{W + Q = U(\beta) - U(\alpha)} \tag{2.10}$$

We can read this as the definition of heat supplied. Of course, when heat is actually extracted from the system, the heat supplied is negative. The **heat capacity at constant volume** for a general system is defined by

$$C_V = \left(\frac{\partial U}{\partial T} \right)_V. \tag{2.11}$$

(The suffix V indicates that the volume is kept constant. If the system is not simple, other thermodynamic variables have to be kept constant as well.)

To emphasize the importance of the fact that the internal energy is a function of state, we show in the following example that this is not the case for the work W performed. This really means that the internal energy is a property of the system itself while the work depends on the process.

EXAMPLE 2.2: *Nonexactness of the work differential.*
Suppose an ideal gas is compressed at constant temperature T from a state α to a state β. Then the work performed on the gas is given by

$$W = - \int_{V_\alpha}^{V_\beta} p \, dV = n R_0 T \ln \frac{V_\alpha}{V_\beta}. \tag{2.12}$$

If, on the other hand, we go from α to β via γ as in figure 2.4, so that the temperature is not constant during the process, then the work performed is

$$W' = p_\alpha (V_\alpha - V_\beta) = n R_0 T \frac{V_\alpha - V_\beta}{V_\alpha} < W. \tag{2.13}$$

The amount of work performed therefore depends on the *path* from α to β in the state space. This means that *there does not exist a function of state* $\mathcal{W}$ such that $W = \mathcal{W}(\beta) - \mathcal{W}(\alpha)$. The reason is, of course, that in the two processes different amounts of heat are extracted from the gas. The notation δW above therefore does not mean 'a change in W', but simply an infinitesimal amount of work. We shall always use the Greek letter δ instead of d in such cases. For example, we also write δQ instead of dQ for an infinitesimal amount of heat.

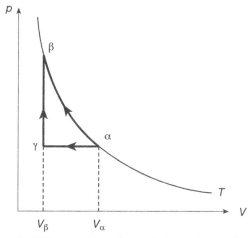

Figure 2.4 The work performed depends on the path.

REMARK 2.1: *Quasistatic processes.*
Note that the state space only applies to *equilibrium* states of the system. The processes considered above are therefore only well-defined if they occur very slowly so that the system is in equilibrium during the entire process. This type of process is called **quasi-static.** If a process is not quasistatic, one cannot draw a curve in the state space representing it. A quasi-static process is an idealization from reality. To produce actual changes, one must always have pressure differences, temperature differences, etc. But by making these differences small enough, one can ensure that the system is sufficiently close to equilibrium. In practice, the process must be slow compared with the relevant relaxation times in order that it may be considered quasi-static. As the system is (nearly) in equilibrium during a quasi-static process such a process is **reversible**, that is, reversing the direction of the process in state space results in another valid process.

A process as that considered by Joule, where the system is thermally insulated so that no heat flows through the walls, is called **adiabatic.** It is only quasi-static if the heating or work is performed sufficiently slowly. We have seen in the above example that an isothermal process is not adiabatic. Indeed, since $T_\alpha = T_\beta$, $U(\alpha) = U(\beta)$ in that process and $Q = -W$. Let us now determine the shape of quasi-static adiabatics in the $p - V$ diagram of an ideal gas. For a small change of volume dV, the work performed on the gas is $\delta W = -p\,dV$. This must equal the change in internal energy: $dU = C_V\,dT$. It follows that

$$\frac{dV}{dT} = -\frac{C_V}{p} = -\frac{C_V\,V}{n\,R_0\,T}. \tag{2.14}$$

Introducing the shorthand

$$\gamma = 1 + \frac{n\,R_0}{C_V}, \tag{2.15}$$

the solution to (2.14) reads

$$V^{\gamma-1}T = \text{constant.} \tag{2.16}$$

By inserting the ideal gas law, we obtain finally

$$\boxed{p\,V^{\gamma} = \text{constant.}} \tag{2.17}$$

As $\gamma > 1$ this means that the adiabatics are steeper than isotherms. It follows from (2.7) that for a monatomic ideal gas, $\gamma = \frac{5}{3}$ and from (2.8) that for a diatomic gas, $\gamma = \frac{7}{5}$.

CHAPTER 3

Differentials

In this chapter we derive some consequences of the first law. The first law is used most often in differential form. This is particularly useful for quasi-static processes, which can be subdivided into small stages during which the change in the various state functions is only small (infinitesimal). In infinitesimal form, the equation (2.10) reads:

$$\delta Q + \delta W = \mathrm{d}U. \tag{3.1}$$

One can give a precise meaning to the differential quantities appearing in this equation by defining the concept of a differential form. We shall discuss this at the end of this chapter, but first let us do some formal calculations with the differential forms appearing in (3.1).

In example 2.2 we already mentioned that the work differential δW cannot be written as a small change in a function of state $\mathcal{W}$. We shall use the notation δ in such a case. On the other hand, the differential $\mathrm{d}U$ does mean: 'a small change in U', and we write d instead of δ. One says that the differential $\mathrm{d}U$ is **exact**. If we assume as basic coordinates of the state space V and T we can write $\mathrm{d}U$ in terms of the coordinate differentials $\mathrm{d}V$ and $\mathrm{d}T$:

$$\mathrm{d}U = \left(\frac{\partial U}{\partial T}\right)_V \mathrm{d}T + \left(\frac{\partial U}{\partial V}\right)_T \mathrm{d}V = C_V \, \mathrm{d}T + \left(\frac{\partial U}{\partial V}\right)_T \mathrm{d}V. \tag{3.2}$$

(The subscript on partial derivatives indicates the variable to be kept fixed.) We have already seen that the work differential can also be written in terms of $\mathrm{d}T$ and $\mathrm{d}V$ (in fact only the latter appears):

$$\delta W = -p \, \mathrm{d}V. \tag{3.3}$$

In general, one can write any arbitrary differential in terms of the coordinate differentials. Denoting an arbitrary differential by ω, one has

$$\omega = f_1(T, V)\mathrm{d}T + f_2(T, V)\mathrm{d}V. \tag{3.4}$$

Using equation (3.1) we have for the heat differential

$$\delta Q = C_V \, dT + \left(p + \left(\frac{\partial U}{\partial V} \right)_T \right) dV. \tag{3.5}$$

Alternatively, one can take T and p as independent variables and write

$$dU = \left(\frac{\partial U}{\partial T} \right)_p dT + \left(\frac{\partial U}{\partial p} \right)_T dp. \tag{3.6}$$

The first coefficient in this equation is *not* the heat capacity at constant pressure! The **heat capacity at constant pressure** is defined as the corresponding coefficient in the expansion of the heat differential in terms of dT and dp. Formally,

$$(\delta Q)_{p\,\text{constant}} = C_p \, dT. \tag{3.7}$$

It is an easy exercise to transform (3.5) to new variables and find

$$\delta Q = C_p \, dT + \left(p + \left(\frac{\partial U}{\partial V} \right)_T \right) \left(\frac{\partial V}{\partial p} \right)_T dp \tag{3.8}$$

where

$$C_p = C_V + \left(p + \left(\frac{\partial U}{\partial V} \right)_T \right) \left(\frac{\partial V}{\partial T} \right)_p. \tag{3.9}$$

For an ideal gas these expressions simplify greatly. Indeed, since $U = C_V T$, where C_V is a constant *independent of the volume*, $(\partial U/\partial V)_T = 0$ and by the ideal gas law, $(\partial V/\partial T)_p = nR_0/p$ and $(\partial V/\partial p)_T = -V/p$ so

$$\delta Q = C_p \, dT - V \, dp \quad \text{(ideal gas)} \tag{3.10}$$

and

$$C_p = C_V + n \, R_0 \quad \text{(ideal gas)}. \tag{3.11}$$

In the following we give a precise mathematical meaning to differentials. This is not essential for the remainder of this book. One can arrive at a formal definition of infinitesimal quantities by simply noting how one calculates with them. The definition is as follows:

> A **differential form of rank 1** or **1-form** on $\mathbb{R}^k$ is a map $\omega : \mathbb{R}^k \times \mathbb{R}^k \to \mathbb{R}$ which depends smoothly on the first argument and is linear in the second. (Here the smoothness condition means that the map $\vec{x} \mapsto \omega(\vec{x}, \vec{v})$ is continuously differentiable for every $\vec{v} \in \mathbb{R}^k$.) We usually write $\omega_{\vec{x}}(\vec{v})$ instead of $\omega(\vec{x}, \vec{v})$.

EXAMPLE 3.1: *Exact differential form.*
If $F : \mathbb{R}^k \to \mathbb{R}$ is $\mathcal{C}^2$, i.e. twice continuously differentiable, then one can define a 1-form $\omega_{\vec{x}} = dF_{\vec{x}}$ by

$$\omega_{\vec{x}}(\vec{v}) = dF_{\vec{x}}(\vec{v}) = \langle \vec{v}, \nabla F(\vec{x}) \rangle. \tag{3.12}$$

Here ∇F is the gradient of F and $\langle \vec{u}, \vec{v} \rangle$ denotes the scalar product: $\langle \vec{u}, \vec{v} \rangle = \sum_{i=1}^{k} u_i v_i$. In the special case $k = 1$ we have $\mathrm{d}F_x = F'(x)\,\mathrm{d}x$. A 1-form is called **exact** if there exists a function F such that $\omega_\alpha = dF_\alpha$.

Every 1-form can be written in the form (cf. equation (3.4))

$$\omega_{\vec{x}} = \sum_{i=1}^{k} f_i(\vec{x})\,\mathrm{d}x_i \tag{3.13}$$

for certain functions $f_1, \ldots, f_k$. The **coordinate forms** $\mathrm{d}x_i$ are given by $\mathrm{d}x_i(\vec{v}) = v_i$. One can view these as unit vectors at the position $\vec{x}$ in the ith coordinate direction. A general $1-$form can then be seen as a vector field giving a vector with coordinates $f_i(\vec{x})$ at every point $\vec{x}$ of space. To prove equation (3.13), put $f_i(\vec{x}) = \omega_{\vec{x}}(\vec{e}_i)$, where $\vec{e}_i$ are the unit vectors in the $i-$th direction. Then equation (3.13) follows from the fact that $\mathrm{d}x_i(\vec{e}_j) = 0$ for $i \neq j$ and $\mathrm{d}x_i(\vec{e}_i) = 1$. Equation (3.12) is a special case of (3.13). Indeed, we can rewrite it as:

$$\mathrm{d}F_{\vec{x}} = \sum_{i=1}^{k} \frac{\partial F(\vec{x})}{\partial x_i}\mathrm{d}x_i. \tag{3.12'}$$

In the following we shall usually omit the subscript $\vec{x}$ from differential forms.

EXAMPLE 3.2. *The vector field corresponding to the work differential.*
The work differential has a special form in the coordinates (T, p) or (V, p). In the latter coordinate system, for example, the corresponding vector field looks like figure 3.1.

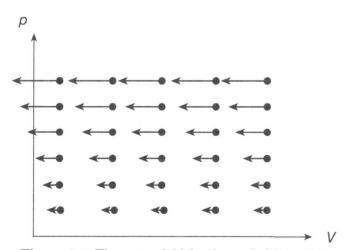

Figure 3.1. The vector field for the work differential.

One can integrate a general $1-$form along a curve Γ in $\mathbb{R}^k$ (figure 3.2). In example 2.2 we in fact already integrated δW along two different paths. In general, the integral is defined analogous to a Riemann integral as follows:

$$\int_\Gamma \omega = \lim_{\Delta_n} \sum_{i=1}^n \omega_{\vec{x}_{i-1}}(\vec{x}_i - \vec{x}_{i-1}),$$

(3.14)

where the limit is taken over finer and finer subdivisions of the curve Γ into intervals $[\vec{x}_{i-1}, \vec{x}_i]$.

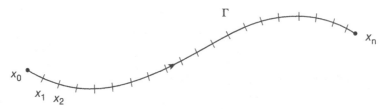

Figure 3.2. Integrating a 1-form along a curve.

In particular, if ω is an exact form, i.e. $\omega = \mathrm{d}F$, then we get

$$\int_\Gamma \mathrm{d}F = \int_\Gamma \nabla F \cdot \mathrm{d}\vec{s} = F(\beta) - F(\alpha),$$

(3.15)

where α and β are the starting point and endpoint of Γ respectively. Note that the result is independent of the path taken from α to β. It follows that the integral of an exact form along a closed path, i.e. a path which returns to its starting point: $\alpha = \beta$, must be zero. The converse is also true.

Theorem 3.1 *A 1-form ω is exact if and only if*

$$\oint_\Gamma \omega = 0$$

for every closed path Γ.

Note that, in the case of the work differential, the integral $\int_\Gamma \delta W$ is the work performed during the quasi-static process from α to β along the curve Γ in the state space. The work differential is not exact because its integral along the path $\alpha \to \beta \to \gamma \to \alpha$ in figure 2.4 is not zero.

An alternative condition for exactness is the following:

Theorem 3.2 *A 1-form $\omega = f_1 \,\mathrm{d}x_1 + f_2 \,\mathrm{d}x_2$ defined on $\mathbb{R}^2$ (or a simply connected region in $\mathbb{R}^2$) is exact if and only if*

$$\frac{\partial f_1}{\partial x_2} = \frac{\partial f_2}{\partial x_1}.$$

(3.16)

This can easily be seen to be a necessary condition because if $\omega = \mathrm{d}F$ then

$$f_1 = \frac{\partial F}{\partial x_1} \text{ and } f_2 = \frac{\partial F}{\partial x_2}$$

and the condition in theorem 3.2 is simply the identity of the mixed second derivatives of F.

CHAPTER 4

The Second Law

The second law of thermodynamics concerns the nonexistence of a perpetuum mobile. This has long been a contentious issue but it is based on one of the most obvious facts of everyday experience, namely that *certain processes are* **irreversible**. This means simply that they cannot be reversed in time. Thus, e.g. the breaking of a glass on the floor is an irreversible process because the reverse process: the *spontaneous* creation of a glass from pieces on the floor simply does not happen. Characteristic of a process like that is that an ordered structure is destroyed which cannot be recreated without "doing something" to the system. It makes sense, therefore, to postulate the existence of a measure for the order in the system. Traditionally, one has chosen to introduce a measure of *disorder* instead, which is called the **entropy.** The second law is then a formulation of the creation of entropy in processes where the system is undisturbed. The first precise formulation of the second law was given by Lord Kelvin and Clausius. Kelvin's formulation reads as follows.

> *There exists no thermodynamic transformation of which the only result is to convert heat from a heat reservoir to work.*

Clausius' formulation is slightly different.

> *There exists no thermodynamic transformation of which the only result is the transfer of heat from a colder heat reservoir to a hotter heat reservoir.*

Clausius deduced from these basic statements the existence of a function of state which can only increase if the system is thermally isolated. This is the entropy which we simply postulate to exist. We then show in chapter 5 that our formulation implies their original statements.

To formulate the law in its entirety we need the following observation:

> *All thermodynamic parameters can be divided into two categories, namely intensive and extensive parameters.* **Intensive parameters** *are thermodynamic parameters that are independent of the*

*size of the system; **extensive parameters** are thermodynamic parameters that are directly proportional to the size of the system.*

This means that, if we double the system the intensive parameters remain the same while the extensive parameters double in value. Doubling a system involves, apart from doubling the volume, also doubling of the number of particles N so we had better include the particle number in our considerations. Examples of intensive parameters are the pressure p and the temperature T; examples of extensive parameters are the internal energy U, the heat capacity C_V, the volume V, the length of a rod, the magnetic moment $\overrightarrow{\mathcal{M}}$ of a magnet, and the number of particles N. Including the number of particles does not actually introduce essential new coordinates because of the fact that all extensive parameters are proportional to N so that we can introduce the corresponding specific quantities, e.g. the **specific volume** $v = V/N$ and the **specific heat** $c_V = C_V/N$. In Part II we argue that it is these quantities that make sense mathematically because the laws of thermodynamics hold only in the thermodynamic limit. (See the introduction. The first law also holds for finite systems of course as it is simply the law of conservation of energy.) We shall discuss this further in chapter 6.

We can now formulate the second law as follows:

> *There exists a function of state S, called the **entropy**, which is an extensive variable and has the following property: During an arbitrary adiabatic process from an equilibrium state α to an equilibrium state β the entropy does not decrease, i.e. $Q_{\alpha \to \beta} = 0 \implies S(\beta) \geqslant S(\alpha)$.*

We first derive some theoretical consequences of this law. In the next chapter we shall discuss an important practical application. A somewhat delicate argument shows that the heat differential is proportional to the entropy differential:

$$\delta Q = \lambda \, dS \tag{4.1}$$

for some function of state (not a constant) λ.

The argument goes roughly as follows: At any point in the state space we can take a coordinate system consisting of the variable S and parameters perpendicular to the surfaces of constant S. Considering a simple system there is only one other parameter, which we can denote X. Then we can write in general (see equation (3.4))

$$\delta Q = \lambda \, dS + \kappa \, dX.$$

Now, for an infinitesimal quasi-static process we have, according to the second law:

$$\delta Q = 0 \implies dS \geqslant 0.$$

However, if $\kappa \neq 0$ we can take an infinitesimal process in the direction given by

$$\frac{\mathrm{d}X}{\mathrm{d}S} = -\frac{\lambda}{\kappa}$$

for which $\delta Q = 0$ but $\mathrm{d}S < 0$ and which therefore violates the second law.

We now show that λ is in fact a function of the temperature only. To see this, consider two systems in thermal contact but thermally isolated from the exterior. By the zeroth law they must have the same temperature T. Assuming that the two systems are simple, the combined system has a three-dimensional state space and can be described by three variables: V_1, V_2, and T, where V_1 and V_2 are the volumes of the two subsystems. By equation (4.1) we have for the heat differentials,

$$\delta Q = \delta Q_1 + \delta Q_2 = \lambda(V_1, V_2, T)\, \mathrm{d}S(V_1, V_2, T)$$

and

$$\delta Q_i = \lambda_i(V_i, T)\, \mathrm{d}S(V_i, T) \quad (i = 1, 2).$$

Hence,

$$\lambda\, \mathrm{d}S = \lambda_1\, \mathrm{d}S_1 + \lambda_2\, \mathrm{d}S_2. \tag{4.2}$$

If we now choose as independent variables S_1, S_2, and T then $S(S_1, S_2, T)$ in fact only depends on T *through S_1 and S_2*. Indeed, $S = S_1 + S_2$. It follows that

$$\frac{\lambda_1(S_1, T)}{\lambda(S_1, S_2, T)} = \left(\frac{\partial S}{\partial S_1}\right)_{S_2} = 1 \text{ and } \frac{\lambda_2(S_2, T)}{\lambda(S_1, S_2, T)} = \left(\frac{\partial S}{\partial S_2}\right)_{S_1} = 1. \tag{4.3}$$

Hence

$$\lambda(S_1, S_2, T) = \lambda_1(S_1, T) = \lambda_2(S_2, T). \tag{4.4}$$

As λ_1 depends only on S_1 and T, and λ_2 depends only on S_2 and T, it follows that $\lambda(T)$ is a *universal function* of the temperature. To determine this function we may gauge the system with a standard thermometer, i.e. a gas thermometer. Now, for an ideal gas,

$$\left(\frac{\partial U}{\partial V}\right)_T = 0$$

by equation (2.6). Hence, by (3.5),

$$\delta Q = \lambda\, \mathrm{d}S = C_V\, \mathrm{d}T + p\, \mathrm{d}V \tag{4.5}$$

and since $\mathrm{d}S$ is an exact differential it follows from theorem 3.2 that

$$\frac{\partial}{\partial T}\left(\frac{p}{\lambda}\right) = \frac{\partial}{\partial V}\left(\frac{C_V}{\lambda}\right) = 0.$$

Inserting the ideal gas law for p we obtain: $\lambda'(T) = \lambda/T$, and hence $\lambda(T) = cT$ for some constant c. Fixing the constant c fixes the measuring scale for S. Setting $c = 1$ we obtain the following important relation which holds for quasi-static processes,

$$\boxed{\delta Q = T\,dS} \tag{4.6}$$

REMARK 4.1: *Definition of absolute temperature.*
The definition of absolute temperature given in chapter 1 is in fact unsatisfactory. Indeed, one does not need the ideal gas to conclude that the absolute zero of temperature exists. To see this, notice that the above argument remains valid if we replace the absolute temperature T by the empirical temperature θ. We conclude in that case that $\lambda(\theta)$ is a universal function of θ. Thus we can *define* the absolute temperature by

$$T(\theta) = C\lambda(\theta). \tag{4.7}$$

The arbitrary constant C can be fixed, as before, by agreeing that the boiling and freezing points of water should differ by 100 degrees. However, as this definition of T does not involve an additive constant, the absolute zero is uniquely determined.

The relation (4.6) holds only for quasi-static processes. However, due to the fact that the entropy is a function of state, one can compute the change in entropy in an arbitrary process between two equilibrium states α and β from

$$\Delta S = S(\beta) - S(\alpha) = \int_\alpha^\beta \frac{\delta Q}{T}, \tag{4.8}$$

where the integral is performed along *any path* in state space from state α to state β.

Inserting (4.6) into (3.5) we obtain the following relation for a simple system:

$$T\,dS = C_V\,dT + \left(p + \left(\frac{\partial U}{\partial V} \right)_T \right) dV. \tag{4.9}$$

Equating the cross derivatives as above we get the **energy equation**

$$\left(\frac{\partial U}{\partial V} \right)_T = T \left(\frac{\partial p}{\partial T} \right)_V - p. \tag{4.10}$$

As an application we can insert this into equation (3.9). Using the general relation

$$\left(\frac{\partial p}{\partial T} \right)_V \left(\frac{\partial V}{\partial p} \right)_T \left(\frac{\partial T}{\partial V} \right)_p = -1, \tag{4.11}$$

we obtain

$$C_p - C_V = \frac{\alpha^2 TV}{\kappa_T}, \tag{4.12}$$

where

$$\alpha = \frac{1}{V}\left(\frac{\partial V}{\partial T}\right)_p \tag{4.13}$$

is the **thermal expansion coefficient** and

$$\kappa_T = -\frac{1}{V}\left(\frac{\partial V}{\partial p}\right)_T \tag{4.14}$$

is the **isothermal compressibility.** For gases one can measure C_V directly but for solids it is easier to measure C_p instead. Relation (4.12) is then very useful for determining C_V.

Another consequence of equations (4.9) and (4.10) is that the entropy change (4.8) for a simple system can be written as

$$\Delta S = \int_{T_\alpha}^{T_\beta} \frac{C_V}{T}\,\mathrm{d}T + \int_{V_\alpha}^{V_\beta}\left(\frac{\partial p}{\partial T}\right)_V \mathrm{d}V. \tag{4.15}$$

REMARK 4.2: *The third law.*
The third law, also called Nernst's law, is a more recent addition to the theory of thermodynamics. Nernst noticed in 1906 that in many chemical reactions the change in entropy seems to become small as $T \to 0$. He therefore formulated the following general law.

> *The change in entropy of a system during a process between two equilibrium states at the same temperature but characterized by different values of another thermodynamic parameter x, tends to zero as the temperature approaches the absolute zero (See figure 4.1).*

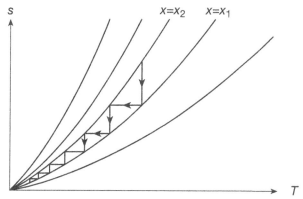

Figure 4.1 A typical cooling process.

Planck gave a more general formulation in 1911:

> *The entropy of all perfect crystals at the absolute zero of tempera-*
> *ture is the same and can therefore be normalized to zero.*

There is considerable experimental evidence for this rule. Moreover, as we shall see in Part II, it can be justified on theoretical grounds. It is important to realize that the crystal (as well as more general systems) must be *in equilibrium* for this law to hold. Many practical systems are *not* in equilibrium but in a **meta-stable state.** Meta-stable states are non-equilibrium states for which the *relaxation time* is large (see the introduction). Glass is a good example of such a system.

An important consequence of the third law is that the absolute zero of temperature cannot be reached with a finite number of quasi-static processes. We shall not prove this fact but it is intuitively easy to understand. Indeed, in order to cool a material one has to extract heat from it. This is achieved by changing its state, that is by changing one of its thermodynamic parameters. However, as the absolute zero of temperature is approached, a given change in this parameter will result in ever smaller changes of entropy and hence a smaller amount of heat being extracted. Eventually, the heat loss due to leakage will equal the amount of heat extracted in the process and the temperature stabilizes. As a practical illustration, figure 4.1 shows an alternating sequence of isothermal and isentropic processes between two values x_1 and x_2 of an independent thermodynamic variable x. (We shall see that adiabatic demagnetization is an example of such a process, where the parameter x is the magnetization of a paramagnetic salt.) As the curves of constant x approach one another for $T \to 0$, it follows that the absolute zero cannot be reached in a finite number of steps.

CHAPTER 5

Thermal Engines and Refrigerators

Thermodynamics was developed in a study of the efficient operation of thermal engines, the steam engine in particular. It is still important for these applications, e.g. the car engine, refrigerators, and steam turbines in electricity plants. The principle of operation of a heat engine is pictured schematically in figure 5.1.

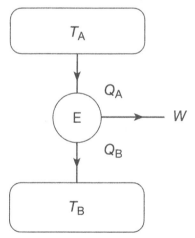

Figure 5.1 Schematic drawing of a heat engine.

Heat is absorbed from a heat reservoir at temperature T_A. The engine E converts part of this heat into useful work W and rejects the surplus heat Q_B to a second heat reservoir at temperature T_B. Clearly,

$$W = Q_A - Q_B. \tag{5.1}$$

The **efficiency** of the engine is defined as

$$\eta = \frac{W}{Q_A}. \tag{5.2}$$

Obviously, $\eta \leqslant 1$, but the second law gives a more stringent bound on the efficiency. Indeed, since

$$\Delta S = -\frac{Q_A}{T_A} + \frac{Q_B}{T_B} \geqslant 0, \tag{5.3}$$

we find

$$\eta \leqslant 1 - \frac{T_B}{T_A} \equiv \eta_C. \tag{5.4}$$

The maximum attainable efficiency η_C is called the **Carnot efficiency.** It also follows that the most efficient engines are reversible, $\Delta S = 0$. An example of a process in which the Carnot efficiency is attained is the **Carnot process.** For an ideal gas this process is represented by the cycle abcd in the following $p - V$ diagram of figure 5.2.

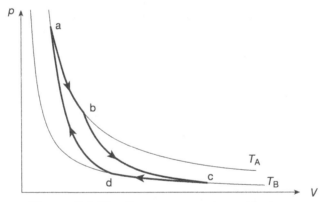

Figure 5.2 The Carnot process in an ideal gas.

ab and cd are isotherms, bc and da are adiabatics. In practice this process is very difficult to carry out. Most practical electricity plants use steam as working fluid. The process can then take place inside the coexistence region of steam and liquid water. This is called *wet steam.* It is then much easier to keep the temperature constant during the transfer of heat as the evaporation occurs at constant temperature. The process is most conveniently represented in a $T - s$ diagram (figure 5.3).

In fact, the actual process used in practical power plants is considerably more complicated. Let us consider two of the most important modifications.

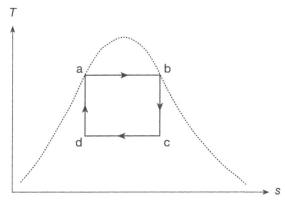

Figure 5.3 Carnot process in the coexistence region of a fluid.

First of all, it is much more convenient to complete the condensation of the steam from c to d' (See figure 5.4) and then use a compressor pump to raise the pressure to that at a in the isentropic process d'a'. This is called the **Rankine cycle**. Its efficiency is lower than the Carnot efficiency but the work done per cycle is larger.

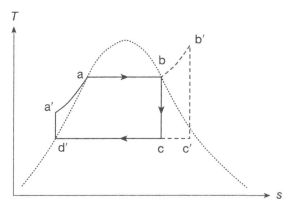

Figure 5.4 The Rankine cycle of a heat engine.

Another modification is to heat the steam further from b to b' in a second set of tubes inside the boiler. This uses some of the waste heat which otherwise escapes with the exhaust fumes from the flue. Moreover, it reduces the amount of condensate in the turbine, which would otherwise cause wear and tear of the turbine blades. This *Rankine process with superheat* is indicated by the broken line in the figure 5.4.

Another example of a heat engine is the car engine. This is in fact an **internal combustion engine**, which means that the fuel is mixed with the working fluid. This has the advantage that higher temperatures can be achieved. It is still a reasonable approximation to neglect the changing fuel mixture and assume that the cycle of the working fluid can be represented in a $p - V$

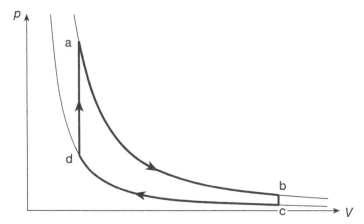

Figure 5.5 The Otto cycle.

diagram. An idealized version of this cycle is the **Otto cycle**. It is depicted in figure 5.5.

ab and cd are adiabatics; bc and da are isochores (constant volume processes). The efficiency of this process is given by

$$\eta_O = 1 - \frac{T_b}{T_a} = 1 - \frac{1}{r_v^{\gamma-1}} < 1 - \frac{T_c}{T_a}, \tag{5.5}$$

where $r_v = V_c/V_d$ is the **compression ratio**.

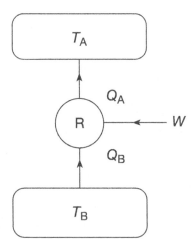

Figure 5.6 Schematic drawing of the principle of a refrigerator.

A **refrigerator** is the converse of a heat engine (figure 5.6): it uses work to extract heat from a cold body, expelling heat to a heat reservoir. In this case, the relevant parameter is the **coefficient of performance** defined by

$$CP = \frac{Q_B}{W}. \tag{5.6}$$

Note that CP can be greater than 1. In the case of a simple reversible re-frigerator, represented by a reversed Carnot process (figure 5.3 with arrows reversed),

$$CP_{rev} = \frac{T_B}{T_A - T_B}. \tag{5.7}$$

We shall discuss the refrigerator in more detail in chapter 11. A refrigerator can also be used as a **heat pump** by using the heat expelled (from d to c in figure 5.3). This is more efficient than an ordinary heater, but it is sadly seldom used except in air conditioners.

CHAPTER 6

The Fundamental Equation

In this chapter, we show that the thermodynamics of a (simple) system is completely determined by a single function: the entropy density as a function of the internal energy density and the specific volume. This function has an important property, namely **concavity**, which enables us to define a number of other thermodynamic functions in chapter 8. Let us recall the definition of convex and concave functions: *A region D in $\mathbb{R}^k$ is called **convex** if for every two points $\vec{x}, \vec{y} \in D$ and every $t \in [0,1]$, also $t\vec{x} + (1-t)\vec{y} \in D$. A function $g : D \to \mathbb{R} \cup \{+\infty\}$ from a convex region $D \subset \mathbb{R}^k$ to the real line united with $+\infty$ is called **convex** if, for every $t \in [0,1]$ and all $\vec{x}, \vec{y} \in D$,*

$$g(t\vec{x} + (1-t)\vec{y}) \leqslant tg(\vec{x}) + (1-t)g(\vec{y}). \tag{6.1}$$

*g is called **concave** if $-g$ is convex.* A region D is convex when it does not have dents. Figures 6.1(a) and 6.1(b) show a convex region and a non-convex region respectively.

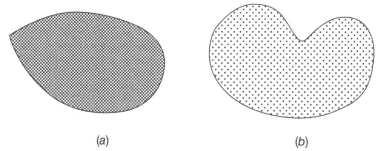

(a)   (b)

Figure 6.1 (a) A convex region (left) and (b) a non-convex region (right).

Similarly, a function is convex if its graph is everywhere bending upwards. In chapter 7, we state and prove various useful facts about convex functions. Here, we first outline the significance for thermodynamic functions.

In the introduction and also in chapter 4 we remarked that, in order to describe a macroscopic system mathematically we have to take the **thermodynamic limit**, i.e. the limit of an infinitely large system with a finite particle number density: $N \to \infty$, $V \to \infty$; $\rho = N/V$ fixed. We also write $v = 1/\rho$ for the **specific volume**. In this limit all *surface effects* disappear and we are left with the pure *bulk properties* of the system. For intensive parameters $X(N,V,T)$ we expect the limit

$$\lim_{N,V \to \infty;\, V/N=v} X(N,V,T) = x(v,T) \tag{6.2}$$

to exist, while for extensive variables $Y(N,V,T)$ we expect the limit

$$\lim_{N,V \to \infty;\, V/N=v} \frac{Y(N,V,T)}{N} = y(v,T) \tag{6.3}$$

to exist. In other words, $X(N,V,T) = x(v,T) + o(1)$ and $Y(N,V,T) = Ny(v,T) + o(N)$ as $N \to \infty$. In fact, in real systems the values of X and Y/N *fluctuate* around their mean values $x(v,T)$ and $y(v,T)$.

Let us now concentrate in particular on the **entropy density** for a simple system defined by

$$s(u,v) = \lim_{N,V \to \infty;\, V/N=v} \frac{1}{N} S(Nu, V, N). \tag{6.4}$$

We shall argue that this function is concave in both its arguments. To show this we consider two systems Σ_1 with parameters (U_1, V_1, N_1) and Σ_2 with parameters (U_2, V_2, N_2), which are brought into thermal and mechanical contact so that they can exchange energy and their relative volumes can adjust. Figure 6.2 gives an impression of this situation.

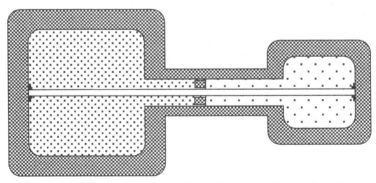

Figure 6.2 Two systems in thermal and mechanical contact.

Denote $v_1 = V_1/N_1$ and $v_2 = V_2/N_2$, and let $c_i = N_i/(N_1+N_2)$ $(i = 1,2)$. Then, by the second law,

$$S(U_1 + U_2, V_1 + V_2, N_1 + N_2) \geqslant S(U_1, V_1, N_1) + S(U_2, V_2, N_2). \qquad (6.5)$$

Taking the thermodynamic limit now means taking the limit where all the parameters $N_1, V_1, N_2, V_2 \to \infty$ but v_1, v_2, c_1, and c_2 are constant. Because U_1 and U_2 are extensive variables the limits $u_1 = \lim_{N_1 \to \infty} (U_1/N_1)$ and $u_2 = \lim_{N_2 \to \infty} (U_2/N_2)$ exist and we obtain

$$s(c_1 u_1 + c_2 u_2, c_1 v_1 + c_2 v_2) \geqslant c_1 s(u_1, v_1) + c_2 s(u_2, v_2). \qquad (6.6)$$

It is a remarkable fact that the thermodynamic functions of an arbitrary system are completely determined by the single function $s(u, v)$.

*The thermodynamics of a simple system is given by a function $s(u, v)$ of two variables, the **entropy density**, with the following properties:*

1. *$s(u, v)$ is a concave function of its arguments;*

2. *$s(u, v)$ is continuously differentiable in both variables, and*

3. *$s(u, v)$ is increasing as a function of u.*

We presently show that the function $s(u, v)$ determines the thermodynamic behaviour of the system completely, which is why the relation between the entropy and the internal energy and the volume is called the **fundamental equation**. To this end we first remark that, because s is increasing in u, we can invert this function and obtain $u(s, v)$ instead. By Property 2 above this is also a continuously differentiable function of its arguments and the differential du is well-defined:

$$du = \left(\frac{\partial u}{\partial s}\right)_v ds + \left(\frac{\partial u}{\partial v}\right)_s dv. \qquad (6.7)$$

Moreover, u is a convex function of s and v. (We leave the proof of this fact as an exercise.) We can then obtain the temperature and the pressure of the system from

$$T = \left(\frac{\partial u}{\partial s}\right)_v \qquad (6.8)$$

and

$$p = -\left(\frac{\partial u}{\partial v}\right)_s \qquad (6.9)$$

respectively. Indeed, by equation (2.5) the work differential δw per particle is given by

$$\delta w = -p \, dv \qquad (6.10)$$

and by equation (4.6) the heat differential δq per particle is given by

$$\delta q = T \, ds. \tag{6.11}$$

Inserting this into equation (3.1), that is, the first law in the case of an infinitesimal quasi-static process,

$$du = \delta w + \delta q, \tag{6.12}$$

yields (6.8) and (6.9). These definitions therefore follow from the laws of thermodynamics for quasi-static processes. In terms of the entropy function $s(u, v)$ we have

$$\left(\frac{\partial s}{\partial u}\right)_v = \frac{1}{T} \tag{6.13}$$

and

$$\left(\frac{\partial s}{\partial v}\right)_u = -\left(\frac{\partial u}{\partial v}\right)_s \left(\frac{\partial s}{\partial u}\right)_v = \frac{p}{T}. \tag{6.14}$$

EXAMPLE 6.1: *The entropy of an ideal gas.*
We can determine the fundamental equation of an ideal gas as follows. By equations (6.13) and (2.6),

$$\frac{\partial s}{\partial u} = \frac{1}{T} = \frac{c_V}{u},$$

where

$$c_V = \frac{C_V}{N} \tag{6.15}$$

is the **specific heat**, and by (6.14) and (1.5),

$$\left(\frac{\partial s}{\partial v}\right)_u = \frac{p}{T} = \frac{nR_0}{V} = \frac{k_B}{v},$$

where **Boltzmann's constant** k_B is given by

$$k_B = R_0/N_0 = 1.3806 \cdot 10^{-23} \text{ J K}^{-1}. \tag{6.16}$$

Integrating, we obtain

$$s(u, v) = c_V \ln u + k_B \ln v + s_0 \tag{6.17}$$

where s_0 is an integration constant. Conversely, both equations (1.5) and (2.6) follow from the fundamental equation (6.17).

REMARK 6.1. *General systems.*
For general, non-simple systems, s is a function of more than two arguments, including the internal energy density u, but the same three axioms still hold.

The other arguments $x_1, \ldots, x_{k-1}$ are called *generalized displacements* and the corresponding conjugate quantities, defined by

$$f_i = -\left(\frac{\partial u}{\partial x_i}\right) \tag{6.18}$$

generalized forces. (k is the dimension of the state space.)

We finish this chapter with some remarks about arbitrary processes. Consider again two systems in equilibrium with respective values (u_1, v_1) and (u_2, v_2) for the internal energy density and the specific volume. Let the relative concentrations be c_1 and c_2 ($c_1, c_2 \geqslant 0$, $c_1 + c_2 = 1$) and the respective entropy functions $s_1(u_1, v_1)$ and $s_2(u_2, v_2)$. If these two systems are brought into **thermodynamic contact** but are otherwise thermally insulated then the total energy and the total volume are constant which implies that $u = c_1 u_1 + c_2 u_2$ and $v = c_1 v_1 + c_2 v_2$ are the same before and after contact. By the second law, the resulting entropy function after equilibrium has been re-established, is maximal given these constraints:

$$s(u, v) = \sup_{\substack{c_1 u_1 + c_2 u_2 = u \\ c_1 v_1 + c_2 v_2 = v}} \{c_1 s_1(u_1, v_1) + c_2 s_2(u_2, v_2)\}. \tag{6.19}$$

Note that this function satisfies the properties 1., 2., and 3. above. The two systems are thus in *thermodynamic equilibrium* if their values for the internal energy density and specific volume are such that the supremum in equation (6.19) is attained. It is an easy calculation to check that this is the case if and only if the partial derivatives of s_1 and s_2 are equal:

$$\frac{\partial s_1}{\partial u_1} = \frac{\partial s_2}{\partial u_2} \quad \text{and} \quad \frac{\partial s_1}{\partial v_1} = \frac{\partial s_2}{\partial v_2}.$$

In view of equations (6.13) and (6.14) these conditions become:

$$T_1 = T_2 \text{ and } p_1 = p_2. \tag{6.20}$$

This is the zeroth law. Note however, that equation (6.19) does not determine the way in which this maximum is attained, that is how u_1 and v_1 (and hence u_2 and v_2) change in the process.

REMARK 6.2: *Partial constraints.*
If one combines two systems only partially, for example by connecting them via a diathermal but immovable separation then the combined system is not a simple system. (This is called **thermal contact.**) The entropy becomes a function of the three variables: u, v_1, and v_2 defined by

$$s(u, v_1, v_2) = \sup_{c_1 u_1 + c_2 u_2 = u} \{c_1 s_1(u_1, v_1) + c_2 s_2(u_2, v_2)\}. \tag{6.21}$$

The same argument as above now applies only to the derivatives w.r.t. u and we obtain as condition for thermal equilibrium: $T_1 = T_2$. Note that in this case it follows easily from the concavity of s that $\partial s/\partial u$ is decreasing in u so that T is increasing in u. Therefore, if initially $T_1 < T_2$ then the final temperature T after contact must lie in between these values: $T_1 < T < T_2$ and, moreover, heat has flowed from system 2 to system 1.

EXAMPLE 6.2: *Free expansion of a gas.*
We considered this example already in the introduction (see figure 2). A container is partitioned into two halves, one of which is evacuated; the other containing a gas. When a valve is opened and the gas allowed to flow into the evacuated half, the gas is said to expand freely. In analogy with equation (6.19), the final state can be described by the formula

$$s(u, v) = \sup_{v_1 \leqslant v} s_1(u, v_1). \tag{6.22}$$

Because the pressure of the gas is positive, the supremum is attained at $v_1 = v$ and $s(u, v) = s_1(u, v)$: the gas spreads over the available volume. If the gas is ideal then the entropy increases by

$$\Delta s = k_B \ln \frac{v}{v_1}. \tag{6.23}$$

The important point to notice here is that u is constant because no energy is supplied. Since there is also no heat flow, the work done is zero as well! Although $\Delta v > 0$, $w = 0$ so equation (2.3) does not hold here. This is generally the case for irreversible processes.

CHAPTER 7

Convexity

The convexity of the internal energy function is an important property directly related to the stability of thermodynamic equilibrium. In this chapter we derive some properties of convex functions which will enable us to develop the theory of thermodynamics further. We state the theorems in general but give proofs for the one-dimensional case only. Proofs in the general k-dimensional case can be found in appendix C. In any case, the proofs are not essential for understanding the subsequent theory though the definitions, concepts, and theorems are.

We start with a definition.

*If $g : \mathbb{R}^k \to \mathbb{R} \cup \{+\infty\}$ is a function which can take the value $+\infty$ then one defines its **essential domain** by*

$$D(g) = \left\{ \vec{x} \in \mathbb{R}^k \,|\, g(\vec{x}) < +\infty \right\}. \tag{7.1}$$

Theorem 7.1 *A convex function $g : \mathbb{R}^k \to \mathbb{R} \cup \{+\infty\}$ is continuous at every interior point of its essential domain.*

Proof. It is clear that the essential domain of g is convex. We give the proof of the theorem in one dimension, so that $D(g) = I$ is an interval. Without loss of generality we may assume that I is an open interval. We shall prove that for all $x_0 \in I$, $g(x_0) \leqslant \liminf_{x \to x_0} g(x)$ and $g(x_0) \geqslant \limsup_{x \to x_0} g(x)$.

To prove that $g(x_0) \leqslant \liminf_{x \to x_0} g(x)$, suppose to the contrary that $g(x_0) > \liminf_{x \to x_0} g(x)$. If there were to exist $x_1 < x_0 < x_2$ such that $g(x_1) < g(x_0)$ and $g(x_2) < g(x_0)$ then

$$g(x_0) \leqslant \frac{x_2 - x_0}{x_2 - x_1} g(x_1) + \frac{x_0 - x_1}{x_2 - x_1} g(x_2) < g(x_0),$$

which is a contradiction. We conclude that $g(x) \geqslant g(x_0)$, either for all $x \geqslant x_0$ or for all $x \leqslant x_0$. Both cases are similar. We consider only the first. We may assume that there exists a sequence $\{x_n\}$ such that $x_n < x_0$ and $g(x_n) \to$

$\liminf_{x \to x_0}[g(x)] < g(x_0)$. Choosing an arbitrary $\tilde{x} > x_0$ we have (see figure 7.1)

$$g(x_0) \leqslant \frac{\tilde{x} - x_0}{\tilde{x} - x_n} g(x_n) + \frac{x_0 - x_n}{\tilde{x} - x_n} g(\tilde{x}).$$

As $x_n \to x_0$ the right hand side tends to $\liminf_{x \to x_0} g(x)$ which contradicts the assumption.

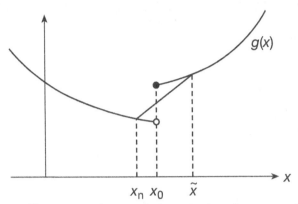

Figure 7.1 Illustrating the proof of continuity of a convex function.

To prove that $g(x_0) \geqslant \limsup_{x \to x_0}[g(x)]$ choose $x_1, x_2 \in I$ with $x_1 < x_0 < x_2$. Then, for $x_0 \leqslant x < x_2$,

$$g(x) \leqslant \frac{x_2 - x}{x_2 - x_0} g(x_0) + \frac{x - x_0}{x_2 - x_0} g(x_2),$$

and for $x_1 < x \leqslant x_0$,

$$g(x) \leqslant \frac{x - x_1}{x_0 - x_1} g(x_0) + \frac{x_0 - x}{x_0 - x_1} g(x_1).$$

Both right hand sides tend to $g(x_0)$ as $x \to x_0$. Hence $\limsup_{x \to x_0}[g(x)] \leqslant g(x_0)$. ∎

We next introduce an important transformation for convex functions.

*Let $g : \mathbb{R}^k \to \mathbb{R} \cup \{+\infty\}$ be an arbitrary function which may take the value $+\infty$. Then we define the **Legendre transform** $g^* : \mathbb{R}^k \to \mathbb{R} \cup \{+\infty\}$ of g by*

$$\boxed{g^*(\vec{t}) = \sup_{\vec{x} \in \mathbb{R}^k} [\langle \vec{x}, \vec{t} \rangle - g(\vec{x})].} \tag{7.2}$$

Lemma 7.1 $g^*(\vec{t})$ *is a convex function of* $\vec{t} \in \mathbb{R}^k$.

Proof. We have, for $\vec{s}, \vec{t} \in \mathbb{R}^k$ and $c \in [0,1]$,

$$
\begin{aligned}
g^*(c\vec{s} + (1-c)\vec{t}) &= \sup_{\vec{x}}\{\langle \vec{x}, (c\vec{s}+(1-c)\vec{t})\rangle - g(\vec{x})\} \\
&= \sup_{\vec{x}}\{c(\langle \vec{x}, \vec{s}\rangle - g(\vec{x})) + (1-c)(\langle \vec{x}, \vec{t}\rangle - g(\vec{x}))\} \\
&\leqslant c\sup_{\vec{x}}\{\langle \vec{x}, \vec{s}\rangle - g(\vec{x})\} + (1-c)\sup_{\vec{x}'}\{\langle \vec{x}', \vec{t}\rangle - g(\vec{x}')\} \\
&= cg^*(\vec{s}) + (1-c)g^*(\vec{t}). \blacksquare
\end{aligned}
$$

The Legendre transform of any function is therefore convex. We presently show that, if g itself is convex then the Legendre transform of its Legendre transform equals g. In fact, we prove a slightly more general result. We need the following lemma.

Lemma 7.2 *Let* $g : \mathbb{R}^k \to \mathbb{R} \cup \{+\infty\}$ *be a function with convex essential domain. Then g is convex if and only if its* **epigraph** *defined by*

$$
\mathrm{epi}(g) = \{(\vec{x}, h) \in D(g) \times \mathbb{R} \mid h \geqslant g(\vec{x})\} \tag{7.3}
$$

is convex.

Proof. This is left as an exercise. $\blacksquare$

REMARK 7.1. *A line intersects the epigraph over an interval.*
Note also that for any line $y = ax + b$ which cuts the graph of $g : \mathbb{R} \to \mathbb{R}$ in two distinct points (x_1, y_1) and (x_2, y_2) with $x_2 > x_1$, the segment between these points lies inside the epigraph, that is above the graph of g, while the remaining parts of the line lie below the graph.

The Legendre transform of a function $g : \mathbb{R} \to \mathbb{R}$ can be interpreted graphically as follows. For any t, $x \mapsto tx$ is a straight line and the Legendre transform $g^*(t)$ at t is the distance over which this line has to be moved down so that it touches the graph of g (or minus the distance over which it has to be moved up).

If $g : \mathbb{R} \to \mathbb{R}$ is such that $g^*(t_0) < \infty$ for at least one point $t_0 \in \mathbb{R}$ then there exists a straight line which lies entirely below the graph of g. For any such line we can consider the closed upper half-plane defined as follows: If the line is the graph of the function $x \mapsto tx - c$ then the closed upper half-plane is given by

$$
H_{t,c} = \{(x, h) \mid h \geqslant tx - c\}. \tag{7.4}
$$

The intersection of all closed half-planes containing the epigraph of g is called the **closed convex hull** of the epigraph of g:

$$
\overline{\mathrm{co}}(\mathrm{epi}(g)) = \bigcap_{\{(t,c)\,\mid\, g(x') \geqslant tx' - c\,\forall x' \in D(g)\}} H_{t,c}. \tag{7.5}
$$

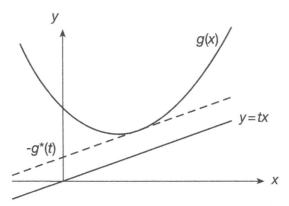

Figure 7.2 Geometric interpretation of the Legendre transform.

Another way to understand this formula is as follows: a point (x, h) belongs to $\overline{co}(\text{epi}(g))$ if and only if it lies above every straight line $y = tx - c$ which lies entirely below the graph of g. In general, the closed convex hull of a subset $A \subset \mathbb{R}^2$ is the smallest closed convex set containing A. This concept can also be defined in higher dimensions.

Lemma 7.3 *The closed convex hull of the epigraph of a function $g : \mathbb{R}^k \to \mathbb{R}$ is convex. Moreover, if g is a convex function, the closed convex hull of* $\text{epi}(g)$ *coincides with the closure of* $\text{epi}(g)$.

Proof. The closed convex hull of the epigraph is convex because the intersection of convex sets is obviously convex and closed half-planes are convex. Similarly, the closed convex hull is also closed. Now suppose that g is a convex function. Then it is clear that the epigraph of g is contained in its closed convex hull. Suppose that the point (x_0, h_0) is not in the closure of the epigraph of g. We shall prove that there is a line separating (x_0, h_0) from the epigraph.

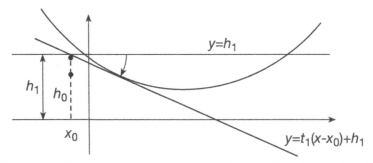

Figure 7.3 Rotating the line $y = h_1$ until it touches the graph of g.

Since $(x_0, h_0) \notin \overline{\text{epi}(g)}$, $h_0 < g(x_0)$. Choose $h_1 \in (h_0, g(x_0))$ and consider the horizontal line through the point (x_0, h_1). It is given by the equation

$y = h_1$. We are going to turn this line until it touches the graph of g (figure 7.3).

It is intuitively obvious that this is possible but to prove it requires some work. We omit this here; but for a complete proof of a more general statement, see appendix C. Let the resulting line be given by the equation $y = t_1(x - x_0) + h_1$. Then the epigraph of g lies in the closed upper half-plane bounded by the line $y = t_1(x - x_0) + h_1$ while the point (x_0, h_0) lies strictly below this line. It follows that (x_0, h_0) does not lie in the closed convex hull of epi(g). ∎

Theorem 7.2 *Let $g : \mathbb{R}^k \to \mathbb{R} \cup \{+\infty\}$ be a function with convex essential domain D. Assume that $g^*(\vec{t}_0) < \infty$ for at least one point $\vec{t}_0 \in \mathbb{R}^k$. Then the epigraph of g^{**} is given by the closed convex hull of the epigraph of g. In particular, if g is convex then $g^{**}(\vec{x}) = g(\vec{x})$ for all $\vec{x} \in$ int(D).*

Proof. We only consider the 1-dimensional case: $k = 1$. The last statement follows from the first and lemma 7.3. To prove the first statement note that, by the definition of $g^*(t)$, $tx_0 - g(x_0) \leqslant g^*(t)$ for all $t \in \mathbb{R}$. This implies $g(x_0) \geqslant tx_0 - g^*(t)$ and hence $g(x_0) \geqslant g^{**}(x_0)$. Thus, epi($g$) $\subset$ epi(g^{**}). To complete the proof it now suffices to prove that if a point (x_0, h_0) is separated from the epigraph of g by a line $y = ax + b$ then it is also separated from the epigraph of g^{**} by that line. But if (x_0, h_0) is separated from epi(g) by the line $y = ax + b$ then $g(x) \geqslant ax + b$ for all $x \in \mathbb{R}$. This implies that $ax - g(x) \leqslant -b$ and hence $g^*(a) \leqslant -b$. This in turn implies that $g^{**}(x) = \sup_{t \in \mathbb{R}}[tx - g^*(t)] \geqslant ax - g^*(a) \geqslant ax + b$ so that the line $y = ax + b$ also separates (x_0, h_0) from epi(g^{**}). ∎

We have seen in theorem 7.1 that convex functions are continuous. It is easy to see that they need not be differentiable; for example the function $f(x) = |x|$ for $x \in \mathbb{R}$ is not differentiable at 0. One-sided directional derivatives do always exist, however.

Theorem 7.3 *Let $D \subset \mathbb{R}^k$ be convex and let $g : D \to \mathbb{R}$ be a convex function. Let $\vec{x} \in D$ and let $\vec{v} \in \mathbb{R}^k$ be a non-zero vector such that $\vec{x} + \lambda\vec{v} \in D$ for $\lambda > 0$ small enough. Then the **directional derivative** of g at $\vec{x}$ in the direction of $\vec{v}$, defined by*

$$\partial_{\vec{v}} g(\vec{x}) = \lim_{\lambda \downarrow 0} \frac{g(\vec{x} + \lambda\vec{v}) - g(\vec{x})}{\lambda} \tag{7.6}$$

exists and it satisfies $-\partial_{-\vec{v}} g(\vec{x}) \leqslant \partial_{\vec{v}} g(\vec{x})$ if $\vec{x} \in$ int(D).

Proof. We first show that $[g(\vec{x} + \lambda\vec{v}) - g(\vec{x})]/\lambda$ is an increasing function of $\lambda > 0$. Indeed, by the definition of convexity, $g(\vec{x} + \lambda_1\vec{v}) \leqslant (1 - \frac{\lambda_1}{\lambda_2})g(\vec{x}) + \frac{\lambda_1}{\lambda_2}g(\vec{x} + \lambda_2\vec{v})$ if $0 < \lambda_1 < \lambda_2$. It follows that

$$\frac{g(\vec{x} + \lambda_1\vec{v}) - g(\vec{x})}{\lambda_1} \leqslant \frac{g(\vec{x} + \lambda_2\vec{v}) - g(\vec{x})}{\lambda_2}. \tag{7.7}$$

It now suffices to show that $[g(\vec{x}+\lambda\vec{v})-g(\vec{x})]/\lambda$ is bounded below to conclude that the limit (7.6) is finite. This need not be the case if $\vec{x}$ is a boundary point, but if $\vec{x} \in \text{int}(D)$ we can choose $\mu > 0$ small enough so that $\vec{x} - \mu\vec{v} \in D$ and we have again, by convexity,

$$-\frac{g(\vec{x} - \mu\vec{v}) - g(\vec{x})}{\mu} \leqslant \frac{g(\vec{x} + \lambda\vec{v}) - g(\vec{x})}{\lambda} \tag{7.8}$$

for all $\lambda > 0$. Taking $\lambda \downarrow 0$ and $\mu \downarrow 0$ we find that $-\partial_{-\vec{v}} g(\vec{x}) \leqslant \partial_{\vec{v}} g(\vec{x})$. ∎

REMARK 7.2: *Right- and left-hand derivatives in one dimension.*
Note that in one dimension the directional derivatives are the left- and right-hand derivatives defined by

$$g'_+(x_0) = \lim_{x \downarrow x_0} \frac{g(x) - g(x_0)}{x - x_0}$$

$$g'_-(x_0) = \lim_{x \uparrow x_0} \frac{g(x) - g(x_0)}{x - x_0}. \tag{7.9}$$

To be precise, $g'_+(x_0) = \partial_{+1} g(x_0)$ and $g'_-(x_0) = -\partial_{-1} g(x_0)$. The last statement of theorem 7.3 therefore says that $g_-(x_0) \leqslant g_+(x_0)$.

We finally give a necessary and sufficient condition for a twice continuously differentiable function to be convex. In one dimension this condition says that the second derivative must be non-negative. In higher dimensions the Hessian must be positive definite. Remember that a matrix A is called **positive definite** if $\langle \vec{v}, A\vec{v} \rangle \geqslant 0$ for all vectors $\vec{v} \in \mathbb{R}^k$.

Theorem 7.4 *Let $D \subset \mathbb{R}^k$ be convex open region and assume that the function $g : D \to \mathbb{R}$ has continuous second derivatives. Then g is convex if and only if the Hessian, that is the matrix of second-order partial derivatives,*

$$\partial^2 g = \begin{pmatrix} \frac{\partial^2 g}{\partial x_1^2} & \cdots & \frac{\partial^2 g}{\partial x_1 \partial x_k} \\ \cdot & & \cdot \\ \cdot & & \cdot \\ \cdot & & \cdot \\ \frac{\partial^2 g}{\partial x_k \partial x_1} & \cdots & \frac{\partial^2 g}{\partial x_k^2} \end{pmatrix} \tag{7.10}$$

is positive definite.

Proof. Assume first that g is convex. Adding the two convexity inequalities

$$g(\vec{x} + \mu\vec{v}) \leqslant \frac{\lambda - \mu}{\lambda} g(\vec{x}) + \frac{\mu}{\lambda} g(\vec{x} + \lambda\vec{v})$$

and

$$g(\vec{x} + \lambda\vec{v}) \leqslant \frac{\mu}{\lambda} g(\vec{x} + \mu\vec{v}) + \frac{\lambda - \mu}{\lambda} g(\vec{x} + (\lambda + \mu)\vec{v})$$

one obtains, for $\lambda > \mu > 0$,

$$g(\vec{x} + \mu\vec{v}) - g(\vec{x}) \leqslant g(\vec{x} + (\lambda + \mu)\vec{v}) - g(\vec{x} + \lambda\vec{v}). \tag{7.11}$$

Dividing by μ and taking $\mu \to 0$ it follows that $\partial_{\vec{v}}g(\vec{x}) \leqslant \partial_{\vec{v}}g(\vec{x} + \lambda\vec{v})$. Thus $\partial_{\vec{v}}g(\vec{x} + \lambda\vec{v})$ is increasing in λ, in other words,

$$\frac{\mathrm{d}^2}{\mathrm{d}\lambda^2}g(\vec{x} + \lambda\vec{v}) \geqslant 0. \tag{7.12}$$

It is a simple calculation to show that the left-hand side equals

$$\frac{\mathrm{d}^2}{\mathrm{d}\lambda^2}g(\vec{x} + \lambda\vec{v}) = \langle \vec{v}, \partial^2 g(\vec{x} + \lambda\vec{v})\vec{v}\rangle.$$

Positivity of the latter for all vectors $\vec{v}$ means that the matrix $\partial^2 g(\vec{x})$ is positive definite. Conversely, if the matrix (7.10) is positive definite then equation (7.12) holds. If $\vec{x}_1, \vec{x}_2 \in D$, take $\vec{v} = \vec{x}_2 - \vec{x}_1$ and $\vec{x} = \vec{x}_1$. By the mean-value theorem, for any given $\lambda > 0$ there exists a $\mu \in (0, \lambda)$ such that $[g(\vec{x} + \lambda\vec{v}) - g(\vec{x})]/\lambda = \partial_{\vec{v}}g(\vec{x} + \mu\vec{v})$. As the right-hand expression is increasing in μ we conclude that $\partial_{\vec{v}}g(\vec{x}) \leqslant [g(\vec{x} + \lambda\vec{v}) - g(\vec{x})]/\lambda \leqslant \partial_{\vec{v}}g(\vec{x} + \lambda\vec{v})$. This implies that the expression $[g(\vec{x} + \lambda\vec{v}) - g(\vec{x})]/\lambda$ is increasing in λ. Indeed, suppose $\lambda' > \lambda$. Then

$$\frac{g(\vec{x} + \lambda\vec{v}) - g(\vec{x})}{\lambda} \leqslant \partial_{\vec{v}}g(\vec{x} + \lambda\vec{v}) \leqslant \frac{g(\vec{x} + \lambda'\vec{v}) - g(\vec{x} + \lambda\vec{v})}{\lambda' - \lambda}$$

which implies

$$\frac{g(\vec{x} + \lambda\vec{v}) - g(\vec{x})}{\lambda} \leqslant \frac{g(\vec{x} + \lambda'\vec{v}) - g(\vec{x})}{\lambda'}.$$

In particular, we have $[g(\vec{x} + \lambda\vec{v}) - g(\vec{x})]/\lambda \leqslant g(\vec{x} + \vec{v}) - g(\vec{x})$ for $\lambda \in (0, 1)$, which is the convexity inequality. ∎

REMARK 7.3: *Positivity of g'' in one dimension.*
In one dimension, theorem 7.4 simply states that a twice differentiable function $g : I \to \mathbb{R}$ on an open interval I is convex if and only if $g''(x) \geqslant 0$ for all $x \in I$.

CHAPTER 8

Thermodynamic Potentials

After the preliminaries about convex functions in chapter 7, we can now define the **free energy density** $f(v,T)$ as minus the Legendre transform of $u(s,v)$ with respect to the variable s:

$$f(v,T) = -\sup_s \{Ts - u(s,v)\}. \tag{8.1}$$

It is easy to see that the function $u(s,v)$ is convex so that we can use theorem 7.2 to invert the Legendre transform and write

$$u(s,v) = \sup_T \{Ts + f(v,T)\}. \tag{8.2}$$

Note that we can also write

$$\boxed{f(v,T) = \inf_u \{u - Ts(u,v)\}} \tag{8.3}$$

that is, $-(1/T)f(v,T)$ is the Legendre transform of $-s(u,v)$. Inverting this relation we obtain

$$s(u,v) = \inf_{T>0} \frac{1}{T}\{u - f(v,T)\}. \tag{8.4}$$

The free energy density is an important quantity in thermodynamics because in many experimental situations it is easier to control the temperature than the internal energy. Note that the supremum in (8.1) is attained at the value of s satisfying (6.8) so that it is legitimate to denote the variable T in the free energy density as the temperature. We can therefore write

$$f(v,T) = u(s(v,T),v) - Ts(v,T), \tag{8.5}$$

where $s(v,T)$ is defined as the solution of (6.8). Alternatively,

$$f(v,T) = u(v,T) - Ts(u(v,T),v) \tag{8.6}$$

where $u(v, T)$ is the solution of

$$\frac{1}{T} = \left(\frac{\partial s}{\partial u}\right)_v. \tag{8.7}$$

Differentiating (8.5) w.r.t. to v we obtain, using equations (6.8) and (6.9),

$$\left(\frac{\partial f}{\partial v}\right)_T (v, T) = \left(\frac{\partial u}{\partial v}\right)_s (s(v, T), v) = -p(v, T). \tag{8.8}$$

Similarly we have

$$\left(\frac{\partial f}{\partial T}\right)_v = -s(v, T), \tag{8.9}$$

so that we can write

$$\boxed{df = -p \, dv - s \, dT} \tag{8.10}$$

Note that this implies in particular that *for an isothermal process* from an equilibrium state α to an equilibrium state β, the work done per particle w equals

$$w = f(\beta) - f(\alpha). \tag{8.11}$$

Thus, the free energy plays a role similar to the potential energy in mechanics: given the work done in an isothermal process, (8.11) determines the new equilibrium state.

In some cases it is convenient to work in terms of the variables s and p. We then take a Legendre transform with respect to v and define the **enthalpy** $h(s, p)$ by

$$\boxed{h(s, p) = \inf_v \{p \, v + u(s, v)\}} \tag{8.12}$$

It is easy to prove that

$$dh = \delta q + v \, dp. \tag{8.13}$$

This potential is useful for processes occurring at constant pressure since in that case the change in h equals the heat absorbed by the system. The enthalpy is therefore especially useful in chemistry since many chemical reactions take place in an open test tube. We shall encounter examples of the use of this thermodynamic potential in chapters 11 and 14. Another important thermodynamic potential is the **Gibbs function** or **Gibbs free energy density** $g(p, T)$ defined as the Legendre transform of $u(s, v)$ with respect to both variables:

$$\boxed{g(p, T) = \inf_{s,v} \{p \, v - T \, s + u(s, v)\}} \tag{8.14}$$

Using again theorem 7.2, now in the case of $k = 2$ that is for functions of two variables, one can also invert this double Legendre transform to obtain $u(s, v)$:

$$u(s, v) = \sup_{p,T} \{T \, s - p \, v + g(p, T)\}. \tag{8.15}$$

The differential of g is given by

$$\boxed{\mathrm{d}g = -s\,\mathrm{d}T + v\,\mathrm{d}p} \tag{8.16}$$

We shall encounter the Gibbs function in chapter 9 when we discuss phase transitions.

Finally, we introduce one other important thermodynamic function. This function will be of importance for systems with a variable number of particles. In that case it is better to normalize all extensive variables with respect to the volume instead of the particle number. Thus, for instance, the internal energy per unit volume is given by $\rho u(v\tilde{s}, v) = \tilde{u}(\tilde{s}, \rho)$ and the free energy per unit volume by $\tilde{f}(\rho, T) = \rho f(\rho^{-1}, T)$. We now define the **grand canonical potential** $\omega(\mu, T)$ by

$$
\begin{aligned}
\omega(\mu, T) &= -\sup_{\rho, \tilde{s}} \left[\rho\,\mu + T\,\tilde{s} - \tilde{u}(\tilde{s}, \rho) \right] \\
&= -\sup_{\rho, s} \left\{ \rho \left[\mu + T\,s - u(s, \rho^{-1}) \right] \right\}.
\end{aligned}
\tag{8.17}
$$

The new variable μ is called the **chemical potential.** Again it is standard to compute the differential of ω:

$$\mathrm{d}\omega = -\tilde{s}\,\mathrm{d}T - \rho\,\mathrm{d}\mu. \tag{8.18}$$

If we analyse the maximization problem in (8.17) in detail we obtain a remarkable result. Indeed, the conditions for a maximum are (6.8) and

$$\mu = u - T\,s - \rho^{-1}\left(\frac{\partial u}{\partial v}\right)_s = u - T\,s + \rho^{-1}p. \tag{8.19}$$

Inserting this into (8.17) we obtain

$$\boxed{\omega(\mu, T) = -p(\rho^{-1}(\mu, T), T)} \tag{8.20}$$

It will turn out that in many cases the function $\omega(\mu, T)$ can be obtained immediately from the microscopic description of a model. (This will be explained at the end of Part II.) It is therefore important to be able to invert the Legendre transform (8.17). This is not entirely trivial because it is not a consequence of the properties of $s(u, v)$ in chapter 6 that the function $\tilde{u}$ is convex as a function of $(\tilde{s}, \rho)$! However, it *is* convex in each of these variables *separately*. This is obvious in the case of $\tilde{s}$ and in the case of ρ it is obtained from the following.

Lemma 8.1 *If* $g : (0, +\infty) \to \mathbb{R}$ *is a convex function, then the function* $\rho \mapsto \rho\,g(\rho^{-1})$ *is convex as a function of* $\rho \in (0, \infty)$.

Proof. Suppose $\rho_1, \rho_2 > 0$ and $\lambda \in [0,1]$. Write $[\lambda\rho_1 + (1-\lambda)\rho_2]^{-1} = \alpha\rho_1^{-1} + (1-\alpha)\rho_2^{-1}$, where $\alpha = \lambda\rho_1/[\lambda\rho_1 + (1-\lambda)\rho_2]$ as can easily be seen. Then, if $\rho = \lambda\rho_1 + (1-\lambda)\rho_2$,

$$
\begin{aligned}
\rho g(\rho^{-1}) &\leqslant \rho\alpha g(\rho_1^{-1}) + \rho(1-\alpha)g(\rho_2^{-1}) \\
&= \lambda\rho_1 g(\rho_1^{-1}) + (1-\lambda)\rho_2 g(\rho_2^{-1}).
\end{aligned}
$$

∎

Lemma 8.2 *The pressure is a non-increasing function of the (specific) volume at constant temperature, that is:*

$$
\left(\frac{\partial p}{\partial v}\right)_T \leqslant 0 \tag{8.21}
$$

whenever this derivative exists.

Proof. This follows immediately from equation (8.8) and the fact that $f(v,T)$ is convex as a function of v. This fact is somewhat tricky to prove. It goes as follows: Suppose $v = cv_1 + (1-c)v_2$ with $c \in (0,1)$. Then we can subdivide s in a similar way:

$$
\begin{aligned}
f(v,T) &= -\sup_s [T\,s - u(s,v)] \\
&= -\sup_{s_1,s_2} \big[T(cs_1 + (1-c)s_2) \\
&\qquad\qquad - u(cs_1 + (1-c)s_2, cv_1 + (1-c)v_2)\big] \\
&\leqslant -\sup_{s_1,s_2} [c(Ts_1 - u(s_1,v_1)) + (1-c)(Ts_2 - u(s_2,v_2))] \\
&= -c\sup_{s_1}[Ts_1 - u(s_1,v_1)] - (1-c)\sup_{s_2}[Ts_2 - u(s_2,v_2)] \\
&= cf(v_1,T) + (1-c)f(v_2,T).
\end{aligned}
$$

(Incidentally, note that f is concave as a function of T!)   ∎

Note that this is indeed reasonable since if $\partial p/\partial v$ were positive for some values of v and T then the system would not be stable against a decrease in volume. Now,

$$
\omega(\mu,T) = -\sup_{\rho\geqslant 0}\{\rho\mu - \rho f(\rho^{-1},T)\}, \tag{8.22}
$$

and by lemmas 8.1 and 8.2, $\rho f(\rho^{-1},T)$ is convex as a function of ρ. We conclude that

$$
\boxed{\rho f(\rho^{-1},T) = \sup_\mu\{\rho\mu + \omega(\mu,T)\}} \tag{8.23}
$$

The internal energy can then be determined from (8.2):

$$u(s,v) = \sup_{\mu,T}\{T\,s + \mu + v\,\omega(\mu,T)\}. \tag{8.24}$$

Alternatively we can determine the entropy density from (8.4):

$$s(u,v) = \inf_{\mu,T} \frac{1}{T}\{u - \mu - v\,\omega(\mu,T)\}. \tag{8.25}$$

CHAPTER 9

Phase Transitions

Consider again the $p - V$ diagram for a general fluid given as figure 1 in the introduction. In the region under the broken curve, liquid and vapour (gas at temperatures below T_c is usually called vapour) coexist. As the volume is increased in this region the pressure remains constant and liquid vaporizes (figure 9.1).

It follows that the pressure at which liquid and vapour coexist is a function of the temperature alone. If the temperature of the liquid is raised while keeping the pressure above the liquid constant, it will no longer be in equilibrium and will evaporate. If we want to keep it in equilibrium, we must increase the pressure. The equilibrium pressure is therefore an increasing function of the temperature. The graph of this function is the **line of coexistence** of liquid and vapour shown in figure 9.2. Above this line, only liquid exists in equilibrium while below the curve only vapour exists in equilibrium. This argument also shows that by lowering the pressure above a liquid its boiling point decreases. This is an important method for obtaining low temperatures (refrigeration). For example, by pumping away the vapour above liquid helium-3 (^{3}He) (a rare isotope of helium), one can reduce the temperature by an order of magnitude, from approximately 3 K to 0.3 K.

Another look at the $p - V$ diagram shows that the coexistence curve must end at a maximum temperature T_c called the **critical temperature.** This point of the coexistence curve $p_{eq}(T)$ corresponds to one single point (p_c, V_c) in the $p-V$ diagram, the **critical point.** At this point remarkable things happen: the so called **critical phenomena.** For example, the isothermal compressibility κ_T (see equation (4.14)) diverges at the critical point. Also the specific heat c_V diverges. This means that the free energy is not twice differentiable at the critical point. One therefore speaks of a **second-order phase transition.** The degree of divergence can be expressed in terms of **critical exponents.** It turns out that κ_T and c_V behave near the critical point asymptotically as

$$\kappa_T \sim \mathcal{K} \, |T - T_c|^{-\gamma} \tag{9.1}$$

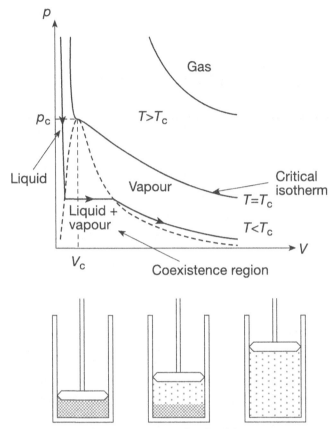

Figure 9.1 Vaporization of a fluid by expansion.

and

$$c_V \sim \mathcal{A}\,|T - T_c|^{-\alpha}. \tag{9.2}$$

The critical exponents γ and α are independent of the substance. This remarkable fact is called **universality.** There are also simple relations between the various critical exponents: so-called **scaling laws.** Although these phenomena are very interesting, their explanation is not easy to understand and we shall not consider them any further here. (They have only recently come to be understood to any degree of satisfaction. The explanation involves the so called **renormalization group.** For an introduction to this subject, see for example Ma (1971), Wilson and Kogut (1974), Pfeuty & Toulouse (1977), or Binney *et al.* (1992).)

Note that for $T > T_c$ the gas cannot be condensed into liquid no matter how high the pressure! This is no problem for water since $T_c(H_2O) = 647\,\mathrm{K} = 374\,^\circ\mathrm{C}$, but it does present a problem for the liquefaction of air for instance: $T_c(O_2) = 154\,\mathrm{K} = -119\,^\circ\mathrm{C}$ and $T_c(N_2) = 126\,\mathrm{K} = -147\,^\circ\mathrm{C}$. The problem is even greater for helium. Normal helium-4 (^{4}He) has a critical temperature of

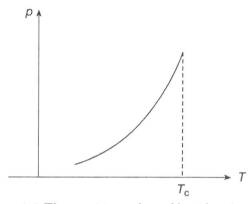

Figure 9.2 The coexistence line of liquid and vapour.

5.2 K and the isotope ^{3}He has an even lower critical temperature of 3.3 K. To liquefy these latter gases, one needs to attain very low temperatures therefore. How this is done will be discussed further in chapter 11.

When the temperature of a liquid is decreased further it eventually solidifies. (Except helium, which stays liquid under normal pressure. It has to be pressurized as well as cooled to solidify: see chapter 12.) The complete picture in a $p - V - T$ diagram looks as in figure 9.3. Note that there is no critical point for the liquid-solid transition. This means that if the temperature is above T_c then the gas solidifies without condensing first into a liquid.

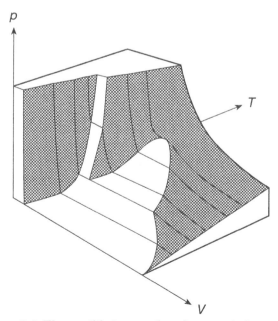

Figure 9.3 The equilibrium surface for a typical material.

A more convenient picture is the $p - T$ diagram with the various coexistence lines. Figure 9.4 shows a typical example. This is called the **phase diagram** for obvious reasons. More complicated phase diagrams can be found in chapter 12.

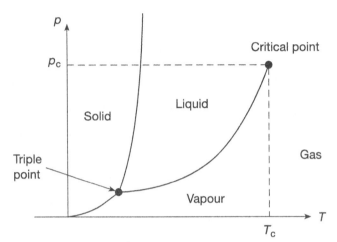

Figure 9.4 The phase diagram of a typical stable material.

Note that there is a unique point where all three phases are in equilibrium; this is the **triple point**. This point marks the lowest pressure at which the liquid can exist. For example, the triple point of carbon dioxide (CO_2) lies at 5.11 bar and 216.8 K. At a normal pressure of 1 bar, therefore, if the temperature is lowered the gas does not liquefy but instead becomes a solid called **dry ice**. The triple point of water lies at 0.006 bar and 273.16 K. It can happen that the humidity of the air is very low so that the *partial pressure* of water is below the triple point pressure. In that case, if the temperature drops below zero (degrees Celsius), the water vapour in the air solidifies. This is called **hoar frost**. Note that the triple point temperature is slightly *above* the normal melting temperature: the melting temperature decreases with increasing pressure (see figure 12.1). This is a peculiarity of water.

During the transition from gas to liquid and from liquid to solid a certain amount of heat is released. The **latent heat** ℓ is the amount of heat needed to melt respectively evaporate a mole of solid respectively liquid. There is an accompanying change of entropy per mole:

$$\Delta s = \frac{\ell}{T}. \tag{9.3}$$

(N.B. The specific entropy is often given per mole (or per unit mass) instead of per molecule.) If g is the Gibbs free energy density then, by equation (8.16),

$$\left(\frac{\partial g}{\partial T}\right)_p = -s \text{ and } \left(\frac{\partial g}{\partial p}\right)_T = v. \tag{9.4}$$

These quantities change discontinuously as we cross the coexistence lines of the previous figure. One therefore speaks of a **first order phase transition.** (At the critical point the second derivatives of g are discontinuous.) We can now use equation (8.16) to derive an equation for the coexistence line. If n_1 moles of liquid coexist with n_2 moles of vapour then

$$g = \frac{n_1}{n_1 + n_2} g_1 + \frac{n_2}{n_1 + n_2} g_2 \tag{9.5}$$

where g_1 is the Gibbs free energy density (per mole) for the liquid and g_2 that for the vapour. In equilibrium, g will attain its minimum value for the given temperature T and pressure p by the definition (8.14). This means that when the liquid and the vapour coexist the values of n_1 and n_2 adjust so that $dg = 0$. This implies with equation (9.5) that

$$g_{\text{liquid}} = g_{\text{vapour}} \tag{9.6}$$

Changing T and p along the coexistence line now leads to

$$\left(\frac{\partial g_\ell}{\partial T} \right)_p dT + \left(\frac{\partial g_\ell}{\partial p} \right)_T dp = \left(\frac{\partial g_v}{\partial T} \right)_p dT + \left(\frac{\partial g_v}{\partial p} \right)_T dp. \tag{9.7}$$

Hence

$$\boxed{\frac{dp}{dT} = \frac{\Delta s}{\Delta v} = \frac{\ell}{T \Delta v}} \tag{9.8}$$

This is the **Clausius-Clapeyron equation** for the coexistence of phases. If we assume that the vapour satisfies the ideal gas law we obtain, since $\Delta v \approx v_{\text{vapour}}$,

$$p = p_0 \exp \left[\frac{\ell}{R_0} \left(\frac{1}{T_0} - \frac{1}{T} \right) \right]. \tag{9.9}$$

The vapour pressure therefore varies exponentially with the inverse temperature.

As remarked before, the laws of thermodynamics do not permit a derivation of the fundamental equation and it is therefore also impossible to derive equations of state from these laws. In order to derive an equation of state, one needs a microscopic description of the system. The first to derive an equation of state in the case of a liquid-gas transition was J. D. van der Waals. He used a very simple argument to include in a rudimentary way the effect of an interaction (force) between the molecules of a gas. Indeed, he argued that in the case of high densities and pressures the ideal gas law needs to be modified to allow for two effects: the finite size of the molecules and the attraction between different molecules. The finite size of molecules means that the volume actually available to a molecule is reduced to $v - b$, where b is the molecular volume. Including this effect into the ideal gas law we obtain $p(v - b) = k_B T$. The attraction between the molecules causes them to decelerate before they hit the container wall. This results in a reduction of the pressure. The force on an individual molecule will be proportional to the density ρ, and since the number of molecules striking the wall per unit time is also proportional to the density we may expect the decrease in pressure to be proportional to ρ^2. This leads to the following modified gas law:

$$p = \frac{k_B T}{v - b} - \frac{a}{v^2} \tag{9.10}$$

This equation, the **van der Waals equation of state**, actually predicts condensation. To see this we draw some typical isotherms in figure 9.5.

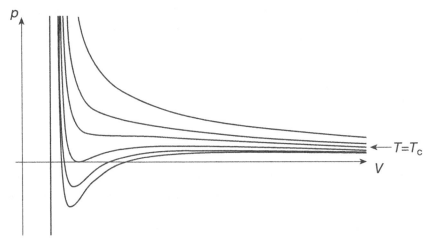

Figure 9.5 Typical isotherms for the van der Waals equation.

It is easy to see that the isotherms of equation (9.10) are not monotonically decreasing for low temperatures, that is, there is a temperature

$$T_c = \frac{8a}{27 \, k_B \, b}, \tag{9.11}$$

such that for $T < T_c$, $(\partial p/\partial v)_T > 0$. This is in disagreement with lemma 8.2, so that the van der Waals equation is unstable. Maxwell proposed the following remedy for this problem. He suggested replacing part of the isotherm by a horizontal line in such a way that the areas A_1 and A_2 are equal (figure 9.6). This horizontal line segment can be interpreted as the transition from vapour to liquid and the temperature T_c as the critical temperature. One can determine a and b from a fit of the equation of state to experimental values for temperatures above T_c or from a comparison with the attractive force between the molecules or atoms of the gas. Inserting these into (9.11) then gives an estimate of the critical temperature. In the case of helium, $a \approx 0.004$ J m^3 mol^{-2} and $b \approx 2.0 \times 10^{-5}$ m^3 mol^{-1}. Inserting these values into equation (9.11) (replacing k_B by R_0 to convert to values per mole) one finds $T \approx 7$ K. This is a reasonable approximation to the actual value $T_c = 5.2$ K, and it gave

Kamerlingh Onnes a good indication as to how far helium had to be cooled when he first succeeded in liquefying it in 1902. (Modern values for a and b are: $a = 0.00346\,\mathrm{J\,m^3\,mol^{-2}}$ and $b = 2.38 \times 10^{-5}\,\mathrm{m^3\,mol^{-1}}$.)

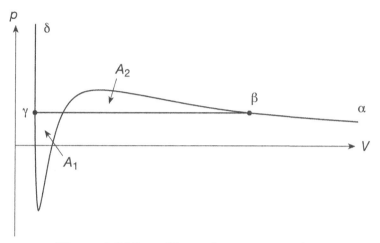

Figure 9.6 Maxwell's equal area construction.

To justify **Maxwell's equal area construction** we consider again the Gibbs free energy density: Integrating equation (8.16) along the isotherm from α to β we get

$$g_\beta - g_\alpha = \int_\alpha^\beta v(p)\,\mathrm{d}p. \tag{9.12}$$

(Note that this integral is with respect to the vertical coordinate in figure 9.6, so that it represents the area to the left of the graph between α and β.) Similarly,

$$g_\gamma - g_\alpha = \int_\alpha^\gamma v(p)\,\mathrm{d}p, \tag{9.13}$$

where again the integral is taken along the van der Waals isotherm from α to γ. In this case, when the graph bends upwards the area must be counted positive but when it bends downwards it must be counted negative. Hence

$$g_\gamma - g_\beta = \int_\beta^\gamma v(p)\,\mathrm{d}p = A_2 - A_1. \tag{9.14}$$

Thus, if $A_2 > A_1$ then β will be the stable equilibrium while if $A_1 > A_2$ then γ will be the stable equilibrium. If $A_1 = A_2$ then the two phases can coexist in equilibrium.

CHAPTER 10

Magnetism

So far, the only example of a thermodynamic system we have encountered has been a gas or liquid, where the thermodynamic variables are the volume and the pressure. Let us now consider another simple system which also exhibits a phase transition in certain cases: a simple magnet. Solids can display various kinds of magnetic behaviour: **diamagnetism, paramagnetism, ferromagnetism, antiferromagnetism**, and **ferrimagnetism**. We shall consider only paramagnetism and ferromagnetism in some detail.

Every solid becomes magnetized when it is placed in a magnetic field $\vec{H}$. This means that the internal microscopic magnetic dipole moments have a tendency to point in a certain direction with respect to the external field $\vec{H}$. The net magnetic moment per unit volume, that is, the vector sum of all internal magnetic moments divided by the volume (see equation (10.6)), is called the **magnetization** $\vec{m}$. If $\vec{m}$ has the same direction as the applied field then the material is called **paramagnetic**, if it is opposite to the field, the material is called **diamagnetic**. The common magnet has the property that it is magnetic ($\vec{m} \neq 0$) even when it is not in a magnetic field; it is **spontaneously magnetized.** This phenomenon is called **ferromagnetism.**

The quantity

$$\vec{B} = \mu_0 \left(\vec{H} + \vec{m} \right) \tag{10.1}$$

determines the force on a hypothetical moving charge in the material; it is called the **magnetic induction.** ($\mu_0 = 4\pi \cdot 10^{-7}\,\mathrm{H\,m^{-1}}$ is an absolute constant.) The quantity χ given by

$$\chi = \frac{\partial m}{\partial H} \tag{10.2}$$

is called the **susceptibility**. (In most cases the magnetization is in the same or opposite direction as the applied field so that χ is a scalar, but in general χ would be a tensor. We disregard this complication here.) The susceptibility depends on the temperature so that a thermodynamical description is needed. Indeed, we shall see that magnetism is a macroscopic cooperative phenomenon so that statistical mechanics is essential for its understanding from a microscopic point of view. Many paramagnetic salts, such as $CoCl_2$, $FeSO_4$, $GdSO_4$, Ga_2O_3, sodium, molybdenum, and cerium magnesium nitrate (CMN) (this material is often used for adiabatic demagnetization, a method for attaining very low temperatures), obey **Curie's Law**:

$$\chi(T) = \frac{C}{T} \tag{10.3}$$

to good precision down to very low temperatures. (Here C is a constant.) This makes these materials very suitable for the measurement of low temperatures.

The zero-field magnetization $m_0(T)$ of a ferromagnet also depends on the temperature (figure 10.1).

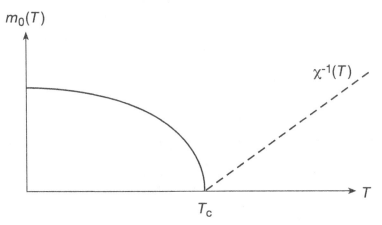

Figure 10.1 The permanent magnetization $m_0(T)$
and the inverse susceptibility of a ferromagnet.

In fact, above a certain critical temperature T_C, the **Curie temperature**, $m_0(T)$ disappears and the material becomes paramagnetic, satisfying the generalized Curie law or **Curie-Weiss law**:

$$\chi(T) = \frac{C}{T - T_C}. \tag{10.4}$$

The generic behaviour of $m_0(T)$ and the Curie-Weiss law (10.4) are combined in figure 10.1.

The isotherms in $m - H$ space are as in figure 10.2. This picture shows a distinct similarity with the $p - V$ diagram of a gas-liquid transition. Indeed, the appearance of a spontaneous magnetization is another example of a phase transition. The magnetic field H is the analogue of the pressure and the magnetization m is the analogue of the density ρ. We shall see that this analogy can be carried further when we consider the statistical mechanics of magnetism.

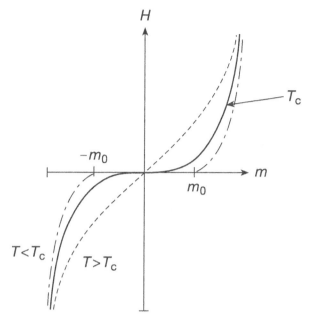

Figure 10.2 Magnetization versus magnetic field
for an ideal ferromagnet (no hysteresis).

REMARK 10.1: *Hysteresis.*
It should be noted that real ferromagnets exhibit a kind of memory called
hysteresis. This means that when a ferromagnetic material is magnetized
in an increasing external magnetic field it follows the magnetization curve of
figure 10.2, but when the external field is then removed the magnetization does
not fall to zero. The remnant magnetization only vanishes when an external
field is applied of opposite direction. This effect is caused by the fact that
a real magnet is not homogeneous but consists of small domains, which are
magnetized independently. See for example the book by Brailsford (1966).

The work differential for a magnetic system is given by

$$\delta W = -\mu_0 \, \vec{M} \cdot \mathrm{d}\vec{H}, \tag{10.5}$$

where

$$\vec{M} = \vec{m} \, V \tag{10.6}$$

is the total magnetic moment. (This follows from the fact that the energy of
a magnetic dipole $\vec{\mu}$ in a magnetic field $\vec{H}$ is given by $-\mu_0 \vec{\mu} \cdot \vec{H}$.)

REMARK 10.2: *Different forms of magnetic work.*
There are in fact two different expressions for the magnetic work differential
in use: (10.5) and

$$\delta W' = \mu_0 \, \vec{H} \cdot \mathrm{d}\vec{M}. \tag{10.7}$$

These expressions are not equivalent but refer to different situations. While equation (10.5) includes the increase in potential energy needed to bring the sample into the external magnetic field $\vec{H}$, equation (10.7) only concerns the work needed to increase the magnetization of the sample.

If we insert (10.5) into (3.1) we obtain

$$dU = \delta Q - \mu_0 \, M \, dH \tag{10.8}$$

and inserting this into (8.1) we find for the free energy density

$$d\tilde{f} = -\tilde{s} \, dT - \mu_0 \, m \, dH. \tag{10.9}$$

(In these formulas we have assumed that $\vec{M}$ and $\vec{H}$ have the same direction and can be considered scalars.) These are the main formulas we shall need in developing the statistical mechanics of magnetism.

CHAPTER 11

The Refrigerator and Liquefaction of Gases

The Refrigerator.

The common refrigerator makes use of the latent heat extracted from the environment when a liquid evaporates. (Clearly, this has to be a volatile liquid. Old refrigerators used the dangerous ammonia but this was replaced by chloro-fluoro carbons which are safe but have now also fallen into disrepute because of their harmfulness to the ozone layer.) The most efficient design of a refrigerator would be a Carnot engine in reverse. In the $p - V$ diagram this process is represented by inverting the arrows in figure 5.2. However, it is more convenient to represent it in a temperature-entropy diagram. Moreover, the working substance is now a liquid in equilibrium with its vapour. (This is usually called a **wet vapour**.) This means that the cycle lies in the coexistence region of the diagram in figure 11.1.

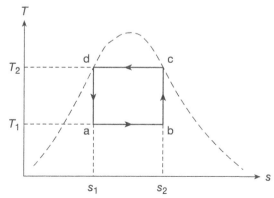

Figure 11.1 An ideal refrigerator cycle.

In this diagram the coexistence curve is indicated by a broken curve. Parts ab and cd of the cycle are isotherms, parts bc and da are adiabatic and quasi-static and therefore isentropic. During ab, some of the fluid evaporates and heat is extracted from the environment, i.e. inside the refrigerator. During bc, the fluid is adiabatically and quasi-statically compressed in a compressor. The temperature rises and the remaining liquid vaporizes to become saturated dry vapour at c. During cd, the fluid condenses in a condenser outside the refrigerator and during da the fluid expands and partially evaporates. The heat extracted per mole of fluid during the evaporation stage ab is given by

$$q_{ab} = T_1 \Delta s, \tag{11.1}$$

where $\Delta s = s_2 - s_1$. Similarly, during the condensation process an amount of heat $q_{cd} = T_2 \Delta s$ is given off. The net flow of heat during the cycle is therefore $q = q_{ab} - q_{cd} = (T_1 - T_2)\Delta s$. In a reversible cyclic process the internal energy does not change, so that by the first law (2.10), $q + w = 0$. The work done on the fluid is therefore $w = (T_2 - T_1)\Delta s$ and the coefficient of performance is

$$CP_{rev} = \frac{T_1}{T_2 - T_1} \tag{11.2}$$

in accordance with equation (5.7).

In fact the Carnot process is not practicable because two pumps are needed: one for compression and one for expansion. One therefore simplifies the design of the refrigerator and replaces the expansion cylinder with a **throttle valve** (figure 11.2). A throttle valve is a porous plug or simply a narrow part in the tube causing a pressure difference over the valve. There is *turbulence* in the liquid so that the throttle process is irreversible. We can nevertheless use thermodynamics to analyse the process provided that we consider the initial and final equilibrium states only. To a good approximation, no heat is exchanged with the surroundings and the pressures on either side of the valve are nearly constant in continuous operation, say p_{in} and p_{out}.

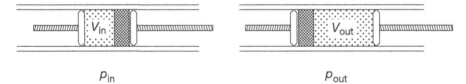

Figure 11.2 A throttle valve.

When a volume V_{in} of fluid is pushed through the valve, an amount of work $W_{in} = p_{in}V_{in}$ must be done to push it into the valve. On the other hand, when it exits on the other side, the fluid expands to a volume V_{out} and does work $W_{out} = p_{out}V_{out}$. As no heat is exchanged, the change in internal energy must equal the net amount of work performed on the fluid:

$$\Delta U = W_{in} - W_{out} = p_{in}V_{in} - p_{out}V_{out}. \tag{11.3}$$

Dividing by the number of moles, it follows that the specific enthalpy defined by equation (8.12) is constant during the process:

$$h = u + pv = \text{ constant.} \tag{11.4}$$

From equation (8.13), we find that there is a change of entropy during the throttling process given by

$$\Delta s = s(B) - s(A) = -\int_{p_A}^{p_B} \frac{v \, dp}{T}. \tag{11.5}$$

Note that in the case of an ideal gas one would have:

$$h_{\text{ideal gas}} = c_V T + R_0 T = c_p T \tag{11.6}$$

so that, for an ideal gas, a throttling process is isothermal and no work is performed. In general, however, work does need to be performed and the temperature drops in the process. The full refrigeration cycle is depicted in figure 11.3.

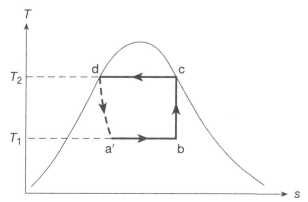

Figure 11.3 Refrigeration cycle with a throttle valve.

da' is the throttling process; it is indicated by a broken curve because it is not quasi-static. The liquid-vapour mixture is circulated by a pump as in figure 11.4, which also indicates the various stages of the process.

Values of the enthalpy and the entropy at saturation, that is on the co-existence curve, are tabulated for various substances. The coefficient of performance is then in practice computed as follows. The heat absorbed by the fluid during evaporation is given by

$$q_{a'b} = h_b - h_{a'} = h_b - h_d. \tag{11.7}$$

This follows immediately from equation (8.13) since during evaporation the pressure is constant. h_b is given by

$$h_b = (1 - x_b)h_\ell + x_b h_g, \tag{11.8}$$

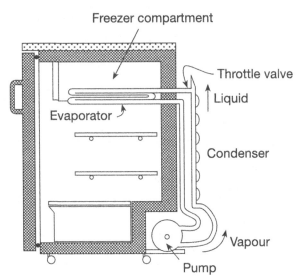

Figure 11.4 Schematic drawing of a refrigerator.

where x_b is the mole fraction of vapour at b and h_ℓ and h_g are the enthalpy densities of saturated liquid and vapour at the same pressure. x_b can be determined from the analogous formula for the entropy

$$s_b = s_c = (1 - x_b)s_\ell + x_b s_g, \qquad (11.9)$$

since s_c is the entropy density of a dry saturated vapour and is tabulated. The work per mole performed on the fluid during the entire process is given by

$$w = h_c - h_b. \qquad (11.10)$$

To see this notice that the total heat absorbed per mole by the fluid during a complete cycle is

$$q = q_{in} - q_{out} = (h_b - h_{a'}) - (h_c - h_d) = h_b - h_c \qquad (11.11)$$

since the throttling process is adiabatic (but *not isentropic*) and $h_{a'} = h_d$. As u is a function of state $\Delta u = 0$ for a cyclic process and hence $q + w = 0$. This yields equation (11.10).

It must finally be remarked that the above process is still not very practical. In fact, the preferred process is sketched in figure 11.5.

Evaporating the liquid completely from a to b' has two advantages. Firstly, the amount of heat extracted in a single cycle is larger and secondly, no liquid enters the compressor pump which leads to a smoother operation of the latter. The line c'c is an isobar: the pressure is constant.

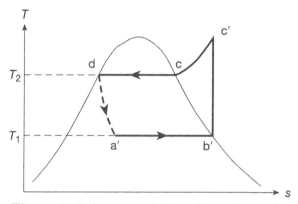

Figure 11.5 A more realistic refrigeration cycle.

Liquefaction of Gases.

We have seen that, if a gas is to be liquefied its temperature must be reduced below the critical temperature. For the final cooling to a wet vapour, a throttling process can be used, but this is not effective for the initial cooling of the gas. To see this, let us consider the throttling process again. If we consider h as a function of the parameters p and T we have a relation analogous to (4.11):

$$\left(\frac{\partial T}{\partial p}\right)_h \left(\frac{\partial p}{\partial h}\right)_T \left(\frac{\partial h}{\partial T}\right)_p = -1. \tag{11.12}$$

Using

$$\left(\frac{\partial h}{\partial T}\right)_p = T\left(\frac{\partial s}{\partial T}\right)_p = c_p \tag{11.13}$$

and equation (8.13) combined with the **Maxwell relation**

$$\left(\frac{\partial s}{\partial p}\right)_T = -\left(\frac{\partial v}{\partial T}\right)_p, \tag{11.14}$$

which follows from equation (8.16), we arrive at

$$\left(\frac{\partial T}{\partial p}\right)_h = \frac{1}{c_p}\left[T\left(\frac{\partial v}{\partial T}\right)_p - v\right]. \tag{11.15}$$

In general, this expression is negative at moderate to high temperatures but positive at low temperatures. (Note that $\Delta p < 0$.) For example, for a van der Waals fluid one obtains

$$\Delta T \approx \frac{1}{c_p}\left(\frac{2a}{k_B T} - b\right)\Delta p. \tag{11.16}$$

The temperature below which $\Delta T < 0$ is called the **inversion tempera-ture**. The lowering of temperature of a gas by throttling is called the **Joule-Thomson effect**. For some gases, in particular helium and hydrogen, the inversion temperature is so low that the gas has to be pre-cooled with another gas in a heat exchanger before it is throttled. The most common design of a liquefaction machine is due to Linde. Figure 11.6 illustrates schematically the operation of this machine.

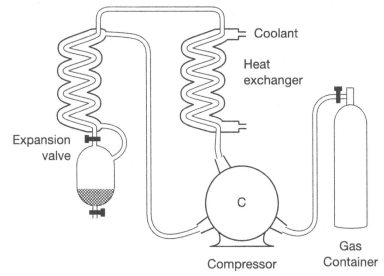

Figure 11.6 Linde's liquefaction apparatus.

The gas to be liquefied is first purified (not shown) and then led to a compressor C. The compressed vapour exits at the top and is pre-cooled in a heat exchanger. It then enters another heat exchanger where it is cooled further by the vapour which remains above the liquid in the liquid container. It is finally throttled by an expansion valve and enters the liquid chamber. After several circulations some of the vapour liquefies. The remaining vapour is recycled through the second heat exchanger and the compressor.

For attaining extremely low temperatures other methods have to be used. One method has already been mentioned: pumping at a ^{3}He bath. The lowest temperatures can be reached by means of adiabatic demagnetization, but for continuous operation one uses a **dilution refrigerator**.

This apparatus is based on the properties of a mixture of ^{3}He and ^{4}He, the two isotopes of helium. The phase diagram of such a mixture is sketched in figure 12.7. Below 0.87 K it separates into two phases: the concentrated phase, which is rich in ^{3}He, and the dilute phase, which is ^{3}He-poor. At very low temperatures, the concentrated phase is almost pure ^{3}He but the dilute

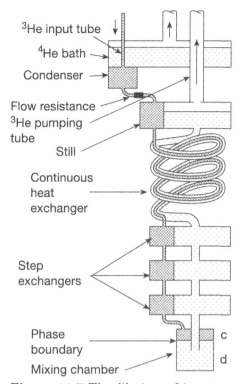

Figure 11.7 The dilution refrigerator.

phase still contains approximately 6.8% ^{3}He. The ^{4}He in the dilute phase is virtually inert so that this phase behaves like a rarefied gas of ^{3}He atoms (a Fermi gas: see chapter 23). In analogy with the cooling of a normal fluid by pumping away the vapour, in the dilution refrigerator lower temperatures are reached by forcing ^{3}He atoms from the concentrated phase to the dilute phase. Temperatures as low as 5 mK can be reached in this way. Figure 11.7 shows a schematic drawing of a dilution refrigerator.

In practice ^{3}He is forced from the concentrated phase, which floats on top because of its lower density, into the dilute phase by pumping on a ^{3}He bath called the still. Incoming ^{3}He gas (figure 11.7) is pre-cooled and liquefied in the condenser attached to a ^{4}He bath, which is maintained at approximately 1 K. The pressure of the ^{3}He is kept sufficiently high by a flow resistance so that condensation will take place. The fluid then enters the heat exchanger with the still, the continuous heat exchangers, and the step exchangers and finally the mixing chamber where the phase separation takes place. After crossing the phase boundary, the ^{3}He atoms return to the still. On their way through the heat exchangers they cool down the incoming fluid. The vapour above the fluid in the still is almost pure ^{3}He as a result of its lower condensation point. So, by pumping on the fluid one obtains a ^{3}He circulation. The pumping causes the fluid in the still to become poor in ^{3}He. This results in an osmotic pressure which forces ^{3}He up from the mixing chamber. Ironically, because the cooling power of the refrigerator is roughly proportional to the number of moles of ^{3}He that cross the phase boundary per second, it is advantageous to heat the still externally to increase the ^{3}He flux.

Examples of Phase Transitions

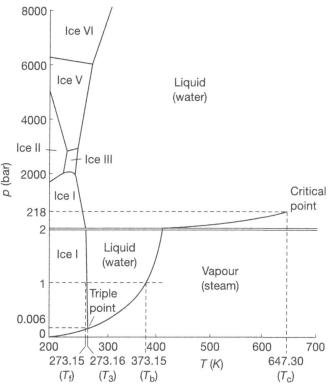

Figure 12.1 The phase diagram of water and ice.

The phase diagrams of most substances are in fact much more complicated than the simple picture in figure 9.2 suggests. In particular, the solid can usually exist in various phases depending on the pressure. These different phases correspond to different arrangements of the molecules with respect to one another. The phase diagram of **water and ice**, for example, is illustrated in figure 12.1.

The usual ice is called ice I, but under high pressures this is unstable and transforms into ice II or ice III, etc. These transformations occur at the bottom of glaciers. Note a peculiarity of water, namely that the melting temperature T_f decreases with increasing pressure.

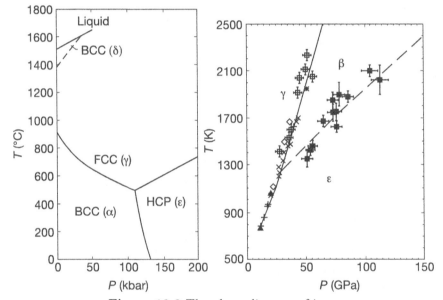

Figure 12.2 The phase diagram of iron.

The phase diagram of **iron** is important for the study of the Earth's core as well as for the manufacture of steel etc. The core of the Earth is thought to be mainly iron. There is an outer core of liquid iron and an inner core of solid iron. There is still no consensus, however, as to the particular solid phase of the inner core. This is due to the fact that the phase diagram for pressures as high as that at the inner core boundary (roughly 3.3 Mbar) cannot be determined experimentally yet, but has to be inferred from measurements at lower pressures and theoretical calculations. In fact, there is considerable uncertainty about the phase diagram even for pressures around 500 kbar. Figure 12.2 shows two parts of the phase diagram: the 'low-pressure' part is well-established; the high-pressure part is an extension proposed by Saxena *et al.* (1993).

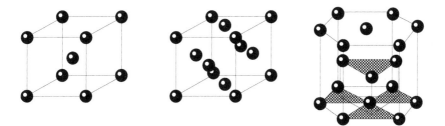

Figure 12.3 Different crystal structures.

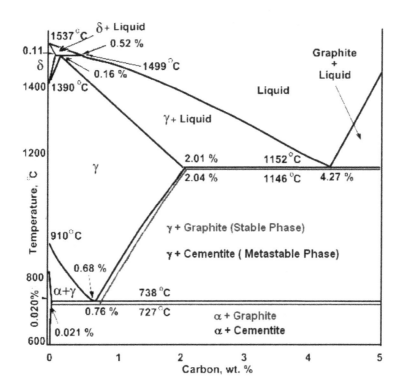

Figure 12.4 The phase diagram of steel.

Of the phases appearing in these diagrams, α-Fe and δ-Fe have a body-centred cubic structure, γ-Fe is face-centred cubic, and ϵ-Fe is hexagonal close-packed. This means that the atoms are arranged in regular arrays with repeating unit cubes as in figure 12.3.

The conjectured β-phase for pressures above 50 GPa is disputed; see for example Söderlind *et al* (1996). In addition, some evidence has been found for another phase transition at a pressure of about 2 Mbar. As the inner-core boundary pressure is approximately 3.3 Mbar, all this means that the structure of the solid iron core of the Earth is still uncertain.

For the manufacture of steel, which is a mixture of iron and carbon, the phase diagram of the mixture is more important. Figure 12.4 shows a simplified version of this diagram. For admixtures of carbon above 2%, the metal is cast iron. It can be seen that it has a lower melting point than steel. This means that it can be easily moulded. On the other hand, it is very hard and brittle. The lowest melting point is at a composition of 4.27% carbon. It is called the *eutectic composition* and melts at 1152 °C. It is this composition that is retained until cold. It leaves the steel making furnace in a state called *pig iron*. The face-centred α-lattice is more spacious than the body-centred γ-lattice and can therefore contain more atoms of carbon. γ-Fe with carbon in solution is called *austenite*, α-Fe with carbon in solution is called *ferrite*. Below 730 °C, austenite cannot exist in equilibrium. Upon slow cooling, steel with a carbon content less than 0.8% first becomes a mixture of austenite and iron and at 730 °C suddenly changes into a combination of ferrite and a structure called *pearlite*, which consists of thin plates of iron and iron carbide (Fe_3C). Austenitic steel can only be manufactured by rapid cooling, a process called **quenching**.

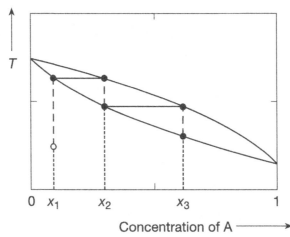

Figure 12.5 Phase diagram of two fully miscible liquids.

The phase diagram of **mixtures** in general is of great importance. For example, consider the phase diagram for a mixture of two fully miscible liquids A and B. We can plot the boiling point as a function of the composition x, the fractional amount of liquid A.

The vapour will be richer in the more volatile component. Assuming that this is A, the composition of the vapour is shifted to the right as indicated by the upper curve in figure 12.5. Now, when the mixture is brought to the boil, starting with a composition x_1 as shown, the vapour will be richer in A; so $x_2 > x_1$. The vapour can be drawn off and condensed. The liquid mixture thus obtained has a higher concentration x_2 of A. By repeating the procedure one gets mixtures of concentration x_3, etc., and eventually almost pure A is obtained. This process is called **fractional distillation.** Not all mixtures have a phase diagram as in figure 12.5, however, so that fractional distillation is not always possible. (For a large collection of phase diagrams for mixtures, see the book by Walas (1985).)

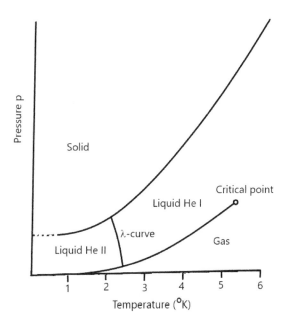

Figure 12.6 The phase diagram of helium (schematic).

A very interesting phase diagram is that of **helium**. It has the peculiarity that there is no solid phase for pressures below 26 atm. Instead, the liquid becomes **superfluid** for temperatures below the λ-**line** or λ-**curve** indicated in figure 12.6. Superfluidity means that the liquid flows without viscosity, that is, it can flow through the smallest opening without experiencing any resistance. Superfluid helium is called **He-II**. It has many extraordinary properties. See for example the book by Wilks (1970). The peculiar behaviour of helium is due to the quantum nature of this fluid. The atoms are so light that their zero-point motion is comparable to the thermal motion at low temperatures. This prevents helium from solidifying at normal pressure. In fact there is an important difference between the two isotopes of helium: ^{3}He and ^{4}He. The former is a Fermi liquid and does not become superfluid until the temperature has reached much lower values; the critical temperature for the superfluid transition is 2.6 mK. We shall consider ^{4}He in more detail in chapter 34.

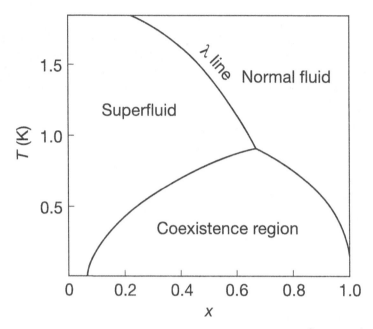

Figure 12.7 The phase diagram for liquid mixtures of ^{3}He and ^{4}He.

The **mixture of ^{3}He and ^{4}He** is also of great interest. Unlike other isotopic mixtures, these two isotopes cannot be mixed arbitrarily at low temperatures: below 0.87 K the mixture separates into a ^{3}He-rich phase and a ^{3}He-poor phase. The former floats on top of the latter due to its lower density. Note that the ^{3}He-poor phase (the dilute phase) still contains a small but non-zero fraction of ^{3}He, even as the temperature approaches the absolute zero. This is the basis of the operation of the **dilution refrigerator** which is commonly used for attaining very low temperatures; see chapter 11.

CHAPTER 13

Equations of State for Solids

The equation of state for a gas is given by the ideal gas law (1.5): $pv = R_0\,T$ if v is the volume per mole of the gas. In the case of solids, the effect of temperature is much less than for gases. In fact, the effect of the temperature is mainly to introduce vibrations in the solid: the individual atoms, ions, or molecules vibrate about their equilibrium positions which are on a regular lattice. (We consider crystalline solids.) Let us therefore write the free energy f as a sum of two terms, the internal energy due to cohesion u_c, and the free energy due to lattice vibrations f_{ph}. (The lattice vibrations are called **phonons**; see chapter 33.)

$$f = u_c + f_{ph}. \tag{13.1}$$

We shall see in Part III, when we discuss a model for phonons due to Debye, that the internal energy per mole of a 'phonon gas' can be written in the form

$$u_{ph} = 3R_0 T\, D\left(\frac{\Theta_D}{T}\right), \tag{13.2}$$

where the **Debye temperature** Θ_D depends only on the volume. Moreover, f_{ph} is also of this form:

$$f_{ph} = T\, f_D\left(\frac{\Theta_D}{T}\right). \tag{13.3}$$

By (8.5) and (8.9),

$$u_{ph} = f_{ph} - T\left(\frac{\partial f_{ph}}{\partial T}\right), \tag{13.4}$$

which implies that

$$f'_D(\lambda) = 3R_0 \frac{D(\lambda)}{\lambda}. \tag{13.5}$$

Now, by equation (8.8),

$$p = -\left(\frac{\partial f}{\partial v}\right)_T = -\frac{du_c}{dv} - f'_D\left(\frac{\Theta_D}{T}\right)\frac{d\Theta_D}{dv} = -\frac{du_c}{dv} - \frac{u_{ph}}{\Theta_D}\frac{d\Theta_D}{dv}. \qquad (13.6)$$

Introducing the **Debye-Grüneisen parameter**

$$\gamma_D = -\frac{d(\ln\Theta_D)}{d(\ln v)} \qquad (13.7)$$

we obtain the **Mie-Grüneisen equation of state**

$$p = -\frac{du_c}{dv} + \gamma_D\frac{u_{ph}}{v}. \qquad (13.8)$$

The cohesion term can be derived from an *Ansatz* (hypothesis) for the internal energy. A much used form for the interaction potential is the **Lennard-Jones potential** given by

$$u_{coh} = -\frac{a}{r^m} + \frac{b}{r^n} \qquad (13.9)$$

where r is the average interatomic distance. It is shown in figure 13.1.

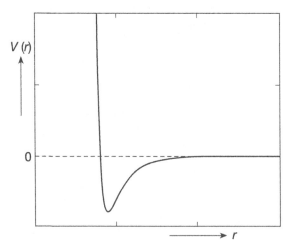

Figure 13.1 The Lennard-Jones potential.

As the volume per atom is proportional to r^3, say $v = \kappa r^3$, we obtain

$$p_{coh} = -\frac{du_{coh}}{dv} = -\frac{ma'}{3}v^{-(m/3+1)} + \frac{nb'}{3}v^{-(n/3+1)} \qquad (13.10)$$

where $a' = a\kappa^{m/3}$ and $b' = b\kappa^{n/3}$. Figure 13.1 shows a graph of the Lennard-Jones potential. It has a minimum at a certain distance r_0 and rises steeply for smaller values of r. For larger values of r, it increases to 0 at large distances.

The attractive part of the interaction is often due to the **van der Waals interaction** between electric dipoles, in which case $m = 6$. Clearly n must be larger than m if the shape of the potential is as in figure 13.1, and one often takes $n = 12$. Other functional forms for the repulsive part are also used, for example the **Morse potential** $u_{\text{repulsive}}(r) = C \exp(-\alpha r)$, where α is a constant of the order of r_0. r_0 is the equilibrium distance at low temperatures and phonons are small vibrations about r_0.

CHAPTER 14

Applications in Chemistry

Most chemical reactions occur at constant pressure. We have seen that in that case the enthalpy is the appropriate thermodynamic potential to use. Just as there is an enthalpy of vaporization as in equation (11.7), there is an enthalpy of ionization of atoms and an enthalpy of reaction for chemical reactions. Examples are

$$H_2O(l) \longrightarrow H_2O(g) \quad \Delta h = +44 \quad \text{kJ mol}^{-1}$$

$$H(g) \longrightarrow H^+(g) + e^- \quad \Delta h = +1312 \quad \text{kJ mol}^{-1}$$

$$Cl(g) + e^- \longrightarrow Cl^-(g) \quad \Delta h = -349 \quad \text{kJ mol}^{-1}$$

$$C_6H_6(l) + \frac{15}{2}O_2(g) \longrightarrow 6CO_2(g) + 3H_2O(l) \quad \Delta h = -3268 \quad \text{kJ mol}^{-1}.$$

When $\Delta h > 0$ the reaction is **endothermic**, that is, energy is absorbed by the reactants; when $\Delta h < 0$ the reaction is **exothermic**, that is, energy is released. Note that it is important to mention the phase of the various reactants. For example, if benzene is burnt in the vapour phase the last equation becomes:

$$C_6H_6(g) + \frac{15}{2}O_2(g) \longrightarrow 6CO_2(g) + 3H_2O(l) \quad \Delta h = -3301 \quad \text{kJmol}^{-1}$$

because it takes 33 kJ mol^{-1} to vaporize benzene. All these values are given at standard temperature and pressure, i.e. 25 °C and 1 bar. Standard chemical reaction enthalpies are tabulated. To obtain enthalpies at other temperatures, one can use relation (11.13). This implies that the enthalpy of each substance should be increased by $c_p \Delta T$. (Normally, c_p is nearly constant over the range of temperatures considered; otherwise one has to use tables for c_p at various intermediate temperatures and interpolate.) The entropy of reaction thus changes to

$$\Delta h' = \Delta h + \Delta c_p \, \Delta T, \tag{14.1}$$

where

$$\Delta c_p = \sum_{\text{products}} n c_p(\text{product}) - \sum_{\text{reactants}} n c_p(\text{reactant}). \qquad (14.2)$$

(n denotes the number of moles in the reaction equation.) This is known as **Kirchhoff's law.**

Whether the reaction is *spontaneous* or not is not determined by whether it is endothermic or exothermic. Instead, this is determined by the Gibbs free energy. Indeed, suppose that the reaction takes place at constant pressure and in thermal contact with the surroundings, which are kept at constant temperature. Then the condition for equilibrium is that g should be minimal; see equation (8.14). We can see this also as follows. By the second law, the total entropy of the system and the surroundings cannot decrease. Thus the condition for a spontaneous reaction is that

$$\Delta s_{\text{tot}} = \Delta s + \Delta s_{\text{surr}} > 0. \qquad (14.3)$$

(If it were zero, the system would already be in equilibrium.) Now, the heat expelled by the system to the surroundings is given by $q = -\Delta h$ (per mole). Hence, by equation (4.6), $\Delta s_{\text{surr}} = -\Delta h/T$ since T is constant. It follows that the condition for spontaneous reaction is

$$\Delta g = \Delta h - T \Delta s = -T \Delta s_{\text{tot}} < 0. \qquad (14.4)$$

If the reaction takes place at constant volume the condition is instead

$$\Delta f = \Delta u - T \Delta s < 0. \qquad (14.5)$$

Note that, if $\Delta h > 0$ the reaction can still proceed spontaneously at high enough temperatures provided that $\Delta s > 0$.

REMARK 14.1: *Nonexpansion work.*
A chemical reaction is not a simple system. The energy liberated in a reaction can be used to do work. Δg is the **maximum nonexpansive work** that the reaction can produce (per mole). To see this, we must add another work term to the expansion work $p\,dv$. An important example is the electrical energy stored in an electrochemical cell (a battery). It is given by

$$W_{\text{electr}} = -n_e F \mathcal{E}, \qquad (14.6)$$

where $\mathcal{E}$ is the electrical potential (at zero current) or **electromotive force** (E.M.F.), n_e is the number of moles of electrons produced and $F = eN_0$ is the charge of a mole of electrons. $F = 96.486\,\text{kC mol}^{-1}$ is called **Faraday's constant.** (C stands for the unit of charge: a coulomb. It is given by $1\,\text{C} = 1\,\text{J V}^{-1}$.) If we add this term to the first law we get

$$du = \delta q + \delta w_{\text{exp}} + \delta w_{\text{electr}} = T\,ds + p\,dv - F\mathcal{E}\,dn_e \qquad (14.7)$$

and hence

$$dg = -s\,dT + v\,dp - F\mathcal{E}\,dn_e. \qquad (14.8)$$

Thus, at constant pressure and volume,

$$\Delta g = -F\mathcal{E}n_{\mathrm{e}}. \tag{14.9}$$

This means that Δg is the maximum electrical energy produced. The maximum can only be reached in a quasi-static process; the current must be infinitesimally small.

CHAPTER 15

Black-Body Radiation

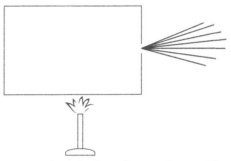

Figure 15.1 Black-body radiation from a black box.

When electromagnetic radiation is in thermodynamic equilibrium with matter at temperature T, one speaks of **black body radiation** because all incident radiation is then perfectly absorbed. The best way to study this situation experimentally is by making a small hole in one side of a closed container (figure 15.1). Experiments by Stefan showed that the **intensity** of the radiation, that is the total energy radiated per unit area per unit time, is given by

$$I(T) = \sigma T^4. \tag{15.1}$$

This result was subsequently derived from thermodynamics by Boltzmann and is now called the **Stefan-Boltzmann Law.** It follows simply from the energy equation (4.10) and the following relation between radiation pressure p and energy density per unit volume $\tilde{u}$:

$$p = \frac{1}{3}\tilde{u}. \tag{15.2}$$

This can be derived as follows. The momentum density of an electromagnetic wave with energy density $\tilde{u}$ is given by $(\tilde{u}/c)\vec{n}$, where c is the speed of light and $\vec{n}$ is a unit vector in the direction of the wave (see for example Landau and Lifshitz (1972)). The perpendicular component of the total momentum of the radiation incident on an area A from a direction in a solid angle $d\omega$ making an angle θ with the normal to A, and in a time dt is therefore

$$d\pi^{\perp} = 2\frac{\tilde{u}}{4\pi c}c\,dt(A\cos\theta)\cos\theta\,d\omega.$$

($\tilde{u}/4\pi$ is the energy density per steradian and there is a factor 2 because of the radiation emitted by the wall. The first factor $\cos\theta$ is due to the fact that a skew column of height $c\,dt$ above the area A at an angle θ has a volume $c\,dt\,A\cos\theta$; the second factor $\cos\theta$ is due to taking the perpendicular component.) Integrating over $d\omega = \sin\theta\,d\theta\,d\phi$ where ϕ ranges from 0 to 2π and θ from 0 to $\pi/2$, yields equation (15.2). (Remember that pressure is force per unit area and force is momentum per unit time by Newton's second law.)

Inserting equation (15.2) into (4.10), we obtain

$$\frac{d\tilde{u}}{dT} = 4\frac{\tilde{u}}{T} \implies \tilde{u} = a\,T^4 \tag{15.3}$$

for some constant a. A similar calculation as above for the pressure leads to equation (15.1). Indeed, the energy of the radiation incident on A during a time interval dt inside a solid angle $d\omega$ is given by

$$dI = \frac{\tilde{u}}{4\pi}\,c\,dt\,A\cos\theta\,d\omega.$$

Integrating over $d\omega$ shows that $\sigma = ca/4$.

If one measures the distribution of the energy radiated at temperature T as a function of the frequency ν one obtains curves for $\tilde{u}(\nu, T)$ with a maximum that shifts to higher values of ν, as T increases (figure 15.2).

Wien succeeded in extending Boltzmann's derivation to show that $\tilde{u}(\nu, T)$ must have the form

$$\tilde{u}(\nu, T) = \nu^3\, f\left(\frac{\nu}{T}\right). \tag{15.4}$$

It follows from this that the value $\nu_{\max}(T)$ of ν at which $\tilde{u}(\nu, T)$ attains its maximum is given by

$$\frac{\nu_{\max}}{T} = \text{constant.} \tag{15.5}$$

This is **Wien's displacement law.**

As in the case of gases, the precise form of the curves $\tilde{u}(\nu, T)$ cannot be determined from thermodynamics. One needs a microscopic description of the radiation. Using the Maxwell laws for this description led to the wrong equations for these curves, however. The resulting expression is called the **Rayleigh-Jeans law:**

$$\tilde{u}(\nu, T) = \frac{8\pi k_{\mathrm{B}}T}{c^3}\nu^2. \tag{15.6}$$

It is in good agreement with experiment for low frequencies but diverges at high frequencies. In 1900, Planck found a remedy for this shortcoming by

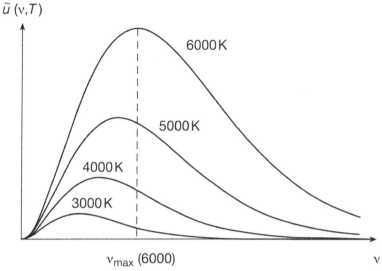

Figure 15.2 Spectrum of black-body radiation.

introducing the bold assumption that the energy comes in **quanta**, i.e. small discrete amounts given by

$$E = h\nu \quad (h = 6.6 \times 10^{-34} \text{Js.}) \tag{15.7}$$

Thus he arrived at the famous expression

$$\tilde{u}(\nu, T) = \frac{8\pi h}{c^3} \frac{\nu^3}{\exp\left(h\nu/k_B T\right) - 1} \tag{15.8}$$

This is **Planck's radiation law.** The assumption (15.7) gave rise to the development of **quantum mechanics.** Integrating over ν yields

$$\tilde{u}(T) = \frac{8\pi h}{c^3} \int_0^\infty \frac{\nu^3}{\exp\left(h\nu/k_B T\right) - 1} d\nu = \frac{8\pi^5 (k_B T)^4}{15(hc)^3} \tag{15.9}$$

by a standard integral formula. The constant a in (15.3) is therefore given by

$$a = \frac{8\pi^5 k_B^4}{15(hc)^3} = 7.56 \times 10^{-16} \text{J m}^{-3} \text{K}^{-4}. \tag{15.10}$$

The corresponding constant σ in equation (15.1) is

$$\sigma = \frac{ca}{4} = 5.67 \times 10^{-8} \text{J m}^{-2} \text{K}^{-4} \text{s}^{-1}. \tag{15.11}$$

The Big Bang

According to the **Big Bang theory of the Universe**, the latter began as a very dense and hot ball of matter and radiation, which has since expanded and hence cooled.

Soon after Einstein formulated his theory of gravity, the **general theory of relativity**, several models of the Universe were constructed. For simplicity these models assumed the Universe to be *homogeneous and isotropic* on a large scale. Einstein himself immediately realized that a stationary Universe is not a solution of Einstein's field equations unless an additional, somewhat artificial term is introduced, known as the **cosmological constant**. In the latter half of the 1920s, Hubble measured the velocity of a number of distant galaxies and found that they all recede from the Earth with speeds which increase with their distance. (The radial velocity of astronomical objects can be measured by the redshift in their spectral lines.) At the same time a Belgian priest, Lemaître, proposed a model of an expanding Universe not involving the cosmological constant. The expansion does not mean that there is a definite centre of expansion but rather that all galaxies move away from each other in a uniform manner so that, as seen from *any* galaxy, all others are flowing outward. The average distance between galaxies $R(t)$ thus depends on time. It satisfies the equation

$$\frac{d^2 R}{dt^2} = -\frac{4}{3}\pi G \left(\rho + 3\frac{p}{c^2} \right) R. \tag{16.1}$$

Here, G is the gravitational constant, ρ the density, c the velocity of light, and p the pressure. If one neglects the pressure term, this equation can in fact also be derived from Newton's theory of gravity. The density ρ includes the density ρ_{mat} of matter and the density ρ_{rad} of radiation. At the present time matter dominates radiation, so that $\rho \approx \rho_{\text{mat}} = \rho(t_0)/R^3(t)$, where $\rho(t_0)$ is the density at the standard time where $R(t_0) = 1$. Neglecting the pressure, equation (16.1) can be integrated to obtain

$$\left(\frac{dR}{dt} \right)^2 = \frac{8\pi}{3} \frac{G\rho(t_0)}{R} - k. \tag{16.2}$$

The constant of integration k is the **curvature of space**. If $k = 0$ this equation can be solved to find

$$R(t) = [6\pi G\rho(t_0)]^{1/3} t^{2/3}. \tag{16.3}$$

This is the **Einstein-de Sitter model**. In this case the Universe would expand forever. If $k > 0$ the Universe is like a 3-dimensional sphere (a balloon in four-dimensional space). It expands until it reaches a maximum size and then reverses into collapse. If $k < 0$ the Universe is infinite in extent and will expand forever. Figure 16.1 illustrates these three possibilities.

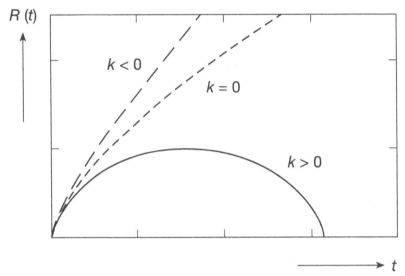

Figure 16.1 Three scenarios for the expansion of the Universe.

For small t, $R(t)$ behaves as $R(t) \propto t^{2/3}$ in all cases. The relative expansion rate

$$H(t) = \frac{1}{R(t)} \frac{\mathrm{d}R(t)}{\mathrm{d}t} \tag{16.4}$$

is called **Hubble's constant**. It is in fact not a constant but varies as the Universe expands. Its present value was estimated by Hubble to be approximately 500 km s^{-1} Mpc^{-1}. The **parsec** (pc) is the usual astronomical unit of distance: 1 pc $= 3.0856 \times 10^{16}$ m. However, this estimate was later found to be erroneous due to systematic errors in his estimates of the distance to galaxies. (Distances are notoriously difficult to measure in astronomy.) The modern estimates of H are

$$H = 72 \pm 10 \text{ km sec}^{-1} \text{ Mpc}^{-1} = 2.4 \pm 0.4 \times 10^{-18} \text{ s}^{-1}. \tag{16.5}$$

The average density of visible matter in the Universe is estimated to be $\rho_{\mathrm{mat}} \approx 10^{-30}$g cm^{-3}. Inserting this into equation (16.2), we find that $k < 0$ so that the Universe would be infinite in extent and forever expanding. However, it is now thought that there is a large amount of invisible matter, so called '*dark matter*' present in the Universe which could turn this conclusion around. On the other hand the cosmological constant is also thought be non-zero and as a result the Universe might still be expanding indefinitely.

In 1946, George Gamov embarked on a bold attempt to extend this theory to explain the abundance of the various chemical elements in the Universe. It was clear from the above dynamical model that the Universe must have been much denser in the distant past. It is then reasonable to assume that it was also much hotter, so hot in fact that nuclear reactions would take place spontaneously. Gamov and co-workers therefore assumed that initially there were only neutrons, which then fused as the Universe cooled to form the elements. From these assumptions Ralph Alpher, Hans Bethe, and Gamov computed the relative abundances of the elements. This theory is known as the $\alpha - \beta - \gamma$ theory. It is flawed in important respects but it did give rise to the idea that there must be a remnant **background radiation**, which is highly isotropic black body radiation. The detection of this radiation in 1964 by Penzias and Wilson was completely accidental but it was a great triumph for the Big Bang theory. It corresponds to a temperature of 2.7 K. In fact, the $\alpha - \beta - \gamma$ theory predicted a temperature in the order of 5 K. This remarkable agreement was rather a lucky coincidence but let us see how they arrived at this conclusion.

To see how the temperature of the radiation decreased as the Universe expanded, we consider the fundamental relations (4.9) and (4.10). With the reasonable assumption that the process is isentropic we obtain using equations (15.2) and (15.3),

$$0 = T \, \mathrm{d}s = \left(\frac{\partial u}{\partial T} \right)_v \mathrm{d}T + T \left(\frac{\partial p}{\partial T} \right)_v \mathrm{d}v$$
$$= 4avT^3 \, \mathrm{d}T + \frac{4}{3} aT^4 \, \mathrm{d}v. \tag{16.6}$$

It follows that $v \, T^3 = $ constant, and

$$\tilde{u} \propto v^{-4/3} \propto R^{-4}, \tag{16.7}$$

where R is the scale factor above. For matter, the energy density also decreases, but more slowly. Indeed, since $\tilde{u}_{\mathrm{mat}} = \rho_{\mathrm{mat}} c^2$ where ρ_{mat} is the matter density,

$$\tilde{u}_{\mathrm{mat}} \propto v^{-1} \propto R^{-3}. \tag{16.8}$$

As long as there is no appreciable conversion of matter into radiation or conversely, therefore, the energy density of radiation decreases more rapidly than the energy density of matter. This means that there must have been a changeover from a *radiation dominated* Universe to a *matter dominated* Universe (provided matter and radiation were decoupled before this moment). In a radiation dominated Universe, the equation (16.2) no longer holds because the pressure cannot be neglected. However, for small t, the curvature term can be neglected and one obtains using equation (15.2) that $R(t) \propto t^{1/2}$. In fact one can do better by using Planck's radiation law (15.8) and (15.9). Indeed, from equation (15.8) it follows that the temperature decreases as $T(t) \propto 1/R(t)$ since the frequency $\nu = c/\lambda$ and the wave length λ increases with the scale factor $R(t)$. Therefore, $\dot{T}/T = -\dot{R}/R$. Moreover, $\rho(t) = a T^4/c^2$ where a is given by equation (15.10). This leads to

$$T = \left(\frac{3c^2}{32\pi a G} \right)^{1/4} t^{-1/2} = \frac{1.5 \times 10^{10}}{t^{1/2}} \, \text{K} \qquad (16.9)$$

for a radiation dominated Universe.

To estimate the present energy density of radiation note that equations (16.7) and (16.8) imply that

$$\rho_{\text{rad}} \rho_{\text{mat}}^{-4/3} = \text{constant}. \qquad (16.10)$$

We have already got an estimate for the present matter density: $\rho_{\text{mat}} \approx 10^{-30}$ g cm^{-3}. The $\alpha - \beta - \gamma$ theory resulted in an estimate for the matter density at the start of the element-forming process of $\rho_{\text{mat}}^{(0)} \sim 10^{-6}$ g cm^{-3}. Assuming a simple neutron addition process and using the known cross-sections for neutron capture of the elements, and taking into account the decay of the neutron into a proton and the expansion of the Universe, it was possible to calculate the present abundance of elements given the initial density. The above value for $\rho_{\text{mat}}^{(0)}$ led to the best fit with the measured abundances. Finally, the initial temperature was estimated to be approximately 10^9 K. This corresponds roughly to $\rho_{\text{rad}}^{(0)} \approx 1$ g cm^{-3}. It then follows that the present radiation density must be around $\rho_{\text{rad}} \sim 10^{-32}$ g cm^{-3} which corresponds to a temperature of 6 K. It must be said, however, that all these estimates have large error bars. More seriously, it was soon found that the original $\alpha - \beta - \gamma$ theory had important flaws. Indeed, it became clear that the initial state could not have consisted of neutrons alone, but rather that there should have been thermal equilibrium between neutrons and protons and various other forms of matter. Moreover, there is a gap at atomic numbers 5 and 8 which cannot be overcome by simple neutron addition. All this meant that elements beyond lithium could not have been formed during the initial stages of the Big Bang. It is now generally agreed that these elements were created inside stars. More and more elaborate calculations were done, including more and more possible nuclear reactions resulting in very good agreement with the measured value of 2.7 K. Other modifications of the theory of the Big Bang were needed; in particular, the assumption of a very rapid exponential expansion in the very early stages. This called **inflation theory**. We cannot go into this here, but see the notes for further reading at the end of this book. Indeed, in recent years there have been major developments in this area. Three subsequent satellites, the COBE satellite (1989), WMAP (2001), and the Planck space observatory (2009) made increasingly accurate measurements of the cosmic microwave background and analysed inhomogeneities in the radiation thought to be due to the initial formation of *structure* (galaxy formation) in the Universe. Moreover, accurate measurement of radiation from distant supernova explosions in particular allowed a more accurate determination of the evolution of the Universe and Hubble's constant, leading to the current model of an open Universe involving a large fraction of *dark matter* as well as the existence of *dark energy* corresponding to a non-zero cosmological constant. All this is obviously well beyond the scope of this book.

CHAPTER 17

The Constitution of Stars

We give here an account of Eddington's original theory of the equilibrium of a star. There are, of course, much more elaborate models known today, but this short account gives a rough idea of the considerations involved. A star is a ball of gas which we shall assume satisfies the ideal gas law

$$p_{\mathrm{G}} = \frac{R_0}{M} \rho\, T. \tag{17.1}$$

Here ρ denotes the mass density and M is the molar mass. The total pressure is composed of gas pressure and radiation pressure:

$$p = p_{\mathrm{G}} + p_{\mathrm{R}}. \tag{17.2}$$

We also assume that

$$\frac{p_{\mathrm{G}}}{p} = \beta \tag{17.3}$$

is approximately constant throughout the star. The pressure p is counterbalanced by the gravitational attraction. Thus we have, inserting equations (15.2) and (15.3),

$$p = \frac{R_0\, \rho\, T}{\beta\, M} = \frac{aT^4}{3(1 - \beta)} \tag{17.4}$$

and hence

$$p = \kappa\, \rho^{4/3}. \tag{17.5}$$

When the pressure is proportional to a power of the density the gas is called **polytropic.** The gravitational potential ϕ satisfies Poisson's equation:

$$\frac{1}{r^2} \frac{\mathrm{d}}{\mathrm{d}r} \left(r^2 \frac{\mathrm{d}\phi}{\mathrm{d}r} \right) = \frac{\mathrm{d}^2\phi}{\mathrm{d}r^2} + \frac{2}{r} \frac{\mathrm{d}\phi}{\mathrm{d}r} = -4\pi\, G\, \rho, \tag{17.6}$$

where the left hand side is the radial part of the Laplace operator in polar coordinates and $G = 6.67 \times 10^{-11} \ \mathrm{N\,m^2\,kg^{-2}}$ is the gravitational constant. This differential equation can be derived as follows: Consider an infinitesimal spherical shell of thickness dr of the star. The inward gravitational force due to this shell is given by $-GM(r)dM(r)/r^2$, where $M(r)$ is the mass inside the ball of radius r:

$$M(r) = 4\pi \int_0^r \rho r^2 \, dr. \tag{17.7}$$

The force per unit area is therefore

$$-g(r)\,\rho(r)\,dr = -\frac{GM(r)\rho(r)dr}{r^2}, \tag{17.8}$$

where $g(r)$ is the gravitational acceleration. This force is sustained by an increase in pressure of

$$dp = -g\,\rho\,dr = \rho\,d\phi, \tag{17.9}$$

where the gravitational potential ϕ is defined by

$$g = -\frac{d\phi}{dr}. \tag{17.10}$$

Differentiating $d\phi/dr = -g$ again with respect to r we obtain equation (17.6). Inserting equation (17.5) into (17.9) we obtain

$$\phi = 4\kappa\,\rho^{1/3} + \text{constant} \tag{17.11}$$

so that

$$\rho = \left(\frac{\phi}{4\kappa}\right)^3 \tag{17.12}$$

and

$$p = \frac{1}{4}\rho\,\phi. \tag{17.13}$$

Inserting this into equation (17.6), we find that ϕ inside the star must satisfy

$$\frac{d^2\phi}{dr^2} + \frac{2}{r}\frac{d\phi}{dr} + \alpha^2\phi^3 = 0 \tag{17.14}$$

with

$$\alpha^2 = \frac{\pi G}{16\kappa^3}. \tag{17.15}$$

This equation can be solved numerically for $\phi(r)$ given the boundary conditions

$$\begin{cases} \phi|_{r=R_*} = 0, \\ \frac{d\phi}{dr}\big|_{r=0} = 0, \end{cases} \tag{17.16}$$

where R_* is the radius of the star. In practice, these boundary conditions are somewhat inconvenient. It is better to transform them into initial conditions as follows. We define

$$u = \frac{\phi}{\phi_0} \text{ and } z = \alpha\,\phi_0\,r. \tag{17.17}$$

Then $u(z)$ satisfies

$$u(0) = 1 \text{ and } \frac{du}{dz}\Big|_{z=0} = 0 \tag{17.18}$$

and

$$\frac{d^2 u}{dz^2} + \frac{2}{z}\frac{du}{dz} + u^3 = 0. \tag{17.19}$$

This is called the **Lane-Emden equation** with index 3. Now, at the boundary of the star, $r = R_*$, the gravitational force is given by

$$g = \frac{G\,M_*}{R_*^2} = -\frac{d\phi}{dr}\Big|_{r=R_*}. \tag{17.20}$$

Transforming this to the variables (17.17) we obtain

$$\alpha = \frac{1}{G\,M_*}\left(-z^2\frac{du}{dz}\right)_{u=0}. \tag{17.21}$$

The solution for $u(z)$ can be read off from table 17.1.

Table 17.1. Solutions for $u(z)$.

z	u	$-z^2\dfrac{du}{dz}$
0.00	1.000 00	0.00
6.80	0.004 17	2.018 237 14
6.85	0.002 00	2.018 237 22
6.89	0.000 29	2.018 237 22
6.90	−0.000 13	2.018 237 22

Solving equation (17.19) with boundary conditions (17.18), we can determine α from (17.21) and hence ϕ_0 from

$$\phi_0 = \frac{1}{\alpha R_*} z\big|_{u=0}. \tag{17.22}$$

The temperature at the centre of the star then follows from

$$T_0 = \frac{\beta\,M}{\rho_0\,R_0}\,p_0 = \frac{\beta\,M}{4\,R_0}\,\phi_0. \tag{17.23}$$

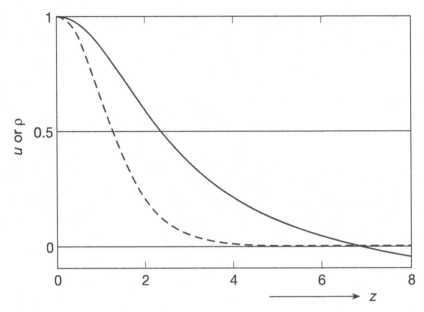

Figure 17.1 The gravitational potential u (——)
and density ρ (- - - -) of a star.

Figure 17.1 shows the normalized gravitational potential u and the density ρ as a function of the normalized radius z.

For the sun we may take $M_\odot = 2 \times 10^{30}$ kg and $R_\odot = 7 \times 10^8$ m and $M \approx M_{\text{Hydrogen}} = 1.672 \times 10^{-27} \times 6.022 \times 10^{23} = 1.0 \times 10^{-3}$. Taking $\beta \approx 1$, this yields eventually $T_\odot = 2 \times 10^7$ K.

Problems to Part I

I-1. Originally, a pressure of 1 atmosphere (atm) was defined as the pressure of a column of mercury with a height of 760 mm. The density of mercury (Hg) is approx. 13.6 g cm^{-3}. Express 1 atm in Pascals. (1 Pa = 1 N m^{-2}.) (Nowadays, 1 atm is defined to be exactly 1.01325 bar, where 1 bar = 10^5 Pa.)

I-2. Show that the volume of 1 mole of an ideal gas at a pressure of 1 atm (= 1.013×10^5 Pa) and a temperature of 0 °C is 22.4 l. This is called the **molar volume at normal temperature and pressure (NTP)**.

I-3. Assume air to be a mixture of 20% oxygen and 80% nitrogen. Show that the gas constant R per unit mass for air is approximately 288 J kg^{-1} K^{-1}. (Note that this does depend on the gas. However, since quantities are often measured in mass units it is often useful to compute this quantity for the given substance.)

I-4. Calculate the internal energy of 1 kg of air occupying 10 l at a pressure of 10 bar. (1 bar = 10^5 Pa) Assume air to be an ideal gas. (Use the result of problem I-3.)

I-5. Oxygen (O_2) is to be stored in a steel vessel at a pressure of 100 bar and a temperature of 20 °C. The capacity of the vessel is 10 l. Calculate the mass of oxygen that can be stored. The vessel has a fusible plug to prevent it from excessive pressure. If the maximal allowable pressure is 120 bar, at what temperature must the fuse melt?

I-6. Air is contained in a container of volume $V = 0.6\,\mathrm{m}^3$ at a pressure of 15 bar and a temperature of 40.5 °C. Some air is allowed to escape and the pressure drops to 14 bar. Calculate the mass of air which has escaped. (Use $R = 288$ J kg^{-1} K^{-1} as computed in problem I-3.)

I-7. Carbon dioxide is contained in a flask at a pressure of 200 bar and a temperature of 20 °C. Some of it escapes to the atmosphere at a pressure of 1 bar. Compute the decrease of the temperature of the escaping gas assuming that the process is adiabatic and quasi-static and that CO_2 is an ideal gas. For CO_2, $\gamma = 1.3$. The boiling point of CO_2 at a pressure of 1 bar is $T_{\mathrm{boil}} = 195$ K. The triple point of CO_2 is at $p = 5$ bar and $T = 217$ K. Argue that the computed temperature would mean that the escaping CO_2 solidifies. (In fact carbon dioxide does not behave as an ideal gas under these circumstances.)

I-8. An iron wire with a length of 1 meter and diameter of 1 mm, is stretched with a force of $F = 157$ N. The extension of the wire is given by Hooke's law:

$$\frac{\Delta\ell}{\ell} = \frac{F}{YA},$$

where $Y \approx 2 \times 10^{11}$ Pa is Young's modulus and A is the cross-sectional area. Show that the work performed on the wire is given by

$$W = \frac{1}{2}\frac{YA}{\ell}(\Delta\ell)^2.$$

The wire is then released and returns adiabatically to its original length. Calculate the increase in temperature, assuming that c_V is constant. (Given: $c_V = 400$ J kg^{-1} K^{-1} and the density of iron is $\rho = 7800$ kg m^{-3}.)

I-9. A bicycle pump has a length of 50 cm (inside) and an inner diameter of 3 cm. The air in the bicycle tyre has pressure of 2 bar. The air in the pump is compressed adiabatically and quasi-statically by pressing down the piston. How far does it have to be lowered before air flows into the tyre? The initial temperature of the air is 17 °C. What is the rise in temperature of the air in this process? (Assume that air satisfies the ideal gas law.)

I-10. Perform the transformation of variables needed to prove equations (3.8) and (3.9) in the text.

I-11. Prove that the 1-form $\omega = (x + y\ln xy)\mathrm{d}x + x\ln xy\,\mathrm{d}y$, defined on the quadrant $x, y > 0$, is exact. Then check that the integral $\oint_\Gamma \omega = 0$ if Γ is the closed path consisting of the line segments from (1,1) to (2,1), from (2,1) to (1,2), and from (1,2) to (1,1).

I-12. Consider the 1-form defined on $\mathbb{R}^2 \setminus \{(0,0)\}$ by

$$\omega = \frac{y}{x^2 + y^2}\mathrm{d}x - \frac{x}{x^2 + y^2}\mathrm{d}y.$$

(i) Prove that

$$\frac{\partial}{\partial y}\frac{y}{x^2 + y^2} = -\frac{\partial}{\partial x}\frac{x}{x^2 + y^2}.$$

(ii) Show that ω is not exact by computing the integral of ω over the unit circle. (This is equation (3.14).)

The apparent contradiction with theorem 3.2 is resolved by remarking that ω is in fact not defined on the whole of $\mathbb{R}^2$.

I-13. Prove the relation

$$\left(\frac{\partial z}{\partial x}\right)_y = -\left(\frac{\partial z}{\partial y}\right)_x\left(\frac{\partial y}{\partial x}\right)_z.$$

for any differentiable function $z(x, y)$ such that the function $y(x, z)$, defined implicitly by $z(x, y(x, z)) = z$, is differentiable. This is the same as equation (4.11) provided all derivatives are well-defined. Hence derive equation (4.12).

I-14. In an air turbine, the air expands from 5 bar and 350 °C to 1 bar and 150 °C. The heat loss from the turbine is negligible. Show that the process is irreversible and compute the change in entropy per kg of air. Draw the process in a $T - s$ diagram. ($\gamma_{\text{air}} = 1.4$.)

I-15. Water cannot be considered an ideal gas. Therefore, in order to analyse its states, one needs to use tables. Many practical engineering processes use steam as working fluid, so these tables are of great importance. Wet steam is water in the region of coexistence of the liquid and the vapour phase (underneath the broken curve in figure 9.1 and figure 1 of the introduction). The **dryness fraction** x is the ratio of the amount of vapour to the total amount of fluid. Consider wet steam at a pressure of 10 bar with a dryness fraction $x = 0.9$ in a cylinder, expanding isothermally and reversibly to a pressure of 1 bar at a temperature of $T = 180\,°C$.
(i) Compute the heat supplied per kg during the process.
(ii) Compute the work done per kg of steam.
(iii) Draw a picture of the process in a $p - V$ diagram.
  Use the following part of the steam table.

	u (kJ kg^{-1})	s (kJ kg^{-1} K^{-1})	h (kJ kg^{-1})
Saturated liquid at 10 bar	762	2.138	763
Saturated vapour at 10 bar	2584	6.586	2778
Steam at 1 bar	2629	7.746	2836

I-16*. Show that the Carnot process in an ideal gas depicted in figure 5.2 has Carnot efficiency given by equation (5.4). Consider a Carnot process with air as the working fluid. Let the maximum temperature be 600 °C and let the minimum temperature be 10 °C. Compute the efficiency. Let, moreover, the maximum and minimum pressures in the cycle be 200 bar and 1 bar respectively. Compute the work ratio, given that $\gamma = 1.4$. (The work ratio is the ratio of the net amount of work done during the cycle and the amount of work done during the expansion part of the cycle (from a to c).)

I-17. Show that the efficiency of the Otto cycle is given by equation (5.5).

I-18. A closed-cycle gas turbine operates approximately according to the **constant pressure cycle** represented by the following diagram. Here, bc and da are isentropes.
(i) Draw the corresponding $T - s$ diagram.
(ii) Show that the efficiency of this cycle is given by $\eta = 1 - r_p^{-(\gamma-1)/\gamma}$, where r_p is the pressure ratio, $r_p = p_a/p_d$.

I-19. Prove that $u(s, v)$ is a convex function of two variables.

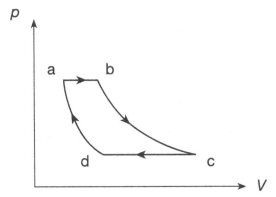

Figure I-18 The constant-pressure cycle.

I-20. Prove that the entropy density (6.17) of the ideal gas is a concave function of u and v.

I-21. Prove lemma 7.2.

I-22. Show that the function $f(x) = |x|$ $(x \in \mathbb{R})$ is convex and compute its Legendre transform.

I-23. Show that the function $f(x) = |1 - x^2|$ $(x \in \mathbb{R})$ is not convex, compute f^* and f^{**}, and verify directly that theorem 7.2 holds in this case.

I-24. Prove equations (8.8) and (8.9) and hence equation (8.10). Similarly, prove also equations (8.13), (8.16), and (8.18).

I-25. Compute the free energy density $f(v, T)$ of an ideal gas and use equation (8.8) to rederive the ideal gas law.

I-26. Prove the Maxwell relations

$$\left(\frac{\partial T}{\partial v}\right)_s = -\left(\frac{\partial p}{\partial s}\right)_v, \quad \left(\frac{\partial T}{\partial p}\right)_s = \left(\frac{\partial v}{\partial s}\right)_p,$$

$$\left(\frac{\partial p}{\partial T}\right)_v = \left(\frac{\partial s}{\partial v}\right)_T, \quad \left(\frac{\partial v}{\partial T}\right)_p = -\left(\frac{\partial s}{\partial p}\right)_T.$$

(Use the expressions for the differentials du, dh, df, and dg.)

I-27. Prove that

$$c_V = -T\left(\frac{\partial^2 f}{\partial T^2}\right)_v$$

and deduce that $c_V \geqslant 0$.

I-28. Derive the identity

$$c_p = -T\left(\frac{\partial^2 g}{\partial T^2}\right)_p$$

and deduce that $c_p \geqslant 0$. A normal solid extracts heat from its surroundings when it melts: the latent heat defined above equation (9.3). Argue that this means that $s_{\text{solid}} < s_{\text{liquid}}$. Now consider the phase diagram of ^{3}He shown in figure I-28.

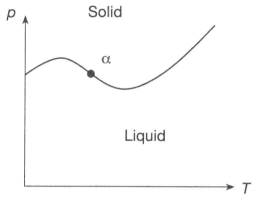

Figure I-28 Low-temperature phase diagram of ^{3}He.

Use the concavity of $g(T, p)$ to show that at the point α, $s_{\text{solid}} > s_{\text{liquid}}$. Argue that this means that an equilibrium mixture of solid and liquid ^{3}He can be cooled by slow compression. This is called **Pomeranchuk cooling**.

I-29. Prove that the isotherms of the van der Waals gas are monotonically decreasing for $T \geqslant T_c$ but not for $T < T_c$, where T_c is given by equation (9.11). Compute also the corresponding critical pressure p_c and the critical specific volume v_c and show that the van der Waals equation can be written in a universal form in terms of the quantities $\tilde{p} = p/p_c$, $\tilde{v} = v/v_c$, and $\tilde{T} = T/T_c$. This is the **law of corresponding states.**

I-30. Show that for a van der Waals gas above the critical temperature,

$$\left(\frac{\partial u}{\partial v}\right)_T = \frac{a}{v^2},$$

and hence that $c_V = c_V(T)$ is independent of the volume. Assuming that c_V is in fact constant, derive an expression for the entropy function $s(u, v)$ for the van der Waals gas.

I-31. Show that for isentropic expansion of a van der Waals gas,

$$T(v - b)^{k_B/c_V} = \text{constant}.$$

Then compute the decrease in temperature in problem I-7 again, but assuming that CO_2 satisfies the van der Waals equation, and notice the difference in volume and final temperature. (For CO_2, $a = 0.3639 \, \text{Pa} \, \text{m}^6 \, \text{mol}^{-2}$ and $b = 0.04267 \, \text{l} \, \text{mol}^{-1}$.)

I-32. Prove that

$$\left(\frac{\partial s}{\partial p}\right)_h < 0,$$

and use this relation to argue that during a throttling process the entropy increases.

I-33*. To compute the change in temperature in filling a bottle from a reservoir, one needs a modification of equation (11.4) for a throttling process because the mass of fluid in the bottle changes in the process. Assuming that the process is adiabatic (no heat is lost to the surroundings), derive the relation

$$h_{\text{source}}(m' - m) = m'u' - mu,$$

where m and m' are the mass of fluid in the bottle before and after the refill, u and u' are the internal energy per unit mass before and after the refill, and h_{source} is the enthalpy per unit mass of the fluid in the reservoir. (Consider a small element of mass δm entering the bottle and write down the energy balance.) Consider for example a rigid bottle of helium at a pressure of 5 bar and a temperature of 15 °C being filled from a large source at 10 bar and 15 °C until the pressure in the bottle has increased to 8 bar. Compute the final temperature of the helium in the bottle assuming the process is adiabatic. ($c_V = 3.12$ kJ kg^{-1} K^{-1}.)

I-34. An ammonia refrigerator operates between an evaporator temperature of -12 °C and a condenser temperature of 32 °C. The vapour enters the condenser in dry saturated condition after isentropic compression. It operates according to the cycle of figure 11.3. Calculate the coefficient of performance. You can use the following table for ammonia.

T (K)	p (bar)	s_{liq}	s_{vap}	h_{liq}	h_{vap}
305	12.37	1.235	4.962	333	1470
261	2.68	0.510	5.504	126	1431

I-35. A ^{3}He cryostat achieves cooling by pumping away the vapour above a ^{3}He bath. Suppose that the ^{3}He vapour is circulated at a rate of 60 cm^3 s^{-1}, and that the minimum temperature reached is 0.3 K. Calculate the maximum sustainable heat flux into the cooling chamber given that the vapour pressure at 0.3 K is 0.2 Pa, and the latent heat is 21 J mol^{-1}, and assuming that the vapour behaves like an ideal gas. Argue that the cooling power is (roughly) proportional to the vapour pressure and hence, using equation (9.9), that the cooling power decreases exponentially with the temperature. Estimate the pressure at 0.1 K using this formula.

I-36*. This problem concerns a more realistic calculation for a power plant. Consider a power plant operating according to a Rankine cycle with superheat (see figure 5.4) and with steam as the working fluid. The boiler pressure is 50 bar, the condensor pressure 0.05 bar. The steam is superheated to 500 °C. Calculate the efficiency of the plant. Use the following part of the steam table.

T (K)	p (bar)	s_{liq}	s_{vap}	h_{liq}	h_{vap}
306	0.05	0.476	8.394	138	2561
537	50.0	2.921	5.973	1155	2794
773	50.0		6.975		3433

(Hint: Show that $w = q_{\text{in}} - q_{\text{out}}$ where $q_{\text{in}} = h_{b'} - h_{a'}$ and $q_{\text{out}} = h_{c'} - h_{d'}$. The work done by the pump is $w_{\text{pump}} = h_{a'} - h_{d'}$. Use equation (8.13) to show that it can be calculated from $w_{\text{pump}} = v\,\Delta p$, where v is the specific volume of water, which is essentially constant. Assume that the pump operates adiabatically.)

I-37. Derive equation (11.15) for the change in temperature in a throttling process.

I-38*. For a van der Waals gas at low density ($v \gg b$ and $a/v \ll k_BT$) derive equation (11.16). Show also that, in general, the inversion temperature for a van der Waals gas is given by

$$\frac{2a}{k_BT}\left(\frac{v-b}{v}\right)^2 = b.$$

Rewrite this as a relation between pressure and temperature and show that the inversion curve in a $p-T$ diagram is a parabola with maximum at $T_{\text{inv}}^{\text{max}} = 8a/9bk_B$. Compute this temperature for CO_2 using the values for a and b given in problem I-31 and compare this value with the experimental value of 1500 K.

I-39. Show that Planck's law (15.8) agrees with the Rayleigh-Jeans law (15.6) for low frequencies ($h\nu \ll k_BT$).

I-40. Show that for Planck's law, $\nu_{max}(T)$ satisfies

$$\frac{h\nu_{max}(T)}{k_BT} = x \text{ where } x = 3(1 - e^{-x}).$$

Numerically, $x = 2.821$. Show also that the maximum energy density per unit wavelength is attained when

$$\frac{hc}{\lambda k_BT} = x' \text{ where } x' = 5(1 - e^{-x'}).$$

Radiation from the Sun is black-body radiation to good approximation. Its maximum energy density per unit wavelength is attained at $\lambda_{max} = 4.8 \times 10^{-7}$ m. Compute the temperature of the photosphere of the Sun.

I-41. When the Universe expands, all wavelengths expand with a common factor: $\lambda(t) \propto R(t)$. Show that if the wavelength is multiplied by a factor f then the energy density $\tilde{u}$ is multiplied by a factor f^{-4} due to two effects: the expansion of the volume and the decrease in energy per photon (see equation (15.7)). Hence derive that black-body radiation remains black-body radiation but with temperature reduced by a factor f.

I-42. Derive equation (16.9).

II

Fundamentals of Statistical Mechanics

As formulated in the introduction, statistical mechanics is the theory of macroscopic systems from a microscopic point of view, that is, it explains the thermodynamics of a large system from the dynamics of its constituent particles. This requires the identification of the macroscopic observables in terms of microscopic quantities. In this part of the book, we formulate this correspondence starting with a simple example: a model of a paramagnetic salt. We have seen in Part I that the thermodynamic behaviour of a macroscopic system is completely determined by the fundamental equation, i.e., the entropy density as a function of the energy density and the specific volume. The latter two quantities are readily identified with the average energy per particle and the average volume per particle, respectively. The main hypothesis of equilibrium statistical mechanics is **Boltzmann's postulate** that the entropy density is given by the logarithm of the number of available states of the entire system of N particles, divided by N. This is formulated more precisely in chapter 19. The available micro-states are determined by the quantum mechanical Hamiltonian. This latter concept is explained in chapter 18; which is a very elementary introduction to quantum mechanics. It is essentially just a compilation of some basic facts, but it suffices for understanding the main part of this book. (Only chapters 33–35 require a more complete understanding of quantum mechanics.) The Hamiltonian determines the possible energy values of the system. It is a sum of a kinetic energy term for each particle and a potential energy term due to the forces between the particles. These forces are usually of a fairly short range and are given by a potential energy function of the form shown in figure 13.1. The forces between the particles cause their individual energies to change during each collision, i.e. every time two particles come within the interaction range of each other. The total energy $E = \epsilon_1 + \cdots + \epsilon_N$ is unchanged, however, provided the system is thermodynamically isolated. In equilibrium, the macroscopic properties of the system are independent of time. We shall identify these macroscopic quantities with averages of corresponding microscopic quantities over all the particles. Thus, for instance, the number of particles with energies ϵ_i in a given range will be independent of time when the system is in equilibrium. Remarkably, although the redistribution of energy over the particles before the attainment of equilibrium is due to the interparticle forces, it turns out that, to a first approximation, the equilibrium state reached can often be described reasonably well by assuming that the particles move independently of one another. We shall therefore first consider models for a system of *free particles*. The simplest such model is that of a paramagnetic salt. It is described in chapter 19 as a system of independent localized spins. In this chapter, the main hypothesis of statistical mechanics is also introduced. Chapter 20 is a mathematical digression about large deviation theory, a special branch of probability theory. This theory is used extensively in the following chapters. It is used in chapter 21 to rederive the thermodynamics of a paramagnet, and in chapter 22 a more general system of independent localized particles is treated in the same way. In chapter 23, this analysis is extended further to quantum gases. This chapter is rather technical and can be omitted if one accepts the general formalism of equilibrium statistical mechanics outlined in chapter 24. The derivation of the general formalism in chapter 24 is not rigorous but should make the basic formulas plausible. These formulas are of course essential for the analysis of all models in equilibrium statistical mechanics. In chapters 25 and 26, a rigorous proof is given of the existence of the free energy density in the thermodynamic limit for classical spin systems (localized particles) and of the large-deviation property for the energy density. These chapters are optional but they put the theory of chapter 24 on a firm footing for these systems.

CHAPTER 18

Quantum Mechanics

At the beginning of the twentieth century, it became clear that microscopic particles do not obey Newton's laws of classical mechanics. Indeed, in order to explain the behaviour of black-body radiation Planck introduced the new postulate that the energy comes in small discrete amounts E called **quanta** proportional to the frequency ν (see chapter 15):

$$E = h\,\nu \text{ where } h = 6.626 \cdot 10^{-34} \text{ J s.} \tag{18.1}$$

The new fundamental constant h is now called **Planck's constant**. It is a very small quantity, which is the reason that the discreteness of the energy is not apparent in everyday phenomena.

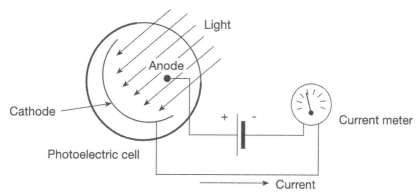

Figure 18.1 A photo-electric cell.

Planck's idea was then used by Einstein to explain the phenomenon of the photoelectric effect. This effect takes place in a photoelectric cell (figure 18.1) which is a vacuum glass bulb with two electrodes, one of which, called the cathode, is illuminated by a light source. The light liberates electrons from the cathode which results in a small current between the electrodes. It was observed experimentally that no current could flow if the wavelength of the light was larger than a certain threshold. Einstein explained this effect by assuming that an electron is liberated by a single light quantum, or **photon**, and that this can only happen if the energy of the photon is high enough. By Planck's relation, therefore, electrons can only be liberated if the frequency is high enough. This idea of light quanta was at odds with the established wisdom that light is a wave phenomenon, of course, but no other explanation of the photoelectric effect seemed possible.

The next step in the development of quantum mechanics was taken by Niels Bohr, who introduced postulates about the discreteness of the energy of the bound electrons in an atom similar to the discreteness of the energy of light quanta. This enabled him to explain the appearance of lines in the radiation spectrum of atoms. His postulates were the following:

1.   Electrons can only exist in certain stable orbits about the nucleus of an atom.
2.   If an electron jumps from an orbit of higher energy E_2 to one of lower energy E_1, it emits a photon with frequency ν given by the relation $h\nu = E_2 - E_1$. Conversely, an electron in the lower orbit can absorb a photon with frequency given by this relation and jump to the higher orbit.
3.   The angular momentum $\vec{l}$ of an electron in one of its stable orbits is always an integral multiple of $\hbar = h/2\pi$.

These initial postulates were later developed into a complete theory for the dynamics of atoms and, more generally, microscopic particles. Modern quantum mechanics was founded by Werner Heisenberg, Max Born, Erwin Schroedinger, Paul Adrian Maurice Dirac, Wolfgang Pauli, Enrico Fermi, and others. It attributes wave *and* particle aspects to all forms of matter. This book is not about quantum mechanics and, indeed, in the first three parts only a few basic facts about the quantum theory will be needed, which will simply be stated without further justification. The meaning of these facts will be clarified with the help of a few examples. (A more complete synopsis of quantum theory can be found in appendix B.)

The most important principle of quantum theory is as follows.

QM 1. *The energy of an isolated system (or single particle) is restricted to a set of allowed energy values. These are given by the eigenvalues E of a certain matrix (operator) $\mathcal{H}$, called the **Hamiltonian**, i.e. the values of E for which there exists a non-zero vector ψ such that*

$$\mathcal{H}\psi = E\,\psi. \tag{18.2}$$

*The eigenvectors ψ are often called **wave functions**. The set of allowed energy values is called the **energy spectrum**.*

The matrix $\mathcal{H}$ is in most cases infinitely large, so that there are infinitely many allowed energy values. Importantly, the energy spectrum of a **bound particle**, i.e. a particle in a potential well, is a **discrete set.** The corresponding energy values are called **energy levels** and the associated eigenvectors are **bound states** because the particle is unable to escape the potential well.

This is the only case of importance for statistical mechanics. It is possible that there are more than one eigenvectors with the same eigenvalue E. In that case the energy level E is called **degenerate.** If a particle has a particular energy E it resides in (one of) the corresponding **eigenstate**(s) defined by the eigenvector ψ.

REMARK 18.1: *Wave mechanics and the probabilistic interpretation.*
The Hamiltonian is usually represented as a differential operator instead of a matrix. This is completely equivalent but allows the following interpretation for the eigenstates ψ. These states are then functions $\psi(x)$ and represent a *wave* associated with the particle. quantum mechanics is therefore sometimes called **wave mechanics.** These waves have the following **probabilistic interpretation.**

> *The square of the modulus $|\psi(x)|^2$ is the probability density of finding the particle at the position x.*

In quantum mechanics, it is in general *fundamentally* impossible to determine the exact outcome of an experiment; one can only determine the **probability** of any given outcome. This fact will not play a major role in our elementary discussions of statistical mechanics, but it is important to realize that this fundamental probability is quite unrelated to the probability due to the unknown random behaviour of large numbers of particles that is essential for statistical mechanics.

Let us now discuss some simple examples. We shall not write down the explicit form of the corresponding Hamiltonian but simply give the energy levels without further explanation. (However, see appendix B for details.)

EXAMPLE 18.1: *Particle on a line in a harmonic well.*
Consider a particle of mass m on a line with position coordinate x attracted to the origin by a central force $F = -k\,x$. Then the energy levels are given by

$$E_n = \hbar\omega\left(n + \tfrac{1}{2}\right), \qquad (n = 0, 1, 2, \dots) \qquad (18.3)$$

where $\omega = \sqrt{k/m}$ and

$$\hbar = \frac{h}{2\pi} = 1.05 \cdot 10^{-34} \text{ J s} \qquad (18.4)$$

is Planck's constant divided by 2π. One can depict this as shown in figure 18.2. The potential well is a parabola. It is the classical potential energy corresponding to the force F:

$$F = -\frac{dV}{dx}; \qquad V(x) = \frac{1}{2}k\,x^2. \qquad (18.5)$$

Note that the lowest possible energy level is not zero; a quantum particle is always moving! This is called **zero-point motion**. In most cases this lowest

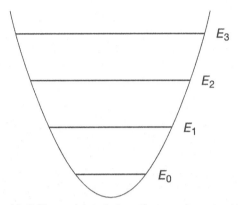

Figure 18.2 Energy levels in a harmonic potential well.

energy level is **non-degenerate**, i.e. there is a unique corresponding eigen-vector, which is called the **ground state.**

EXAMPLE 18.2: *Free particle in a box.*
Consider first a free particle moving on a line confined to the interval 0 to L. In that case the energy levels are given by

$$E_n = \frac{\pi^2 \hbar^2}{2mL^2} n^2 \quad (n = 1, 2, 3, \ldots). \tag{18.6}$$

(These are the energy levels in the case of **Dirichlet boundary conditions**, which means roughly that the particle bounces elastically off the walls.) Note that the lowest energy level is again non-zero. We have the picture in figure 18.3.

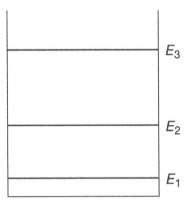

Figure 18.3 Energy levels of a particle confined to an interval.

The energy well is now a square well: the infinite potential prevents the particle from crossing the walls.

Next consider a free particle in a three-dimensional box with side L. This example will be of importance to us later because an ideal gas can be modelled as an assembly of independent free particles in a box (container). The three degrees of freedom can be treated as completely independent so that the energy levels are given by a sum of three levels of the type (18.6):

$$E(n_1, n_2, n_3) = \frac{\pi^2 \hbar^2}{2mL^2}(n_1^2 + n_2^2 + n_3^2), \tag{18.7}$$

where $n_1, n_2, n_3 = 1, 2, 3, \ldots$. Again, the ground state energy $E_0 = E(1, 1, 1)$ is positive and the ground state is non-degenerate. However, the higher, so-called **excited energy levels** are **degenerate**: there are more than one combinations of integers n_1, n_2 and n_3 with the same $n_1^2 + n_2^2 + n_3^2$.

EXAMPLE 18.3: *Charged particle in a central Coulomb potential.*
For atomic physics, the hydrogen atom has proved to be of special importance because it is the simplest atom and the only atom for which the energy levels can be calculated *exactly*! Basically, the hydrogen atom consists of a **proton** p with positive charge $+e$ and relatively large mass $M_p \approx 1.67 \times 10^{-27}$ kg, and an **electron** e$^-$ with equal but opposite charge $-e$ and with a much smaller mass $m_e \approx 9.1 \times 10^{-31}$ kg; so $M_p \approx 1800\, m_e$. The proton charge is $e = 1.602 \times 10^{-16}$ C.

Opposite charges attract each other by the **Coulomb force** (figure 18-4)

$$\vec{F}_{\text{Coulomb}} = -\frac{e^2}{4\pi\epsilon_0 r^3}\vec{r}. \tag{18.8}$$

(The constant $4\pi\epsilon_0 = 1.112 \times 10^{-10}$ C^2 J^{-1} m^{-1}.)

p $\longrightarrow$ F   F $\longleftarrow$ e$^-$

Figure 18.4 Coulomb attraction between electron and proton.

Because $M_p \gg m_e$ we can neglect the motion of the proton. (This is an unnecessary simplification.) The problem then reduces to an electron e$^-$ moving in a central force field given by equation (18.8). The corresponding **electrical potential** is given by

$$\phi_{\text{Coulomb}}(r) = \frac{e}{4\pi\epsilon_0 r}. \tag{18.9}$$

This means that the potential energy $V(r)$ is given by

$$V(r) = -e\,\phi(r). \tag{18.10}$$

($-e$ is the charge of the electron.) The energy levels for the bound states are

$$E_n = -\frac{m\,e^4}{8\epsilon_0^2 h^2 n^2} \qquad (n = 1, 2, \cdots).\qquad (18.11)$$

Here $m = m_{\rm e}$. Figure 18.5 gives an impression of these energy levels.

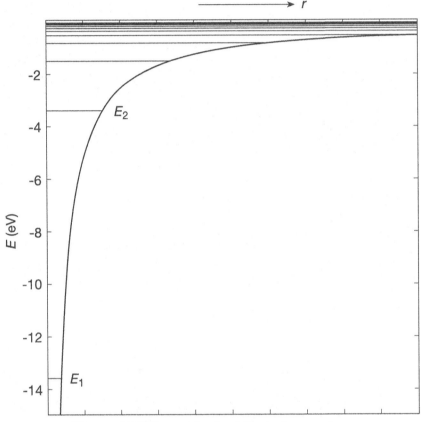

Figure 18.5 Energy levels of the hydrogen atom.

Again there is a lowest energy level or ground state. The ground state energy is of course not positive because the energy well is not positive. However, the ground state energy is elevated above the bottom of the well. This means that the electron cannot fall into the infinitely deep well owing to its zero-point motion. This in fact solves one of the main problems of classical mechanics applied to atoms. A charged electron would, according to classical mechanics, radiate electromagnetic waves and loose energy in the process. A classical atom would therefore be unstable. Note that in this case the energy well does not go to $+\infty$ as $r \to \infty$. Instead, $V(r) \to 0$. As a result the energy levels do not have wider and wider spacing as $n \to \infty$ but instead tend to zero. Actually, positive energies $E > 0$ are also allowed. The electron is then free or **unbound** to the proton and the hydrogen atom is **ionized.** The kinetic energy of the electron is then too large for it to be captured by the proton.

REMARK 18.2: *The hydrogen spectrum according to Bohr.*
The spectrum (18.11) also follows from Bohr's postulates. To see this remember that the angular momentum is given by $\vec{L} = m\vec{v} \times \vec{r}$, where $\vec{r}$ is the position vector and $\vec{v}$ the velocity of the electron. Using the third postulate, this gives one relation for the unknowns r and v. Another relation is obtained by equating the Coulomb force (18.8) and the centripetal force:

$$-\frac{e^2}{4\pi\epsilon_0 r^2} = -\frac{mv^2}{r}.$$

The sum of the kinetic energy $\frac{1}{2}mv^2$ and the potential energy $V(r)$ then yields (18.11). Note that by Bohr's postulate 2, the hydrogen atom can emit photons with frequencies $(E_{n_2} - E_{n_1})/h$. These correspond to lines in the spectrum of hydrogen. For $n_1 = 1$ they are called **Lyman lines**, for $n_1 = 2$ they are called **Balmer lines**, and for $n_1 = 3$ they are called **Paschen lines**. Although Bohr's postulates yield the correct spectrum for hydrogen, they are only an approximation to quantum mechanics in general. In fact, the orbit of an electron is not well-defined; the position of an electron is given by a probability distribution as described above. Moreover, Bohr's postulate 3 is incorrect; there are only two independent scalar angular momentum parameters l and m determining the eigenstates corresponding to a given energy value E_n. (In addition, there is a spin angular momentum s; see below.)

The level scheme for other atoms has a similar appearance although more complicated. This has important consequences for the radiation spectrum of an atom. Normally, the atom will be in the ground state. However, when enough energy is supplied an outer electron can be excited to a higher energy level. This can be done, for instance, by electromagnetic radiation (light) or by heating. A particle of light or **photon** can knock the electron into an excited state. The energy of a photon of frequency ν is given by equation (18.1). The frequency of visible light is in the range between $4 \cdot 10^{14}$ Hz and $7.5 \cdot 10^{14}$ Hz. This corresponds to an energy range between 1.7 eV and 3 eV. A hydrogen atom in its ground state can therefore not be brought into an excited state by means of visible light; one needs ultraviolet light. Outer electrons of some other atoms can be excited more easily, however. When an electron has been excited it will fall back spontaneously to a lower energy level and light of a very precisely defined characteristic frequency is emitted. This gives rise to **spectral lines**: colour lines in the spectrum of light emitted by atoms.

It should be mentioned here that the energy spectrum of a hydrogen atom is in fact not as simple as it was presented here. This is due to several small effects. The energy levels in figure 18.5 are $2n^2$−fold *degenerate*. (The angular momentum variables mentioned in the remark above can have the values $l = 0, 1, \ldots, n - 1$ and $m = -l, -l + 1, \ldots, l$ and the spin angular momentum has two possible values.) These small perturbing effects cause the degeneracy to be lifted and the spectral lines to split up into closely spaced separate lines (figure 18.6).

Figure 18.6 Splitting of degenerate energy levels due to perturbations.

The main effects are:

(a)   magnetic interactions between the magnetic moment due to the orbital motion of the electron and the electron spin and between the electron spin and the proton spin; and

(b)   relativistic corrections due to the fact that the speed of the electron in its orbit is very high.

EXAMPLE 18.4: *Electron spin.*
The **electron spin** mentioned as a small complication to the spectrum of the hydrogen atom is of great importance for the phenomena of ferromagnetism and paramagnetism. Roughly speaking one can picture the electron as a tiny rotating sphere (figure 18.7).

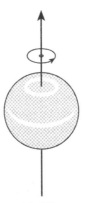

Figure 18.7 Electron spin.

The rotating charge of the electron then induces a magnetic dipole moment μ_e. According to quantum mechanics, the electron cannot spin at just any rate. It can only spin at one specific angular frequency in either direction. The corresponding angular momentum is **quantized**: it takes on only two discrete values: $\pm S$ with

$$S = \tfrac{1}{2}\hbar. \tag{18.12}$$

Remember that, in general, the angular momentum L of a rotating body is given by $L = I\omega$, where I is the moment of inertia and ω is the angular velocity. If the body is a solid ball with radius r then $I = \tfrac{2}{5}mr^2$ as opposed to

mr^2 for a particle in orbit. However, the radius r of the electron is not known. The angular frequency ω can therefore not be determined. In fact, the picture of a spinning ball is incorrect; the quantum theory of the electromagnetic field and its interactions with charged particles, **quantum electrodynamics**, or **QED**, assumes that $r = 0$! Nevertheless, the electron does have a spin angular momentum given by equation (18.12). The induced magnetic moment is given by $\mu_e = \mp \mu_B$, where

$$\mu_B = \frac{e\hbar}{2m} \tag{18.13}$$

is called the **Bohr magneton**. When the electron is subjected to a magnetic field $\vec{H}$, this field will exert a force on the electron trying to align $\vec{\mu}_e$ and $\vec{H}$. As a result the electron has a lower energy when $\vec{\mu}_e$ and $\vec{H}$ are aligned than when they are anti-aligned. (According to quantum mechanics there are *only* these two possibilities: the direction of $\vec{\mu}_e$ is also quantized.) When $\vec{\mu}_e$ and $\vec{H}$ are parallel the total energy of the electron is reduced by an amount

$$\epsilon = \mu_0 \, \mu_B \, H, \tag{18.14}$$

while if they are anti-parallel the total energy is increased by this amount. We thus have two magnetic energy levels for the electron (figure 18.8).

Figure 18.8 Magnetic energy levels of an electron.

This makes magnetic systems particularly easy to analyse, and in the following chapter we shall therefore start by considering a macroscopic system of non-interacting electrons with fixed positions in a magnetic field $\vec{H}$.

However, first we have to introduce two more principles of quantum mechanics.

QM 2. *The energy levels of a composite system of non-interacting parts are given by the sums of energy levels for each of its parts.*

Consider as an example two electrons with energies $E \pm \epsilon$ in a magnetic field. Then the total energy has the following possible values: $2(E + \epsilon)$, $2E$, and $2(E - \epsilon)$, where the level $2E$ is twofold degenerate.

The third principle is not so easy to grasp.

QM 3. *All (elementary) particles can be subdivided into two groups:* **bosons** *and* **fermions**. *Bosons are particles whose spin is an entire multiple of $\hbar$, fermions have spin $\frac{1}{2}m\hbar$, where m is an odd integer. Bosons obey* **Bose-Einstein statistics**; *fermions obey* **Fermi-Dirac statistics** *and the* **Pauli exclusion principle**.

We shall see later what **statistics** means. It has to do with the degeneracy of energy levels of an assembly of identical particles and is therefore of great importance to statistical mechanics. The **Pauli principle** says that at most $2s + 1$ fermions can occupy the same energy state if $s = S/\hbar$. In particular, at most *two electrons* can occupy the same eigenstate. Alternatively, one can say that only one electron can occupy the same state, but every energy level has an extra two-fold degeneracy due to the electron spin. This degeneracy is lifted when a magnetic field is applied giving rise to the magnetic levels of figure 18.8.

The Pauli principle is the reason behind the periodic table of the elements. This can be understood as follows. Atoms other than hydrogen have more than one electron in orbit around the nucleus. To a first approximation, all these electrons move in a central force field due to the charge of $+Ze$ of the nucleus. This gives rise to a similar but deeper potential well to the one in figure 18.5. Now, each energy level can accommodate only as many electrons as its degeneracy (including a factor 2 for spin degeneracy). This means that the ground state of the atom is formed by filling up the energy levels from the bottom upwards. The chemical properties of the elements are completely determined by their detailed energy level structure, and in particular the energy of the highest filled levels. The energy levels can again be labelled by the quantum numbers n, l, m, and s (see remark 18.2), which take on the values $n = 1, 2, 3, \ldots$, $l = 0, 1, 2, \ldots, n-1$, $m = -l, -l+1, \ldots, l$, and $s = \pm\frac{1}{2}$. Traditionally, the different l-values are indicated by letters: s for $l = 0$, p for $l = 1$, d for $l = 2$, f for $l = 3$, and then alphabetically upwards. Electrons with the same n-value are said to be in the same **shell**. The number of electrons in a given shell and with a given value of l is indicated as an index. Thus, the ground-state configuration of silicon (Si) is indicated as follows:

$$1s^2\, 2s^2\, 2p^6\, 3s^2\, 3p^2.$$

This means that Si has a total of 14 electrons ($Z = 14$), two in the 1s state (with opposite spin), two in the 2s state, six in the 2p state (with $m = -1, 0, 1$ and $s = \pm\frac{1}{2}$), two in the 3s state, and two in the 3p state. Note that the highest shell is only partially filled.

Elements with few electrons in the highest shell are called **metals**. They have many chemical properties in common because the chemical properties are almost entirely determined by the highest non-empty shell. For example, sodium (Na) has the configuration

$$1s^2\, 2s^2\, 2p^6\, 3s^1.$$

The behaviour is not quite so straightforward for heavier elements because the order of the energy levels is mixed up. For example, the 4s levels fill up before the 3d levels and elements with partially filled 3d levels behave like metals; these are the **transition metals.** Iron is an example; it has the configuration

$$1s^2 \, 2s^2 \, 2p^6 \, 3s^2 \, 3p^6 \, 4s^2 \, 3d^6.$$

(The 3d shell can take a maximum of 10 electrons.) Elements with a completely filled highest shell are chemically inactive; they are the **inert gases** helium (He), neon (Ne), argon (Ar), etc.

CHAPTER 19

Non-Interacting Localized Spins

Let us now consider a simple model of a macroscopic system consisting of a large number N of *identical particles* with 'spin-$\frac{1}{2}$', i.e. $S = \frac{1}{2}\hbar$, pinned to fixed positions and immersed in a magnetic field H. We shall see that with the help of one simple postulate we can derive the whole thermodynamics of this system. As explained in the previous chapter, the external field H causes the energy levels ϵ_i of each particle to split up into two levels $\epsilon_i \pm \epsilon$, where $\epsilon = \mu_0 \mu H$. (See equation (18.14); μ is the magnetic moment of the particle, $\mu_0 = 4\pi \cdot 10^{-7}$ H m^{-1} is an absolute constant.) Since all particles are pinned to fixed positions, we can disregard excited states and consider only the ground state of each particle with energy ϵ_0. Thus each particle has effectively only two available energy levels: $\epsilon_0 - \epsilon$ and $\epsilon_0 + \epsilon$. (If the particles are fixed then it costs a lot of energy to move them, which means that the excited states are much higher than $\epsilon_0 + \epsilon$.) For simplicity we shall put $\epsilon_0 = 0$ in the following. This does not affect the final result. Since there are N particles in all, this means that there are 2^N **microstates** available to the system. If the system is isolated it cannot exchange energy with the surroundings so that the total energy E is fixed. This means that only those microstates are actually accessible to the system for which

$$M\epsilon - (N - M)\epsilon = E, \qquad (19.1)$$

where M is the number of particles in the upper energy level. The total number of microstates for given M ($0 \leqslant M \leqslant N$) is

$$\Omega(E) = \binom{N}{M}. \qquad (19.2)$$

The basic postulate of statistical mechanics is **Boltzmann's postulate**:

SM. *The entropy density of a localized system in equilibrium with given energy per particle u is*

$$s(u) = \lim_{N \to \infty} \frac{k_B}{N} \ln \Omega(Nu) \tag{19.3}$$

where $\Omega(E)$ is the number of microstates accessible to the system at energy E.

Remember that the fundamental function $s(u, v)$ determines the thermodynamics of a simple system completely. Here the volume plays no role because all the particles are fixed to their positions. To evaluate the limit (19.3) we need Stirling's formula for $\ln N!$ which we shall first derive from Laplace's **principle of steepest descent**.

Theorem 19.1 (Laplace) *Let $G : I \to \mathbb{R}$ be a continuous function on an interval I, which is bounded above, that is $G(x) \leqslant K$ for some constant K and all $x \in I$. Assume that there exists $\alpha > 0$ such that for large $|x|$, $G(x) < -\alpha|x|$. Then*

$$\lim_{N \to \infty} \frac{1}{N} \ln \int_I e^{N G(x)} dx = \sup_{x \in I} G(x). \tag{19.4}$$

Proof. Let L be so large that $G(x) < -\alpha|x|$ for $|x| \geqslant L$ and $-\alpha L < \sup_{x \in I} G(x)$. Then

$$\int_I e^{N G(x)} dx \leqslant 2L \exp\left[N \sup_{x \in I} G(x)\right] + 2 \int_L^\infty e^{-\alpha N x} dx$$

$$< \left(2L + \frac{2}{\alpha N}\right) \exp\left[N \sup_{x \in I} G(x)\right].$$

Taking N to infinity we find that

$$\limsup_{N \to \infty} \frac{1}{N} \ln \int_I e^{N G(x)} dx \leqslant \sup_{x \in I} G(x). \tag{19.5}$$

It remains to prove the inequality

$$\liminf_{N \to \infty} \frac{1}{N} \ln \int_I e^{N G(x)} dx \geqslant \sup_{x \in I} G(x). \tag{19.6}$$

Let $K = \sup_{x \in I} G(x)$ and choose $\epsilon > 0$ arbitrarily small. Because G is continuous there exist $x_0 \in I$ and $\delta > 0$ such that, if $|x - x_0| < \delta$ then $G(x) > K - \epsilon$. Hence

$$\int_I e^{N G(x)} dx > \int_{(x_0 - \delta, x_0 + \delta) \cap I} e^{N G(x)} dx > \delta \, e^{N(K - \epsilon)}$$

so that

$$\frac{1}{N} \ln \int_I e^{N G(x)} dx > K - \epsilon + \frac{1}{N} \ln \delta$$

and, taking $N \to \infty$,

$$\liminf_{N \to \infty} \frac{1}{N} \ln \int_I e^{N\,G(x)} \mathrm{d}x \geqslant K - \epsilon.$$

As this holds for arbitrary $\epsilon > 0$, it implies equation (19.6). ∎

As an application we have the following.

Theorem 19.2 (Stirling's formula) *The following asymptotic formula for $N!$ holds:*

$$\lim_{N \to \infty} \left(\frac{1}{N} \ln (N!) - \ln N \right) = -1. \tag{19.7}$$

Proof. We have

$$N! = \int_0^\infty e^{-x} x^N \, \mathrm{d}x = \int_0^\infty e^{N \ln x - x} \, \mathrm{d}x. \tag{19.8}$$

Putting $x = Nt$ we obtain

$$\frac{1}{N} \ln(N!) - \ln N = \frac{1}{N} \ln N + \frac{1}{N} \ln \int_0^\infty e^{N(\ln t - t)} \mathrm{d}t.$$

Taking $N \to \infty$ we find, using theorem 19.1 with $G(t) = \ln t - t$ on $(0, +\infty)$,

$$\lim_{N \to \infty} \left(\frac{1}{N} \ln(N!) - \ln N \right) = \sup_{t \geqslant 0} (\ln t - t) = -1.$$

∎

REMARK 19.1: *Traditional form of Stirling's formula.*
Traditionally, Stirling's formula is written in asymptotic form as follows:

$$N! \approx N^N \, e^{-N} \, \sqrt{2\pi N}.$$

The first two factors in this formula are covered by equation (19.7). The last factor is obtained by expanding $\ln t$ around the point $t = 1$ where $G(t)$ is maximal: $\ln t \approx t - 1 - \frac{1}{2}(t - 1)^2$. After replacing the lower bound in the resulting Gaussian integral by $-\infty$ it can be calculated. The resulting factor $\sqrt{2\pi N}$ is not important for the following, however.

Let us now apply this result to our model. Inserting equation (19.2) into (19.3) and using theorem 19.2 we obtain

$$\begin{aligned}
s(u) &= k_{\mathrm{B}} \lim_{N \to \infty} \frac{1}{N} \ln \binom{N}{M} \\
&= k_{\mathrm{B}} \lim_{N \to \infty} \left[\frac{1}{N} \ln(N!) - \ln N - \frac{M}{N} \left(\frac{1}{M} \ln(M!) - \ln M \right) \right. \\
&\quad - \frac{N - M}{N} \left(\frac{1}{N - M} \ln(N - M)! - \ln(N - M) \right) \\
&\quad \left. + \ln N - \frac{M}{N} \ln M - \frac{N - M}{N} \ln(N - M) \right]
\end{aligned} \tag{19.9}$$

By equation (19.1), $u = \dfrac{2M - N}{N}\epsilon$, so that we arrive at the formula

$$s(u) = -k_{\mathrm{B}} \left[\frac{1 + u/\epsilon}{2} \ln \frac{1 + u/\epsilon}{2} + \frac{1 - u/\epsilon}{2} \ln \frac{1 - u/\epsilon}{2} \right]. \tag{19.10}$$

The temperature of the system now follows from this fundamental formula by means of equation (6.13):

$$\frac{1}{T} = \frac{ds}{du} = \frac{k_{\mathrm{B}}}{2\epsilon} \ln \left(\frac{\epsilon - u}{\epsilon + u} \right). \tag{19.11}$$

Inverting this relation we obtain

$$u = -\epsilon \tanh \left(\frac{\epsilon}{k_{\mathrm{B}}T} \right). \tag{19.12}$$

The magnetization m follows from equation (10.6):

$$m = -\frac{(2M - N)\mu}{V} = -\frac{\mu}{\epsilon v} u, \tag{19.13}$$

which is

$$m = \frac{\mu}{v} \tanh \left(\frac{\epsilon}{k_{\mathrm{B}}T} \right). \tag{19.14}$$

Introducing the **magnetic induction in vacuum** given by $B_0 = \mu_0 H$, we have $\epsilon = \mu B_0$. Some graphs of B_0 as a function of m have been sketched in figure 19.1.

For small B_0, that is $\mu B_0 \ll k_{\mathrm{B}}T$,

$$m \approx \frac{\mu^2}{v\, k_{\mathrm{B}}T} B_0 \tag{19.15}$$

and the susceptibility χ follows from equation (10.2):

$$\chi \approx \frac{\mu^2 \mu_0}{v\, k_{\mathrm{B}}T}. \tag{19.16}$$

This is Curie's law (10.3)! The system is therefore a simple model for a paramagnet.

The magnetic spins in a paramagnet are usually d-shell electrons in magnetic ions of a paramagnetic salt. (This model does not account for paramagnetism in metals because there the conduction electrons are not localized.) Other materials are paramagnetic due to their nuclear spin. In that case the magnetic moment μ tends to be much smaller. We have assumed that the interaction between the spins is negligible. Later, we shall see that in some cases where the interaction is not negligible the system can behave like a ferromagnet. For some materials Curie's law holds very accurately down to very

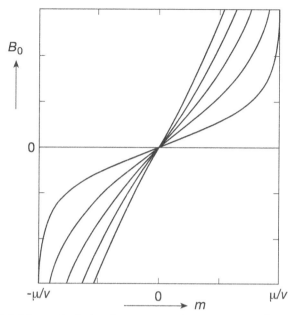

Figure 19.1 Magnetic induction versus magnetization of a paramagnet.

low temperatures. This makes them very suited for the measurement of low temperatures as well as for the attainment of low temperatures. The method of using paramagnetic salts for attaining low temperatures is called **adiabatic demagnetization**. It works as indicated in figure 4.1 where the parameter x is the magnetic field H. Measurement of low temperatures can be done using these salts down to temperatures in the millikelvin range. For even lower temperatures, one uses nuclear spins instead of electron spins. These have even smaller interaction and the Curie law thus holds down to extremely low temperatures.

REMARK 19.2: *About SI units.*
The magnetic induction B_0 is measured in *Tesla* (T): $1 \text{ T} = 1 \text{ J A}^{-1} \text{ m}^{-2}$. Magnetic field strength is measured in amperes per metre (A m^{-1}) and the proportionality constant μ_0 is given by $\mu_0 = 4\pi \cdot 10^{-7} \text{ H m}^{-1}$, where $1 \text{ H} = 1 \text{ J A}^{-2}$. Finally, μ is measured in A m^2 so that $\mu^2 \mu_0$ is measured in J m^3, which are also the units of $v k_B T$.

CHAPTER 20

Large Deviation Theory

We want to generalize the theory of the previous chapter to particles with more than two energy levels. To this end we shall first generalize Laplace's theorem to a result more suited to the general situation. This generalization is called **large-deviation theory**. In chapter 21 we apply this theory in re-deriving equation (19.10) for the entropy of a paramagnet. In chapter 22 we consider the more general situation. Large-deviation theory also has important applications in other areas of mathematics, for example information theory, dynamical systems, and queueing theory. See problems II-19 to II-23.

The main theorem of large-deviation theory is a generalization of theorem 19.1, where the integral over x is replaced by an integral with respect to the distribution function F_N of a general random variable X_N. Remember that, if X_N is a random variable then its distribution function F_N is defined by $F_N(x) = \mathbb{P}(X_N \in (-\infty, x])$, where $\mathbb{P}$ denotes the probability. Similarly, if X_N is a random variable with values in $\mathbb{R}^d$ then

$$F_N(x_1, \ldots, x_d) = \mathbb{P}\left(X_N \in (-\infty, x_1] \times \cdots \times (-\infty, x_d]\right). \tag{20.1}$$

To formulate the theorem we first need some more definitions.

Definition 20.1 *A function* $I : \mathbb{R}^d \to [0, +\infty]$ *(which may also take the value* $+\infty$*) is called a* **rate function** *if, for every* $B \in [0, +\infty)$*, the set* $I^{-1}(B) = \{x \in \mathbb{R}^d \mid I(x) \leqslant B\}$ *is compact.*

Definition 20.2 *A sequence of random variables* X_N $(N \in \mathbb{N})$ *with values in* $\mathbb{R}^d$ *satisfies the* **large-deviation principle (LDP)** *if there exists a rate function* $I : \mathbb{R}^d \to [0, \infty]$ *such that the following hold:*

LD 1. *For all closed subsets* $F \subset \mathbb{R}^d$,

$$\limsup_{N \to \infty} \frac{1}{N} \ln \mathbb{P}\left(X_N \in F\right) \leqslant -\inf_{x \in F} I(x) \tag{20.2}$$

LD 2. *For all open subsets* $G \subset \mathbb{R}^d$,

$$\liminf_{N \to \infty} \frac{1}{N} \ln \mathbb{P}\left(X_N \in G\right) \geqslant -\inf_{x \in G} I(x). \tag{20.3}$$

REMARK 20.1: *The one-dimensional case.*
In the one-dimensional case $(d = 1)$ which we shall mostly consider, it suffices to prove LD 1 and LD 2 for closed and open intervals respectively. Moreover, if F_N is continuous, in particular if X_N has a density,

$$\mathbb{P}\left(X_N \in (a, b)\right) = \mathbb{P}\left(X_N \in [a, b]\right) = F_N(b) - F_N(a). \qquad (20.4)$$

If $I(x)$ is also continuous, then the two inequalities (20.2) and (20.3) combine and become

$$\lim_{N \to \infty} \frac{1}{N} \ln \mathbb{P}\left(X_N \in (a, b)\right) = - \inf_{x \in (a,b)} I(x). \qquad (20.5)$$

These definitions are rather abstract, so let us consider some simple examples.

EXAMPLE 20.1: *Distribution function satisfying the LDP.*
Define distribution functions F_N by

$$F_N(x) = \frac{N}{2} \int_{-\infty}^{x} e^{-N|u|} \, du.$$

If X_N is a random variable with distribution function F_N then the sequence $\{X_N\}_{N=1,2,\ldots}$ satisfies the LDP with rate function $I(x) = |x|$. To see this we compute $F_N(x)$:

$$F_N(x) = \begin{cases} \frac{1}{2} e^{-N|x|}, & \text{if } x < 0 \\ 1 - \frac{1}{2} e^{-Nx}, & \text{if } x \geqslant 0. \end{cases}$$

Since X_N has a density function, and $I(x)$ is continuous it follows by the remark above that we must prove the identity (20.5). Now if $a < b < 0$ then $F_N(b) - F_N(a) = \frac{1}{2} \left(e^{-N|b|} - e^{-N|a|}\right)$ so that

$$\frac{1}{N} \ln \left(F_N(b) - F_N(a)\right) \to -|b| = - \inf_{x \in (a,b)} I(x) \text{ as } N \to \infty.$$

If $a \leqslant 0 \leqslant b$ then $F_N(b) - F_N(a) = 1 - \frac{1}{2} \left(e^{-Nb} + e^{-N|a|}\right)$ and

$$\frac{1}{N} \ln \left(F_N(b) - F_N(a)\right) \to 0 = - \inf_{x \in (a,b)} I(x).$$

Finally, if $0 \leqslant a < b$, $F_N(b) - F_N(a) = \frac{1}{2}(e^{-Na} - e^{-Nb})$ and

$$\frac{1}{N} \ln \left(F_N(b) - F_N(a)\right) \to -a = - \inf_{x \in (a,b)} I(x).$$

The situation is illustrated in figure 20.1, which shows the probability density $f_N(x) = \frac{1}{2} N e^{-N|x|}$ for a number of values of N. For $x \neq 0$, this tends to zero exponentially fast with rate given by $I(x) = |x|$.

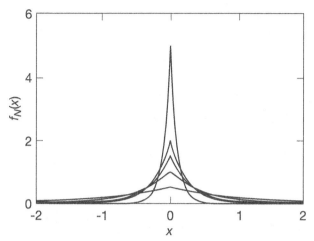

Figure 20.1 Probability density for a sequence of random variables satisfying the LDP.

EXAMPLE 20.2: *A discontinuous rate function.*
If X_N is the random variable with distribution function defined by

$$F_N(x) = \begin{cases} N \int_0^x e^{-Nx}\,\mathrm{d}x & \text{for } x \geqslant 0 \\ 0 & \text{for } x < 0, \end{cases}$$

then the sequence $\{X_N\}_{n=1,2,\dots}$ satisfies the LDP with rate function

$$I(x) = \begin{cases} +\infty & \text{if } x < 0 \\ x & \text{if } x \geq 0. \end{cases}$$

Note that $I(x)$ is not continuous but does satisfy the condition for a rate function: $\{x \mid I(x) \leq B\} = [0, B]$. The proof of the LDP is similar to example 20.1.

EXAMPLE 20.3: *A discrete version of example 20.2.*
Let X_N be the random variable with distribution function given by

$$F_N(x) = \begin{cases} 1 - \exp\{-N[x]\}, & \text{if } x > 0, \\ 0 & \text{if } x \leqslant 0, \end{cases}$$

where $[x]$ denotes the integer part of x. Then $\{X_N\}_{N=1,2,\dots}$ satisfies the LDP with rate function

$$I(x) = \begin{cases} +\infty & \text{if } x \notin \{1, 2, \dots\}, \\ x - 1 & \text{if } x \in \{1, 2, \dots\}. \end{cases}$$

To prove this, consider first the upper bound. We can distinguish between two cases: $[a, b] \cap \{1, 2, \dots\} = \emptyset$ and $[a, b] \cap \{1, 2, \dots\} \neq \emptyset$. In the first case $\mathbb{P}(X_N \in [a, b]) = 0$ and the upper bound is trivial. In the second case, let $k_{\min}$ be the minimum positive integer in $[a, b]$ and $k_{\max}$ the maximum integer in $[a, b]$. Then $\mathbb{P}(X_N \in [a, b]) = e^{-N(k_{\min}-1)} - e^{-Nk_{\max}}$ and hence

$$\frac{1}{N} \ln \mathbb{P}(X_N \in [a, b]) \leqslant -(k_{\min} - 1) = - \inf_{x \in [a,b]} I(x).$$

For the lower bound we consider the same two cases. If $(a, b) \cap \{1, 2, \dots\} = \emptyset$ then $\inf_{x \in (a,b)} I(x) = +\infty$ and the bound is again trivial. Otherwise, $\mathbb{P}(X_N \in (a, b)) = e^{-N(k_{\min}-1)} - e^{-Nk_{\max}} > e^{-N(k_{\min}-1)}(1 - e^{-N})$ and therefore

$$\liminf_{N \to \infty} \frac{1}{N} \ln \mathbb{P}(X_N \in (a, b)) \geqslant -(k_{\min} - 1) = \inf_{x \in (a,b)} I(x).$$

EXAMPLE 20.4: *A degenerate example of LDP.*
Let X_N be the random variable with distribution function given by

$$F_N(x) = \begin{cases} 0 & \text{if } x \leqslant -1/N, \\ \frac{1}{2}(Nx + 1) & \text{if } -1/N \leqslant x \leqslant 1/N, \\ 1 & \text{if } x \geqslant 1/N. \end{cases}$$

Then $\{X_N\}_{N=1,2,\dots}$ satisfies the LDP with rate function

$$I(x) = \begin{cases} 0 & \text{if } x = 0, \\ +\infty & \text{if } x \neq 0. \end{cases}$$

Indeed, if $x < 0$ then $F_N(x) = 0$ for N large enough and if $x > 0$ then $F_N(x) = 1$ for N large enough.

> The large-deviation principle says that if $I(x) \neq 0$ in a (small) interval (a, b) then the probability $\mathbb{P}(X_N \in (a, b))$ that X_N takes its value in this interval tends to zero exponentially fast with rate given by the infimum of $I(x)$ over this interval.

In the present case $\mathbb{P}(X_N \in (a, b))$ is equal zero for N large and hence tends to zero faster than exponentially.

EXAMPLE 20.5: *More difficult bounds.*
Put

$$F_N(x) = \sqrt{\frac{N}{2\pi}} \int_{-\infty}^{x} e^{-Nu^2/2} \, du.$$

Then the sequence of random variables X_N with distribution functions F_N satisfy the LDP with rate function $I(x) = x^2/2$. To prove this we must show that equation (20.5) holds. We consider again the three cases $a < b \leqslant 0$, $a \leqslant 0 \leqslant b$ and $0 \leqslant a < b$. In the first case we have

$$F_N(b) - F_N(a) = \sqrt{\frac{N}{2\pi}} \int_a^b e^{-Nx^2/2} \, dx \leqslant \sqrt{\frac{N}{2\pi}} (b - a) \, e^{-Nb^2/2},$$

and hence

$$\limsup_{N \to \infty} \frac{1}{N} \ln \mathbb{P}\left(X_N \in [a, b]\right)$$

$$\leqslant \lim_{N \to \infty} \left[\frac{1}{N} \ln \sqrt{N} + \frac{1}{N} \ln \left(\frac{b - a}{\sqrt{2\pi}} \right) - \frac{1}{2} b^2 \right]$$

$$= -\frac{1}{2} b^2 = - \inf_{x \in [a,b]} I(x).$$

On the other hand, if $\epsilon > 0$ is small enough,

$$F_N(b) - F_N(a) > \sqrt{\frac{N}{2\pi}} \int_{b-\epsilon}^b e^{-Nx^2/2} \, dx > \epsilon \sqrt{\frac{N}{2\pi}} e^{-N(b-\epsilon)^2/2}$$

and hence

$$\liminf_{N \to \infty} \frac{1}{N} \ln \mathbb{P}\left(X_N \in (a, b)\right) > -\frac{1}{2}(b - \epsilon)^2$$

for arbitrarily small $\epsilon > 0$. The right hand side tends to $-b^2/2 = - \inf_{x \in (a,b)} I(x)$ as $\epsilon \to 0$. The other two cases are similar and will be left as an exercise.

Theorem 19.1 can be generalized to integrals with respect to a sequence of distribution functions F_N and even to higher dimensions. For this we need a lemma.

Lemma 20.1 *Suppose that* $\left(a_N^{(j)}\right)_{N=1,2,\dots}$ *with* $j = 1, 2, \dots, r$ *are* r *sequences of non-negative numbers. Then*

$$\limsup_{N \to \infty} \frac{1}{N} \ln \sum_{j=1}^r a_N^{(j)} \leqslant \bigvee_{j=1}^r \limsup_{N \to \infty} \frac{1}{N} \ln a_N^{(j)}. \tag{20.6}$$

(Here $\bigvee$ *denotes 'the maximum of'.)*

  Proof. Clearly,

$$\sum_{j=1}^r a_N^{(j)} \leqslant r \bigvee_{j=1}^r a_N^{(j)}.$$

It follows that

$$\ln \sum_{j=1}^r a_N^{(j)} \leqslant \ln \bigvee_{j=1}^r a_N^{(j)} + \ln r = \bigvee_{j=1}^r \ln a_N^{(j)} + \ln r$$

and hence

$$\limsup_{N\to\infty} \frac{1}{N} \ln \sum_{j=1}^{r} a_N^{(j)} \leqslant \limsup_{N\to\infty} \bigvee_{j=1}^{r} \frac{1}{N} \ln a_N^{(j)}$$

$$\leqslant \bigvee_{j=1}^{r} \limsup_{N\to\infty} \frac{1}{N} \ln a_N^{(j)}.$$

∎

Theorem 20.1 (Varadhan) *Let* $\left(X_N^{(1)}, \ldots, X_N^{(d)}\right)_{N=1}^{\infty}$ *be a sequence of d-dimensional random variables (that is, d-tuples of random variables) with corresponding distribution functions* $(F_N)_{N=1,2,\ldots}$ *satisfying the LDP with rate function* $I : \mathbb{R}^d \to [0, +\infty]$.
Then, if $G : \mathbb{R}^d \to \mathbb{R}$ *is continuous and bounded above,*

$$\lim_{N\to\infty} \frac{1}{N} \ln \int_{\mathbb{R}^d} e^{N\,G(\vec{x})} \, dF_N(x_1, \ldots, x_d) = \sup_{\vec{x}\in\mathbb{R}^d} [G(\vec{x}) - I(\vec{x})]. \qquad (20.7)$$

Proof. We only consider the one-dimensional case. The general case is proved in an analogous fashion. The proof uses the same strategy as that of theorem 19.1, finding an upper bound for the lim sup and a lower bound for the lim inf. Fix $A > 0$ large and $\epsilon > 0$ small. Put $K_A = \{x \mid I(x) \leqslant A\}$ and let L_A be so large that $K_A \subset (-L_A, L_A)$. We can cover this interval by a finite number n of smaller intervals $(a_i - \delta_i, a_i + \delta_i)$ $(i = 1, 2, \ldots, n)$ such that for $x \in (a_i - \delta_i, a_i + \delta_i)$ the following hold:

$$|G(x) - G(a_i)| < \epsilon \text{ and } I(x) > I(a_i) - \epsilon.$$

This is possible because G is continuous and $\{x \mid I(x) \leqslant I(a_i) - \epsilon\}$ is closed.

To see this remember that by the Heine-Borel theorem every open cover of a compact set has a finite subcover. Now, for any $a \in [-L_A, L_A]$, there is a $\delta_a > 0$ such that $|G(x) - G(a)| < \epsilon$ and $I(x) > I(a) - \epsilon$ for any $x \in (a - \delta_a, a + \delta_a)$. The open intervals $(a - \delta_a, a + \delta_a)$ with $a \in [-L_A, L_A]$ obviously cover $[-L_A, L_A]$ and therefore there exist finitely many points $a_1 < a_2 < \cdots < a_n \in [-L_A, L_A]$ such that $[-L_A, L_A] \subset \bigcup_{i=1}^{n} (a_i - \delta_{a_i}, a_i + \delta_{a_i})$. Finally, put $\delta_i = \delta_{a_i}$.

We now have

$$\int_{-\infty}^{\infty} e^{N\,G(x)} \, dF_N(x) \leqslant e^{NM} \mathbb{P}\left[|X_N| \geqslant L_A\right]$$

$$+ \sum_{i=1}^{n} e^{N(G(a_i)+\epsilon)} \mathbb{P}\left[X_N \in [a_i - \delta_i, a_i + \delta_i]\right],$$

where $M = \sup_{x \in \mathbb{R}} G(x)$. Using the lemma above we find that

$$\limsup_{N \to \infty} \frac{1}{N} \ln \int_{-\infty}^{\infty} e^{N\, G(x)} \, \mathrm{d}F_N(x)$$

$$\leqslant \max \left(M - A, \bigvee_{i=1}^{n} [(G(a_i) + \epsilon) - (I(a_i) - \epsilon)] \right)$$

$$\leqslant \max \left(M - A, \sup_{x \in \mathbb{R}} [G(x) - I(x)] + 2\epsilon \right)$$

$$= \sup_{x \in \mathbb{R}} [G(x) - I(x)] + 2\epsilon$$

if we choose $A > M - \sup_{x \in \mathbb{R}} [G(x) - I(x)]$. Since ϵ is arbitrary we get

$$\limsup_{N \to \infty} \frac{1}{N} \ln \int_{-\infty}^{\infty} e^{N\, G(x)} \, \mathrm{d}F_N(x) \leqslant \sup_{x \in \mathbb{R}} [G(x) - I(x)], \qquad (20.8)$$

which is the analogue of equation (19.5).

On the other hand, to obtain the analogue of equation (19.6), suppose the supremum is attained at x_0: $G(x) - I(x) \leqslant G(x_0) - I(x_0)$ for all $x \in \mathbb{R}$. Then, given $\epsilon > 0$, there exists $\delta > 0$ such that $|x - x_0| < \delta \implies |G(x) - G(x_0)| < \epsilon$. It follows that

$$\int_{-\infty}^{\infty} e^{N\, G(x)} \, \mathrm{d}F_N(x) \geqslant \int_{x_0 - \delta}^{x_0 + \delta} e^{N\, G(x)} \, \mathrm{d}F_N(x)$$

$$\geqslant e^{N(G(x_0) - \epsilon)} \, \mathbb{P}\left(X_N \in (x_0 - \delta, x_0 + \delta) \right).$$

Hence

$$\liminf_{N \to \infty} \frac{1}{N} \ln \int_{-\infty}^{\infty} e^{N\, G(x)} \, \mathrm{d}F_N(x)$$

$$\geqslant G(x_0) - \epsilon - \inf_{x \in (x_0 - \delta, x_0 + \delta)} I(x)$$

$$\geqslant G(x_0) - I(x_0) - \epsilon.$$

Taking $\epsilon \to 0$ we obtain

$$\liminf_{N \to \infty} \frac{1}{N} \ln \int_{-\infty}^{\infty} e^{N\, G(x)} \, \mathrm{d}F_N(x) \geqslant \sup_{x \in \mathbb{R}} [G(x) - I(x)]. \qquad (20.9)$$

The equations (20.8) and (20.9) imply (20.7). ∎

In the examples above we have seen some sequences of random variables satisfying the LDP. In those cases it was easy to guess the rate function. In general it is not so straightforward to find a candidate for the rate function. The following old result provides us with a general formula for the rate function for sums of independent random variables:

Theorem 20.2 (Cramér) *Let* $\left(X_k^{(1)}, \ldots, X_k^{(d)}\right)_{k=1,2,\ldots}$ *be a sequence of independent d-dimensional random variables with identical distribution functions* $F_k = F$, *and assume that the* **cumulant generating function** $C(t)$ *defined by*

$$C(t) = \ln \int_{\mathbb{R}^d} e^{\langle \vec{t}, \vec{x} \rangle} \, \mathrm{d}F(x_1, \ldots, x_d), \tag{20.10}$$

where $\langle \vec{t}, \vec{x} \rangle = \sum_{i=1}^{d} t_i x_i$ *is the scalar product, is finite for all* $\vec{t} \in \mathbb{R}^d$. *Then the sequence of d-dimensional random variables*

$$\left(\bar{X}_N^{(1)}, \ldots, \bar{X}_N^{(d)}\right) = \left(\frac{1}{N} \sum_{k=1}^{N} X_k^{(1)}, \ldots, \frac{1}{N} \sum_{k=1}^{N} X_k^{(d)}\right) \tag{20.11}$$

satisfies the LDP with rate function I *given by*

$$I(x) = \sup_{\vec{t} \in \mathbb{R}^d} \left\{ \langle \vec{t}, \vec{x} \rangle - C(\vec{t}) \right\}. \tag{20.12}$$

REMARK 20.2: *Derivation of the rate function.*
For simplicity consider the one-dimensional case. Note that if we take $G(x) = tx$ in equation (20.7) then we obtain

$$\lim_{N \to \infty} \frac{1}{N} \ln \int_{-\infty}^{\infty} e^{Ntx} \, \mathrm{d}F_N(x) =$$

$$= \lim_{N \to \infty} \frac{1}{N} \ln \left(\int_{-\infty}^{\infty} e^{tx} \, \mathrm{d}F(x) \right)^N$$

$$= C(t)$$

$$= \sup_{x \in \mathbb{R}} \left[tx - I(x) \right] \tag{20.13}$$

Inverting this Legendre transform assuming that $I(x)$ is convex we obtain equation (20.12) as a *candidate* for the rate function. Note that this is not a proof of theorem 20.2 because the conditions of theorem 20.1 are not satisfied: the function $G(x) = tx$ is not bounded above. Notwithstanding this, the result (20.13) is in fact correct. This follows from theorem 20.2 and the fact that the function $C(t)$ is convex: see lemma 20.3. To prove this, first remark that $C(t)$ is differentiable.

Lemma 20.2 *The function* $C : \mathbb{R}^d \to \mathbb{R}$ *is infinitely differentiable.*

Proof. We want to differentiate underneath the integral sign. This is allowed provided the integrand $e^{\langle \vec{t}, \vec{x} \rangle}$ and its derivatives $\left(\prod_{i=1}^{d} x_i^{k_i}\right) e^{\langle \vec{t}, \vec{x} \rangle}$ are uniformly integrable on neighbourhoods of any point t. This is the case as can easily be verified. For example, in one dimension, suppose $t \in (t_-, t_+)$ and let $t_m = |t_-| \vee |t_+|$. Then since $|x^k| \leqslant k! \, e^{|x|}$,

$$\left| x^k e^{tx} \right| \leqslant k! \left(e^{(t_m+1)x} + e^{-(t_m+1)x} \right)$$

and the integral on the right-hand side is finite because $C(t)$ is finite for all t.

∎

Lemma 20.3 *Suppose that for all $\vec{t} \in \mathbb{R}^d$, $C(\vec{t}) \in \mathbb{R}$. Then $C(\vec{t})$ is convex for all $\vec{t} \in \mathbb{R}^d$.*

Proof. We use theorem 7.4. In the one-dimensional case this simply means that the second derivative must be non-negative, which is easily checked:

$$C''(t) = \frac{\int x^2 e^{tx} \, dF(x)}{\int e^{tx} \, dF(x)} - \left(\frac{\int x e^{tx} \, dF(x)}{\int e^{tx} \, dF(x)} \right)^2$$
$$= \mathbb{E}_t[(X - \mathbb{E}_t(X))^2] \geqslant 0 \qquad (20.14)$$

where we have defined

$$\mathbb{E}_t[g(X)] = \frac{\int_{-\infty}^{\infty} g(x) e^{tx} \, dF(x)}{\int_{-\infty}^{\infty} e^{tx} \, dF(x)}. \qquad (20.15)$$

More generally, in d dimensions,

$$\frac{\partial^2 C}{\partial t_i \partial t_j} = \mathbb{E}_t \left[(X_i - \mathbb{E}_t(X_i))(X_j - \mathbb{E}_t(X_j)) \right] \qquad (20.16)$$

and this matrix is positive definite:

$$\langle \vec{v}, (\partial^2 C)\vec{v} \rangle = \mathbb{E}_t \left[\left(\sum_{i=1}^{d} (X_i - \mathbb{E}_t(X_i)) v_i \right)^2 \right].$$

∎

We finish this chapter with a proof of theorem 20.2 in the one-dimensional case. This is somewhat more difficult than the rest of the text. It uses the **central limit theorem** which is stated in appendix A, but will not be needed anywhere else in this book. This proof is not essential for the understanding of the following chapters. In fact we shall freely use a variety of generalizations of Cramér's theorem which will not be proved. These generalizations have similar proofs, although technically more involved. The main point is that one first obtains a candidate for the rate function in the same way as above.

Proof of theorem 20.2 We first show that $I(x)$ is a rate function. Clearly, $C(0) = 0$ so that $I(x) \geqslant 0$. In order to prove that the level sets $\{x \in \mathbb{R} | I(x) \leqslant B\}$ are bounded we must show that $I(x)$ tends to $+\infty$ as $|x| \to \infty$. We consider only the case $x \to +\infty$. As $x \to +\infty$ the supremum will be attained at increasingly higher values of t. We choose $t_0 > 0$ fixed and estimate $I(x)$ as follows: for $t' > t_0$,

$$C(t') \leqslant C(t_0) + C'(t')(t' - t_0)$$

so that if $x \geqslant C'(t')$ then

$$I(x) \geqslant t' x - C(t') \geqslant t' x - [C(t_0) + C'(t')(t' - t_0)] \geqslant t_0 x - C(t_0),$$

which tends to $+\infty$ as $x \to +\infty$.

In order to show that the level sets are closed suppose that $x_n \to x$ and $I(x_n) \leqslant b$. We have to show that $I(x) \leqslant b$. Now, since $I(x_n) \leqslant b$, $t\, x_n - C(t) \leqslant b$ for all t. Taking the limit $n \to \infty$ we get $t\, x - C(t) \leqslant b$ for all t and hence $I(x) \leqslant b$.

Now note that

$$C'(0) = \int_{-\infty}^{\infty} x \, \mathrm{d}F(x) = x_0 \tag{20.17}$$

and hence by convexity

$$C(t) \geqslant t\, x_0 \tag{20.18}$$

for arbitrary t. This implies that $I(x_0) = 0$. Let $[x_-, x_+] = \{x \in \mathbb{R} \mid I(x) = 0\}$. Then $I(x)$ is decreasing for $x \leqslant x_-$ and increasing for $x \geqslant x_+$. We now start the proof of equation (20.2). Suppose that $a > x_+$. Then $\inf_{x \in [a,b]} I(x) = I(a)$ and since for $t \geqslant 0$ and $x \geqslant a$, $e^{Nt(x-a)} \geqslant 1$ we obtain

$$\mathbb{P}\left(X_N \in [a,b]\right) \leqslant \int_{-\infty}^{\infty} e^{Nt(x-a)} \, \mathrm{d}F_N(x) = \exp\left[-Nta + NC(t)\right] \tag{20.19}$$

and hence

$$\frac{1}{N} \ln \mathbb{P}\left(X_N \in [a,b]\right) \leqslant -ta + C(t).$$

But $I(a) = t_a\, a - C(t_a)$ where $a = C'(t_a) > x_0$ and hence $t_a > 0$ since C' is increasing. Taking $t = t_a$, we find

$$\frac{1}{N} \ln \mathbb{P}\left(X_N \in [a,b]\right) \leqslant -I(a).$$

Similarly, if $b < x_-$ then

$$\frac{1}{N} \ln \mathbb{P}\left(X_N \in [a,b]\right) \leqslant -I(b) = - \inf_{x \in [a,b]} I(x).$$

Finally, if $[a,b] \cap [x_-, x_+] \neq \emptyset$ then $\inf_{x \in [a,b]} I(x) = 0$ and equation (20.2) is trivial.

The proof of equation (20.3) is a bit more difficult. We shall use the central limit theorem. Choose $\epsilon > 0$ arbitrary and let $z_0 \in (a,b)$ be such that $I(x) > I(z_0) - \epsilon$ for all $x \in (a,b)$. Assume first that we can choose z_0 so that there exists $\tau \in \mathbb{R}$ such that $z_0 = C'(\tau)$. We define new random variables $Z^{(k)}$ with distribution functions given by

$$G(z) = \int_{-\infty}^{z} e^{\tau x - C(\tau)} \, \mathrm{d}F(x). \tag{20.20}$$

Taking $\delta > 0$ so small that $(z_0 - \delta, z_0 + \delta) \subset (a, b)$ we have

$$\mathbb{P}(X_N \in (a, b)) \geqslant \int_{z_0-\delta}^{z_0+\delta} e^{N(C(\tau)-\tau z)} \, \mathrm{d}G_N(z)$$

$$\geqslant e^{N(C(\tau)-\tau z_0 -|\tau|\delta)} \, \mathbb{P}\left(\sum_{k=1}^{N} Z^{(k)} \in (N(z_0 - \delta), N(z_0 + \delta))\right).$$

$$(20.21)$$

The probability in the right-hand side expression can be written as

$$\mathbb{P}\left(\frac{1}{\sqrt{N}} \sum_{k=1}^{N} Z^{(k)} \in (\sqrt{N}(z_0 - \delta), \sqrt{N}(z_0 + \delta))\right)$$

which, by the central limit theorem, tends to

$$\int_{-\infty}^{\infty} e^{-(z-z_0)^2/2\sigma^2} \frac{\mathrm{d}z}{\sqrt{2\pi}\sigma} = 1$$

where $\sigma^2 = \int (z - z_0)^2 \mathrm{d}G(z)$ is the variance (see theorem A.9). Now taking logarithms in equation (20.21) we obtain

$$\liminf_{N\to\infty} \frac{1}{N} \ln \mathbb{P}(X_N \in (a, b)) \geqslant C(\tau) - \tau z_0 - \delta|\tau| = -I(z_0) - \delta|\tau|. \quad (20.22)$$

This proves equation (20.3) as $\epsilon > 0$ and $\delta > 0$ are arbitrarily small.

It remains to consider the case that $z_0 > C'(t)$ for all $t \in \mathbb{R}$. (The case $z_0 < C'(t)$ is similar.) In that case $C'(t) \to c \leqslant z_0$ as $t \to \infty$. If $z_0 > c$ there is nothing to prove since $I(z_0) = +\infty$; so we may assume $z_0 = c$. Since I is convex we have, as in the proof of theorem 7.1,

$$I(c) \geqslant \liminf_{x\to c} I(x). \quad (20.23)$$

To see this, assume the contrary and pick $\tilde{x} < c$. Then, for any $x \in (\tilde{x}, c)$,

$$I(x) \leqslant \frac{c-x}{c-\tilde{x}} I(\tilde{x}) + \frac{x-\tilde{x}}{c-\tilde{x}} I(c).$$

As x tends to c the right-hand side tends to $I(c)$ proving equation (20.23). The inequality (20.23) implies that we can replace $z_0 = c$ by a slightly smaller z_0 so that $I(x) > I(z_0) - 2\epsilon$ for all $x \in (a, b)$ and the previous argument applies. ∎

CHAPTER 21

The Paramagnet Revisited

Let us now apply large-deviation theory to the problem of chapter 19. We introduce the following probability distribution for the random variable $u_N = E/N$:

$$F_N(x) = \frac{\sum_{M:E_M/N \leqslant x} \Omega(E_M)}{\sum_{M=0}^{N} \Omega(E_M)}, \tag{21.1}$$

where E_M is defined by equation (19.1). Note that the normalization constant

$$\sum_{M=0}^{N} \Omega(E_M) = 2^N \tag{21.2}$$

is the total number of microstates. The probability distribution (21.1) is the uniform distribution giving equal probabilities to every possible microstate. We shall use this distribution in the large deviation principle to derive the entropy density (19.10). Before embarking on this, note that we have been slightly inaccurate in the definition of $s(u)$ in chapter 19: the limit (19.3) can only be defined for $u = E/N$ with E given by equation (19.1), i.e. if u/ϵ is a rational number! Clearly this is an undesirable artefact of the definition (19.3) so we replace it with the more accurate

$$\boxed{s(u) = \lim_{\Delta \to 0} \lim_{N \to \infty} \frac{k_B}{N} \ln \Omega_\Delta(Nu)} \tag{21.3}$$

where

$$\Omega_\Delta(E) = \sum_{-N\Delta < \delta E < N\Delta} \Omega(E + \delta E). \tag{21.4}$$

This is well defined for all values of u and we now prove that it is the same as equation (19.10). For this we use the fact that

$$\Omega_\Delta(E) = 2^N \mathbb{P}\left(u_N \in \left(\frac{E}{N} - \Delta, \frac{E}{N} + \Delta\right)\right). \tag{21.5}$$

Since the spins are all independent we can write

$$u_N = \frac{1}{N} \epsilon \sum_{k=1}^{N} s^{(k)}$$

where the $s^{(k)}$ are independent random variables, equal to ± 1 with probability $\frac{1}{2}$. By theorem 20.2, u_N satisfies the LDP with rate function given by (20.12) and

$$C(t) = \ln \int_{-\infty}^{\infty} e^{t\epsilon x} \, dF(x), \tag{21.6}$$

where $F(x)$ is the distribution function for $s^{(k)}$, that is

$$F(x) = \begin{cases} 0 & \text{if } x < -1, \\ \frac{1}{2} & \text{if } -1 \leqslant x < 1, \\ 1 & \text{if } x \geqslant 1. \end{cases} \tag{21.7}$$

$C(t)$ can be computed as follows:

$$
\begin{aligned}
e^{C(t)} &= \lim_{L \to \infty} \int_{-L}^{L} e^{t\epsilon x} \, dF(x) \\
&= \lim_{L \to \infty} \left(e^{t\epsilon x} F(x)|_{-L}^{L} - \epsilon t \int_{-L}^{L} e^{\epsilon t x} F(x) \, dx \right) \\
&= \lim_{L \to \infty} \left(e^{\epsilon t L} - \frac{1}{2} \epsilon t \int_{-1}^{1} e^{t\epsilon x} \, dx - \epsilon t \int_{1}^{L} e^{\epsilon t x} \, dx \right) \\
&= \lim_{L \to \infty} \left(e^{\epsilon t L} - \frac{1}{2} (e^{\epsilon t} - e^{-\epsilon t}) - (e^{\epsilon t L} - e^{\epsilon t}) \right) \\
&= \frac{1}{2} (e^{\epsilon t} + e^{-\epsilon t}) = \cosh(\epsilon t).
\end{aligned}
$$

Hence we conclude that

$$C(t) = \ln \cosh(\epsilon t) \tag{21.8}$$

and therefore

$$I(x) = \sup_{t \in \mathbb{R}} [t\, x - \ln \cosh(\epsilon t)]. \tag{21.9}$$

The maximum is attained when

$$x = \epsilon \tanh(\epsilon t) \tag{21.10}$$

and inverting this we have

$$t = \frac{1}{2\epsilon} \ln \left(\frac{\epsilon + x}{\epsilon - x} \right) \quad (-\epsilon < x < \epsilon). \tag{21.11}$$

Inserting this into equation (21.9) we find

$$I(x) = \frac{1}{2}\left(1 + \frac{x}{\epsilon}\right)\ln\left(1 + \frac{x}{\epsilon}\right) + \frac{1}{2}\left(1 - \frac{x}{\epsilon}\right)\ln\left(1 - \frac{x}{\epsilon}\right). \tag{21.12}$$

Using equation (21.5) in (21.3), we obtain finally

$$\begin{aligned}
s(u) &= \lim_{\Delta\to 0}\lim_{N\to\infty}\frac{k_B}{N}\ln\Omega_\Delta(E)\\
&= k_B\ln 2 + \lim_{\Delta\to 0}\lim_{N\to\infty}\frac{k_B}{N}\ln\mathbb{P}\left(u_N \in (u - \Delta, u + \Delta)\right)\\
&= k_B\ln 2 - k_B I(u)\\
&= \frac{k_B}{2}\left[2\ln 2 - \left(1 + \frac{u}{\epsilon}\right)\ln\left(1 + \frac{u}{\epsilon}\right) - \left(1 - \frac{u}{\epsilon}\right)\ln\left(1 - \frac{u}{\epsilon}\right)\right] \tag{21.13}
\end{aligned}$$

in accordance with equation (19.10).

Figure 21.1 shows the graph of $s(u)$ and figure 21.2 shows s as a function of the temperature.

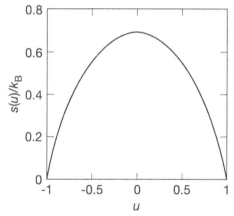

Figure 21.1 The entropy of a paramagnet.

REMARK 21.1: *Negative temperatures.*
Note that the temperature computed in equation (19.11) can be negative. This happens if $u > 0$, that is, if the majority of the spins is in the upper energy state $+\epsilon$. This is of course an idealization. In fact all physical systems have an unbounded spectrum, that is there is no maximum energy level, and this means that the temperature can only be positive. Nevertheless, this model can be a useful representation even in the case of negative temperatures because the situation described by negative temperatures can actually be achieved. It is called **population inversion** and is the mechanism according to which **lasers** operate. In a way it is better to consider the negative temperature region as a continuation of the high-temperature region: as the temperature increases the number of spins in the upper energy state increases and at $T = +\infty$ it equals

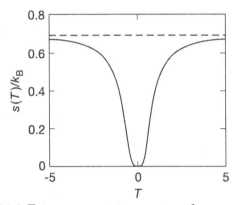

Figure 21.2 Entropy versus temperature for a paramagnet.

the number in the lower energy state; for still higher temperatures, i.e. for negative temperatures, the number of spins in the upper state exceeds that in the lower energy state. In practice this situation cannot be attained by simply increasing the temperature. Instead, one has to immerse the spins in a bath of radiation with a frequency corresponding to the energy difference 2ϵ.

CHAPTER 22

Identical Localized Free Particles

In general, a particle in an energy well will have infinitely many energy levels (See for instance example 18.1). In this chapter, we extend the analysis of the previous chapter to this case. We shall again assume that the particles are *localized* so that the volume does not play a role in the analysis. We also assume again that there is no interaction between the particles. In chapter 24 we shall see that this analysis forms the basis of the analysis of general statistical mechanical systems, even when there is interaction! In chapter 23 we consider the case of non-localized particles without interaction, i.e. gases.

Let us denote the energy levels of a single particle by $\epsilon_0 < \epsilon_1 < \epsilon_2 < \ldots$, each with a certain degeneracy $g_0, g_1, g_2, \ldots$. Again we define the entropy density $s(u)$ by

$$\boxed{s(u) = \lim_{\Delta \to 0} \lim_{N \to \infty} \frac{k_B}{N} \ln \Omega_\Delta(N, u)} \tag{22.1}$$

where $\Omega_\Delta(N, u)$ is defined as in equation (21.4):

$$\Omega_\Delta(N, u) = \sum_{-N\Delta < \delta E < N\Delta} \Omega(N, Nu + \delta E) \tag{22.2}$$

and $\Omega(N, E)$ is the total number of microstates with total energy E. The total number of microstates can be evaluated as follows. Let n_i denote the number of particles in a state with energy ϵ_i. Since the total number of particles is N and we have fixed the total energy E, the numbers n_i must satisfy the conditions

$$\sum_{i=0}^{\infty} n_i = N \text{ and } \sum_{i=0}^{\infty} n_i \epsilon_i = E \tag{22.3}$$

if a microstate with the given values of n_i is to be allowed. The number of microstates with *given* $n_0, n_1, \ldots$ is:

$$\Omega\left(N, \{n_i\}_{i=0}^{\infty}\right) = \frac{N!}{\prod_{i=0}^{\infty} n_i!} \prod_{i=0}^{\infty} g_i^{n_i}. \tag{22.4}$$

$(N!/(\prod n_i!))$ is the number of ways to choose which particles go into which energy level ϵ_i and $g_i^{n_i}$ is the number of ways the n_i particles in level i can be subdivided over the g_i degenerate states.)

The total number of allowed states with total energy E is therefore

$$\Omega(E, N) = \sum_{\{n_i\}}^{*} \Omega(N, \{n_i\}) \tag{22.5}$$

where the * indicates that the sum is to be restricted to those sequences $\{n_i\}$ satisfying equation (22.3). It is now not so easy to compute the limit (22.1) directly using Stirling's formula. Instead, we use the methods of large-deviation theory. This also leads to a difficulty, however. If we assign equal probabilities to all microstates as we did in chapter 21 (see equation (21.1)) then the probability of each individual microstate will be zero because there are infinitely many microstates. To avoid this problem we employ a trick: we give decreasing weights $e^{-\beta E}$ to higher energy states and compensate for this factor in the expression for Ω_Δ. We choose therefore an arbitrary constant $\beta > 0$ and define random variables X_N^β with distribution functions

$$F_N(x) = \frac{1}{Z_N(\beta)} \sum_{\{n_i\}: \sum n_i = N, \, E/N \leqslant x} \Omega(N, \{n_i\}) \, e^{-\beta E} \tag{22.6}$$

where $E = \sum_{i=0}^{\infty} n_i \, \epsilon_i$ and

$$Z_N(\beta) = \sum_{\{n_i\}: \sum n_i = N} \Omega(N, \{n_i\}) \, e^{-\beta E}$$

$$= \left(\sum_{i=0}^{\infty} g_i \, e^{-\beta \epsilon_i}\right)^N = (Z_\beta)^N \tag{22.7}$$

is a normalizing factor called the **partition function.** We have the following inequalities:

$$\begin{cases} \Omega_\Delta(N, u) > e^{\beta N(u-\Delta)} Z_N(\beta) \, \mathbb{P}_N^\beta(u - \Delta, u + \Delta) \\ \Omega_\Delta(N, u) < e^{\beta N(u+\Delta)} Z_N(\beta) \mathbb{P}_N^\beta(u - \Delta, u + \Delta) \end{cases} \tag{22.8}$$

where we have written $\mathbb{P}_N^\beta(a, b)$ instead of $\mathbb{P}\left(X_N^\beta \in (a, b)\right)$. Note that, since $e^{-\beta E} = \prod e^{-\beta \epsilon_i}$, the variable X_N^β is an average of independent random variables

$$X_N^\beta = \frac{1}{N} \sum_{k=1}^{N} X_\beta^{(k)}, \tag{22.9}$$

where $X_\beta^{(k)}$ has the distribution

$$F_\beta(x) = \frac{1}{Z_\beta} \sum_{i:\, \epsilon_i \leqslant x} g_i \, e^{-\beta\,\epsilon_i}. \tag{22.10}$$

By Cramér's theorem, X_N^β satisfies the LDP with rate function given by

$$I_\beta(x) = \sup_{t\in\mathbb{R}} [tx - C_\beta(t)]. \tag{22.11}$$

The cumulant generating function can be computed as follows:

$$\begin{aligned} C_\beta(t) &= \ln \int_{-\infty}^\infty e^{tx} \, \mathrm{d}F_\beta(x) \\ &= \ln\left(\frac{\sum_{i=0}^\infty g_i \, e^{(t-\beta)\epsilon_i}}{\sum_{i=0}^\infty g_i \, e^{-\beta\epsilon_i}} \right) \\ &= \ln Z_{\beta-t} - \ln Z_\beta \end{aligned} \tag{22.12}$$

for $t < \beta$ while $C_\beta(t) = +\infty$ for $t \geqslant \beta$. Therefore,

$$I_\beta(x) = \ln Z_\beta + \sup_{t<\beta} [tx - \ln Z_{\beta-t}]. \tag{22.13}$$

We now use the same reasoning as in equation (21.13) together with the bounds (22.8) to obtain an expression for $s(u)$:

$$\begin{aligned} \limsup_{N\to\infty} \frac{1}{N} &\ln \Omega_\Delta(N, u) \\ &< \beta(u + \Delta) + \ln Z_\beta + \limsup_{N\to\infty} \frac{1}{N} \ln \mathbb{P}_N^\beta(u - \Delta, u + \Delta) \\ &\leqslant \beta(u + \Delta) + \ln Z_\beta - \inf_{x\in[u-\Delta, u+\Delta]} I_\beta(x) \end{aligned} \tag{22.14}$$

and

$$\begin{aligned} \liminf_{N\to\infty} \frac{1}{N} &\ln \Omega_\Delta(N, u) \\ &> \beta(u - \Delta) + \ln Z_\beta + \liminf_{N\to\infty} \frac{1}{N} \ln \mathbb{P}_N^\beta(u - \Delta, u + \Delta) \\ &\geqslant \beta(u - \Delta) + \ln Z_\beta - \inf_{x\in(u-\Delta, u+\Delta)} I_\beta(x). \end{aligned} \tag{22.15}$$

Assuming that I_β is continuous we find

$$\begin{aligned} s(u) &= k_{\mathrm{B}}(\beta u + \ln Z_\beta) - k_{\mathrm{B}} \ln I_\beta(u) \\ &= k_{\mathrm{B}} \inf_{\beta>0} [\beta u + \ln Z_\beta] \end{aligned} \tag{22.16}$$

by a simple change of variables. Of course, this formula is not as explicit as equation (21.13) or (19.10) because we have considered a general set of energy levels ϵ_i. However, the formalism for the statistical mechanics of the system remains the same. In particular, we can derive the temperature of the system from the relation (6.13) (cf. equation (19.11)):

$$\frac{1}{T} = \left(\frac{\partial s}{\partial u}\right)_v = k_{\mathrm{B}}\beta(u) \tag{22.17}$$

where $\beta(u)$ is the solution of:

$$u = \frac{\sum_{i=0}^{\infty} \epsilon_i\, g_i\, e^{-\beta(u)\,\epsilon_i}}{\sum_{i=0}^{\infty} g_i\, e^{-\beta(u)\,\epsilon_i}} = \mathbb{E}\left(X_{\beta(u)}\right). \tag{22.18}$$

We can interpret this formula as follows.

> *If, instead of the total energy E of the system, we fix its temperature T and hence the parameter β according to (22.17), the particles of the system are distributed over the available energy levels* $\epsilon_0, \epsilon_1, \ldots$ *according to the distribution (22.10) with mean energy given by equation (22.18).*

In chapter 24 we shall see that the above formalism is very general and applies equally well to general systems of particles where interaction between the particles is present.

REMARK 22.1: *Equilibrium population of energy levels.*
The distribution of particles over the available energy levels can also be studied using large-deviation theory. One then defines an entropy function $s(\underline{x})$ of infinitely many variables $x_0, x_1, x_2, \ldots$ representing the occupancy of the energy levels $\epsilon_0, \epsilon_1, \epsilon_2, \ldots$, that is $x_i = n_i/N$. It can be written as follows:

$$s(\underline{x}) = \lim_{\Delta\to 0}\lim_{N\to\infty} \frac{k_{\mathrm{B}}}{N} \ln \Omega_\Delta(\underline{x}), \tag{22.19}$$

where $\Omega_\Delta(\underline{x})$ is the number of microstates $\{n_i\}_{i=0,1,\ldots}$ with $\|\underline{x} - \{n_i/N\}\| < \Delta$ for some norm on the infinite-dimensional space of sequences. As remarked in chapter 20, large-deviation theory can be extended to infinite-dimensional spaces. We will not go into that here but simply remark that it leads to a similar formula to (22.16) for $s(\underline{x})$. Moreover, it then easily follows that

$$s(u) = \sup_{\underline{x}:\, \sum x_i \epsilon_i = u} s(\underline{x}). \tag{22.20}$$

The maximizer in this expression is given by the distribution (22.10), i.e.

$$x_i = \frac{1}{Z_\beta} g_i e^{-\beta(u)\epsilon_i} \tag{22.21}$$

where $\beta(u)$ is determined by equation (22.18).

CHAPTER 23

Quantum Gases

In chapter 22 we considered a system of identical, localized particles. These particles can therefore be distinguished according to their position in space. *When identical particles are not localized they can no longer be distinguished.* This situation occurs in the case of a *gas* since the particles of a gas are unconstrained in their motion except by the walls of the container. It also occurs when the particles are so close together that their wave functions overlap, that is when they are free to exchange positions. The indistinguishability of particles has an important consequence for the counting of microstates:

> *The microstates of indistinguishable particles are completely determined by the number of particles in each individual particle eigenstate; one says that the **statistics** of indistinguishable particles is different from that of localized particles.*

It should be stressed that this is a typical **quantum-mechanical phenomenon**, formulated as QM 3 in chapter 18. Classical particles are always distinguishable because one can *in principle* follow each particle along its trajectory. According to quantum mechanics this is impossible even in principle. According to principle QM 3, there are in fact two different kinds of statistics depending on the type of particle: **Bose-Einstein statistics** for **bosons** and **Fermi-Dirac statistics** for **fermions**. This is related to the fact that fermions satisfy the **Pauli exclusion principle**; there can be no more than one particle in each eigenstate. (As explained in chapter 18, the degeneracy of each energy level contains a factor of $(2s + 1)$ due to the spin degrees of freedom; each energy level can then contain as many particles as its degeneracy.)

For **bosons** then, the number of particles $n_{i,\alpha}$ in each energy state is in principle unbounded except that at fixed total energy E and total particle number N the restrictions (22.3) have to be satisfied:

$$\sum_{i=0}^{\infty} \sum_{\alpha=1}^{g_i} n_{i,\alpha} = N \text{ and } \sum_{i=0}^{\infty} \sum_{\alpha=1}^{g_i} n_{i,\alpha}\, \epsilon_i = E. \qquad (23.1)$$

A microstate is completely determined by the set of numbers $\{n_{i,\alpha}\}$ since individual particles cannot be distinguished. In defining the entropy density we must now take the **thermodynamic limit** in the proper sense: $V \to \infty$ but $N/V \to \rho$, where the particle number density ρ is *given*. Indeed, while for localized particles the limits $N \to \infty$ and $V \to \infty$ are the same thing, for gaseous systems the number of particles, and hence the density, is a variable independent of V. We therefore define the entropy density (now per unit volume; see the definition (8.17) of the grand potential) $\tilde{s}(\tilde{u}, \rho)$ as follows:

$$\boxed{\tilde{s}(\tilde{u}, \rho) = \lim_{\Delta, \delta \to 0} \lim_{V \to \infty} \frac{k_{\mathrm{B}}}{V} \ln \Omega_{\Delta, \delta}(\tilde{u}, \rho)} \qquad (23.2)$$

where $\Omega_{\Delta, \delta}(\tilde{u}, \rho)$ is the number of microstates with density $N/V \in (\rho - \delta, \rho + \delta)$ and energy density $E/V \in (\tilde{u} - \Delta, \tilde{u} + \Delta)$.

To evaluate the limit (23.2), we need to use the large-deviation theory of chapter 20 for random variables with values in $\mathbb{R}^2$. We define *two-dimensional random variables*

$$X_V^{\beta, \lambda} = \left(\frac{1}{V} \sum_{i,\alpha} n_{i,\alpha}, \; \frac{1}{V} \sum_{i,\alpha} n_{i,\alpha} \epsilon_i \right) \qquad (23.3)$$

with distribution function

$$F_V^{\beta, \lambda}(x, y) = \frac{1}{\mathcal{Z}_V(\beta, \lambda)} \sum_{\{n_{i,\alpha}\}: \, N/V \leqslant x, \, E/V \leqslant y} e^{-\lambda N - \beta E} \qquad (23.4)$$

where N and E are defined by equation (23.1). The random variables $X_V^{\beta, \lambda}$ are averages of independent random variables $X_{i,\alpha} = (n_{i,\alpha}, n_{i,\alpha}\epsilon_i)$ with distribution functions given by

$$F_i^{\beta, \lambda}(x, y) = \frac{1}{z(\lambda + \beta \epsilon_i)} \sum_{n: \, n \leqslant x, \, n \epsilon_i \leqslant y} e^{-\lambda n - \beta \epsilon_i n} \qquad (23.5)$$

where

$$z(\lambda) = \sum_{n=0}^{\infty} e^{-\lambda n} = (1 - e^{-\lambda})^{-1}. \qquad (23.6)$$

Note that the analogue of Cramér's theorem (theorem 20.2) does not apply here because the distribution functions $F_i^{\beta, \lambda}$ depend on i. Nevertheless, we may still expect that the LDP holds if the limit of the cumulant generating functions exists. If we put $G(x, y) = t_1 x + t_2 y$ in equation (20.7) as we did for the one-dimensional case in equation (20.13), and we replace N in that formula by a sequence of volumes V_l ($l = 1, 2, \dots$) tending to infinity, we get

$$
\begin{aligned}
C^{\beta,\lambda}(t_1, t_2) &= \lim_{l \to \infty} \frac{1}{V_l} \ln \int_{-\infty}^{\infty} \int_{-\infty}^{\infty} e^{V_l(t_1 x + t_2 y)} \mathrm{d}F_{V_l}^{\beta,\lambda}(x, y) \\
&= \lim_{l \to \infty} \frac{1}{V_l} \ln \left[\prod_{i=0}^{\infty} \left(\int_{-\infty}^{\infty} \int_{-\infty}^{\infty} e^{t_1 x + t_2 y} \mathrm{d}F_i^{\beta,\lambda}(x, y) \right)^{g_i} \right] \\
&= \lim_{l \to \infty} \left[\frac{1}{V_l} \sum_{i=0}^{\infty} g_i \ln \left(\sum_{n=0}^{\infty} e^{(t_1 - \lambda)n + (t_2 - \beta)\epsilon_i n} \right) - \frac{1}{V_l} \sum_{i=0}^{\infty} g_i \ln z(\lambda + \beta \epsilon_i) \right] \\
&= \lim_{l \to \infty} \frac{1}{V_l} \sum_{i=0}^{\infty} g_i \left[\ln z(\lambda - t_1 + (\beta - t_2)\epsilon_i) - \ln z(\lambda + \beta \epsilon_i) \right].
\end{aligned}
\tag{23.7}
$$

It is clear that this limit can only exist if the sum in the final expression is of order V_l. We must therefore investigate the dependence of ϵ_i on V. Let us consider the example of free particles in a cubic box discussed in example 18.2. In one dimension we have $V = L$ and the energy levels are given by equation (18.6). Inserting this in equation (23.7), we get

$$
C^{\beta,\lambda}(t_1, t_2) = \gamma(\lambda - t_1, \beta - t_2) - \gamma(\lambda, \beta) \tag{23.8}
$$

where

$$
\gamma(\lambda, \beta) = \lim_{L \to \infty} \frac{1}{L} \sum_{i=1}^{\infty} \ln \left\{ \left[1 - \exp \left(-\lambda - \beta \frac{\pi^2 \hbar^2}{2mL^2} i^2 \right) \right]^{-1} \right\}. \tag{23.9}
$$

This is a Riemann integral! Writing $k = \delta i$ with $\delta = \pi / L$ we have

$$
\begin{aligned}
\gamma(\lambda, \beta) &= \lim_{\delta \to 0} \frac{1}{\pi} \sum_{i=1}^{\infty} \delta \ln z(\lambda + \beta \epsilon(\delta i)) \\
&= \frac{1}{\pi} \int_0^{\infty} \ln z(\lambda + \beta \epsilon(k)) \, \mathrm{d}k
\end{aligned}
\tag{23.10}
$$

where

$$
\epsilon(k) = \frac{\hbar^2}{2m} k^2. \tag{23.11}
$$

In 3 dimensions we get similarly

$$
\gamma(\lambda, \beta) = \lim_{l \to \infty} \frac{1}{V_l} \ln \mathcal{Z}_{V_l}(\beta, \lambda) = \frac{1}{(2\pi)^3} \int_{\mathbb{R}^3} \ln z(\lambda + \beta \epsilon(k)) \, \mathrm{d}^3 k. \tag{23.12}
$$

As before, the rate function is given by

$$
I^{\beta,\lambda}(x, y) = \sup_{t_1 < \lambda, \, t_2 < \beta} \{ t_1 x + t_2 y - C^{\beta,\lambda}(t_1, t_2) \} \tag{23.13}
$$

where $C^{\beta,\lambda}(t_1, t_2)$ is given by equation (23.8); the specific entropy is

$$
\begin{aligned}
\tilde{s}(e, \rho) &= \lim_{\Delta,\delta\to 0} \lim_{V\to\infty} \frac{k_B}{V} \ln \mathbb{P}_{\Delta,\delta}(\rho, e) \\
&\quad + k_B \lambda\rho + k_B \beta e + k_B \gamma(\lambda, \beta) \\
&= -k_B I^{\beta,\lambda}(\rho, e) + k_B \gamma(\lambda, \beta) + k_B \lambda\rho + k_B \beta e \\
&= k_B \inf_{\lambda,\beta>0} [\lambda\rho + \beta e + \gamma(\lambda, \beta)].
\end{aligned}
\tag{23.14}
$$

We have written e instead of $\tilde{u}$. Again, we can determine the temperature from

$$
\frac{1}{T} = \left(\frac{\partial s}{\partial u}\right)_v = \left(\frac{\partial \tilde{s}}{\partial e}\right)_\rho = k_B \beta(e, \rho),
\tag{23.15}
$$

where $\beta(e, \rho)$ is the solution of the simultaneous equations

$$
e = -\frac{\partial}{\partial \beta}\gamma(\lambda, \beta)\Big|_{\substack{\beta=\beta(e,\rho)\\\lambda=\lambda(e,\rho)}} = \frac{1}{(2\pi)^3} \int_{\mathbb{R}^3} \frac{\epsilon(k)\, d^3k}{e^{\lambda(e,\rho)+\beta(e,\rho)\epsilon(k)} - 1}
\tag{23.16}
$$

and

$$
\rho = -\frac{\partial}{\partial \lambda}\gamma(\lambda, \beta)\Big|_{\substack{\beta=\beta(e,\rho)\\\lambda=\lambda(e,\rho)}} = \frac{1}{(2\pi)^3} \int_{\mathbb{R}^3} \frac{d^3k}{e^{\lambda(e,\rho)+\beta(e,\rho)\epsilon(k)} - 1}.
\tag{23.17}
$$

Similarly, we can determine the chemical potential for fixed values of e and ρ. Indeed, from equation (8.17) for the grand potential we can obtain the maximum condition

$$
\mu = \left(\frac{\partial \tilde{u}}{\partial \rho}\right)_{\tilde{s}} = -\frac{(\partial \tilde{s}/\partial \rho)_e}{(\partial \tilde{s}/\partial e)_\rho}.
\tag{23.18}
$$

In the present case this becomes

$$
\mu = -\frac{\lambda(e, \rho)}{\beta(e, \rho)}.
\tag{23.19}
$$

From the expression for $\tilde{s}(e, \rho)$ we can deduce all the thermodynamics of the system. Let us first write it in terms of the variables μ and T:

$$
\tilde{s}(e, \rho) = \inf_{\mu,T} \left[-\frac{1}{T}\rho\mu + \frac{1}{T}e + k_B\gamma\left(-\frac{\mu}{k_B T}, \frac{1}{k_B T}\right)\right].
\tag{23.20}
$$

From this we obtain the entropy per particle

$$
s(u, v) = \inf_{\mu,T} \left[\frac{1}{T}\left\{u - \mu + k_B T v \gamma\left(-\frac{\mu}{k_B T}, \frac{1}{k_B T}\right)\right\}\right].
\tag{23.21}
$$

Inverting the Legendre transform with respect to T we find (compare equation (8.3))

$$
f(v, T) = \sup_{\mu} \left[\mu - k_B T v \gamma\left(-\frac{\mu}{k_B T}, \frac{1}{k_B T}\right)\right]
\tag{23.22}
$$

and inverting the Legendre transform with respect to μ,

$$\omega(\mu, T) = -k_B T \, \gamma\left(-\frac{\mu}{k_B T}, \frac{1}{k_B T}\right). \tag{23.23}$$

Comparing this with equation (23.12), we may conclude that the grand potential $\omega(\mu, T)$ is directly related to the **grand-canonical partition function** $\mathcal{Z}_V(\beta, \mu)$ which is normally defined as a function of β and μ rather than λ. Indeed, the parameter λ is unconventional and was introduced here only to make the formalism symmetric in N and E. We have

$$\omega(\mu, T) = -k_B T \lim_{V \to \infty} \frac{1}{V} \ln \mathcal{Z}_V\left(\frac{1}{k_B T}, \mu\right)$$
$$= \frac{k_B T}{(2\pi)^3} \int_{\mathbb{R}^3} \ln\left(1 - e^{(\mu - \epsilon(k))/k_B T}\right) \mathrm{d}^3 k. \tag{23.24}$$

Note that the pressure of the gas is given by equation (8.20):

$$p(v, T) = -\omega(\mu(v, T), T) = \frac{k_B T}{(2\pi)^3} \int_{\mathbb{R}^3} \ln\left[\left(1 - e^{(\mu(v,T) - \epsilon(k))/k_B T}\right)^{-1}\right] \mathrm{d}^3 k \tag{23.25}$$

where $\mu(v, T)$ is given implicitly by equation (23.17) with fixed $\beta = 1/k_B T$:

$$\frac{1}{v} = \frac{1}{(2\pi)^3} \int_{\mathbb{R}^3} \frac{\mathrm{d}^3 k}{e^{(\epsilon(k) - \mu(v,T))/k_B T} - 1}. \tag{23.26}$$

REMARK 23.1: *Bose-Einstein condensation.*
We have been somewhat careless in the above analysis. In equation (23.13) for the rate function, it was assumed that $C^{\beta, \lambda}(t_1, t_2)$ is infinite for $t_1 \geqslant \lambda$ or $t_2 \geqslant \beta$. It can be seen from equation (23.12) that this is almost but not quite true. In fact $\gamma(0, \beta) < +\infty$ if $\beta > 0$. This does not affect (23.13) as such, but the same phenomenon also occurs in equation (23.17) for the density at the point $\lambda = 0$ (which corresponds to $\mu = 0$ in equation (23.26)). Indeed, it is not difficult to see that

$$\rho_c = \frac{1}{(2\pi)^3} \int_{\mathbb{R}^3} \frac{\mathrm{d}^3 k}{e^{\beta \epsilon(k)} - 1} < +\infty. \tag{23.27}$$

As ρ is increasing in μ and not defined for $\mu > 0$, one may wonder if this means that the density cannot exceed ρ_c. This is not the case. If we fix $\rho > \rho_c$ in equation (23.14) the infimum is simply attained at $\lambda = 0$. We shall study this peculiarity further in chapter 34. It turns out that the surplus of particles $\rho - \rho_c$ resides in the minimum energy state corresponding to $k = 0$ (see equation (23.11)). This can be interpreted as **condensation** of particles in the ground state and indeed the free boson gas has a phase transition at $\rho = \rho_c$. This so-called **Bose-Einstein phase transition** was first proposed by Bose and Einstein in 1925 but has only very recently been observed in a real physical system. (See M. H. Anderson *et al.* (1995) who first demonstrated Bose-Einstein condensation experimentally.)

In the case of a **fermion gas** the analysis is very similar. The only change needed is in the formula (23.6) for $z(\lambda)$, where we must restrict the sum to $n = 0, 1$ because there can be no more than one particle in each eigenstate. This gives

$$z(\lambda) = \sum_{n=0,1} e^{-\lambda n} = 1 + e^{-\lambda}. \tag{23.28}$$

Equation (23.7) for $C^{\beta,\lambda}$ remains unchanged. This leads again to equation (23.8) where, for particles in a three-dimensional box, $\gamma(\lambda, \beta)$ is given by equation (23.12). The end result is that $\tilde{s}(e, \rho)$ is again given by (23.14). Equations (23.15) and (23.18) are also unchanged but $\beta(e, \rho)$ and $\mu(e, \rho) = -\lambda(e, \rho)/\beta(e, \rho)$ now satisfy:

$$e = \frac{1}{(2\pi)^3} \int_{\mathbb{R}^3} \frac{\epsilon(k)\, d^3 k}{e^{\beta(e,\rho)(\epsilon(k)-\mu(e,\rho))} + 1} \tag{23.29}$$

and

$$\rho = \frac{1}{v} = \frac{1}{(2\pi)^3} \int_{\mathbb{R}^3} \frac{d^3 k}{e^{\beta(e,\rho)(\epsilon(k)-\mu(e,\rho))} + 1}. \tag{23.30}$$

The pressure of a free fermion gas is given by

$$p(v, T) = \frac{k_{\mathrm{B}} T}{(2\pi)^3} \int_{\mathbb{R}^3} \ln(1 + e^{(\epsilon(k)-\mu(v,T))/k_{\mathrm{B}} T})\, d^3 k, \tag{23.31}$$

where the chemical potential is given implicitly by equation (23.30).

Equations (23.25) and (23.31) are the **equations of state** for the free boson gas and the free fermion gas, respectively. The implicit equations for $\mu(v, T)$, unfortunately, cannot be solved exactly. However, let us examine the behaviour of the pressure in the case of high temperatures. If we keep v *fixed* in equation (23.26) or equation (23.30) and let the temperature become high, then $\mu/k_{\mathrm{B}} T$ must be large negative. The same is also true if we keep T fixed but let ρ become small, i.e. for rare gases. To see what this implies for the pressure we write the integrals in terms of the dimensionless variable $x = \lambda_T k/2\sqrt{\pi}$ where λ_T is the so-called **thermal wavelength** defined by

$$\lambda_T = \frac{h}{\sqrt{2\pi m\, k_{\mathrm{B}} T}}. \tag{23.32}$$

We have

$$\rho(\mu, T) = \frac{4}{\sqrt{\pi} \lambda_T^3} \int_0^\infty \frac{x^2\, dx}{e^{x^2 - \mu/k_{\mathrm{B}} T} \pm 1} \tag{23.33}$$

and

$$p(v, T) = \frac{4 k_{\mathrm{B}} T}{\sqrt{\pi} \lambda_T^3} \int_0^\infty x^2 \ln\left[z\left(x^2 - \frac{\mu}{k_{\mathrm{B}} T}\right)\right] dx. \tag{23.34}$$

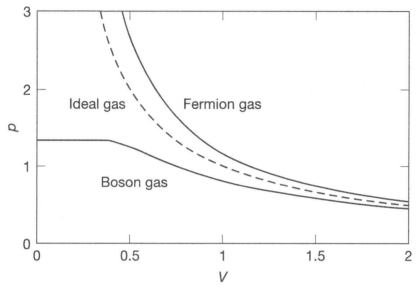

Figure 23.1 Pressure as a function of the volume for the free gases.

For large negative μ we have according to equations (23.6) and (23.28),

$$\ln\left[z\left(x^2 - \frac{\mu}{k_B T}\right)\right] \approx e^{\mu/k_B T - x^2} \tag{23.35}$$

for fermions as well as bosons. The limit in which equation (23.35) holds is called the **classical limit**. The integrals for $\rho(\mu, T)$ and $p(v, T)$ can be evaluated in this limit with the result

$$\rho = \frac{1}{\lambda_T^3} e^{\mu/k_B T} \tag{23.36}$$

and

$$p(v, T) = \frac{k_B T}{\lambda_T^3} e^{\mu(v,T)/k_B T} = \rho\, k_B T. \tag{23.37}$$

This, of course, is the *ideal gas law*! It is also easily seen that in this classical limit

$$e(v, T) = \frac{3 k_B T}{2v} \tag{23.38}$$

which corresponds to equation (2.6) with $c_V = \frac{3}{2} k_B$ as for a monatomic gas. Applications of the free quantum gases will be discussed in Part III. Here we conclude by comparing their pressures graphically with the classical ideal gas (figure 23.1).

Note that in case of the boson gas the pressure is constant for small values of v ($v \leqslant 1/\rho_c(T)$). This corresponds to the phenomenon of Bose-Einstein condensation.

CHAPTER 24

Ensembles

In chapter 22 we considered a system of non-interacting localized particles with general energy levels $\epsilon_0 < \epsilon_1 < \epsilon_2 < \ldots$. As first argued by Gibbs, this analysis can in fact be applied to general macroscopic systems if each particle is replaced by an entire macroscopic system! Thus one obtains a large collection of identical macroscopic systems, which is called an **ensemble**. The energy levels ϵ_i must be replaced by the energy levels of an individual macroscopic system, which we denote by E_i. One can argue that the behaviour of a given macroscopic system, averaged over a long time span compared with the motion of the individual particles, is given by the average over an ensemble of identical copies of this system. This idea of replacing the average over time with an instantaneous average over copies of the same system is called **ergodicity.** The theory of **ergodic systems** has developed into a whole new branch of mathematics which we shall not discuss further here. However, to make the concept clear, let us be a bit more specific about the definition of ergodicity. Let $X^{(k)}$ be an observable of system number k ($k = 1, 2, \ldots, M$). For instance, $X^{(k)}$ could be the pressure (force per unit area) exerted on the walls. Over a short time span this observable will depend on time: $X^{(k)}(t)$, but if we measure it over a sufficiently long time, the average will be constant provided the system is in equilibrium. The system is said to be **ergodic** if this average equals an average over an ensemble, more precisely,

$$\lim_{T \to \infty} \frac{1}{T} \int_0^T X^{(k)}(t)\, \mathrm{d}t = \lim_{M \to \infty} \mathbb{E}(X_M), \tag{24.1}$$

where $X_M = (1/M) \sum_{k=1}^{M} X^{(k)}(t)$ for an arbitrary point t in time ! The idea of ergodicity has come under severe criticism and it is now known that 'most'mechanical (i.e. Hamiltonian) systems are *not* ergodic.

One can look at the problem from a different viewpoint, however. Let us consider a single very large system and subdivide it into a large number M of cells which, by themselves, can still be considered macroscopic. Let us first assume that the particles of the system are localized, so that they do not wander from cell to cell. Because the cells are still very large, the energy of interaction between particles of one cell and those of all other cells will be negligibly small compared with the energy of the particles within a cell, that is,

$$E_k \gg \sum_{i \neq k} E_{ik}, \tag{24.2}$$

where E_k is the total energy of the particles in cell k and E_{ik} is the total energy of interaction between particles of cell i and particles of cell k. We can therefore neglect the energy of interaction between cells and assume that the cells are independent and localized, provided they are so large that equation (24.2) holds. This means that we are almost in the situation of independent identical macroscopic systems as imagined by Gibbs. The corresponding ensemble is called the **canonical ensemble**. If E_i $(i = 0, 1, 2, \dots)$ are the energy levels of a single cell then, in the limit $M \to \infty$, at a given temperature T the distribution of the cells over the energy levels will be given by

$$\mathbb{P}_\beta(E = E_i) = \frac{1}{Z(\beta)} g_i \, e^{-\beta E_i} \tag{24.3}$$

where

$$Z(\beta) = \sum_{i=0}^{\infty} g_i \, e^{-\beta E_i} \tag{24.4}$$

is called the **canonical partition function**. This follows immediately from the discussion of systems of localized free particles in chapter 22. As remarked below equation (22.18), the distribution is just (22.10) and the internal energy *per cell* is given by

$$U(\beta) = \frac{1}{Z(\beta)} \sum_{i=0}^{\infty} E_i \, g_i \, e^{-\beta E_i} = -\frac{\mathrm{d}}{\mathrm{d}\beta} \ln Z(\beta). \tag{24.5}$$

The entropy of an entire cell is given by the analogue of equation (22.16):

$$S(U) = k_{\mathrm{B}} \inf_{\beta > 0} \{\beta U + \ln Z(\beta)\} \tag{24.6}$$

and the free energy of a cell is given by

$$F(\beta) = \inf_{U} \left(U - \frac{1}{k_{\mathrm{B}}\beta} S(U) \right) = -\frac{1}{\beta} \ln Z(\beta). \tag{24.7}$$

This analysis, however only applies in the limit of infinitely large cells so that the interaction between cells is strictly negligible. In equation (24.7), we must therefore take the limit of infinitely large cells and define a free energy *per particle* by

$$\boxed{f(\beta) = -\frac{1}{\beta} \lim_{N \to \infty} \frac{1}{N} \ln Z_N(\beta)} \tag{24.8}$$

where $Z_N(\beta)$ is still given by equation (24.4) except that the energy levels E_i depend on the number of particles N per cell:

$$Z_N(\beta) = \sum_{i=0}^{\infty} g_i(N) e^{-\beta E_i(N)} \tag{24.9}$$

(We often consider f as a function of $\beta = 1/k_B T$ rather than T.) In fact, assuming that the limit (24.8) exists, one can prove directly that the thermodynamics of the system is well-defined by introducing the random variables u_N with distribution given by $\mathbb{P}_\beta(u = E_i/N) = \mathbb{P}_\beta(E = E_i)$. The corresponding cumulant generating function exists and is given by

$$C(t) = \lim_{N \to \infty} \frac{1}{N} \ln \int e^{Ntx} \, dF_N(x) = -\beta[f(\beta - t) - f(\beta)].$$

The rate function is obtained by taking the Legendre transform of $C(t)$ and one eventually obtains the following formula for the entropy density:

$$s(u) = k_B \inf_{\beta > 0} [\beta u - \beta f(\beta)] \tag{24.10}$$

The internal energy density then follows from the minimizing condition:

$$u(\beta) = \frac{d}{d\beta}[\beta f(\beta)]. \tag{24.11}$$

Equations (24.8)–(24.11) form the basic formalism of **statistical mechanics in the canonical ensemble**, which applies when the system consists of localized particles. (In fact it applies also for general systems because the number of particles crossing the boundary is small compared to the total number of particles in a cell. In practice, however, the grand-canonical formalism developed below is more convenient in that case.) In particular, we can re-derive the formulas of chapter 21 using this formalism. In Part III we consider some other models which do involve interaction between the particles (spins), and analyse these with the help of the canonical formalism. In chapter 25, we prove the existence of the limit (24.8) for a general spin system with interaction. The proof is based on equation (24.2). We give a proper proof of the LDP for u_N and hence equation (24.10) in chapter 26.

In the above discussion we have assumed that the number of particles in each cell is fixed. If, instead, we assume that the cells can exchange particles then the energy levels of each cell depend on the number of particles N_j in this cell (labelled j): $E(N_j, i)$. In this case one speaks of the **grand-canonical ensemble**. As in chapter 23 we must then introduce two-dimensional random variables

$$X_{M,V} = \left(\frac{1}{MV} \sum_{N,i} N \, n(N, i), \frac{1}{MV} \sum_{N,i} E_V(N, i) \, n(N, i) \right), \tag{24.12}$$

where M is the number of cells, V is the volume of each cell and $n(N, i)$ is the number of cells j with $N_j = N$ and residing in a state with energy $E_V(N, i)$. Note that the situation is different from that in chapter 23 in that the individual cells are *distinguishable*. Nevertheless, one can still use large-deviation theory with the following results analogous to the corresponding formulas in chapter 23.

$$\omega(\beta, \mu) = -\frac{1}{\beta} \lim_{V \to \infty} \frac{1}{V} \ln \mathcal{Z}_V(\beta, \mu) \tag{24.13}$$

where $\mathcal{Z}_V(\beta, \mu)$ is the **grand-canonical partition function** defined by

$$\mathcal{Z}_V(\beta, \mu) = \sum_{N=0}^{\infty} e^{\beta \mu N} \sum_{i=0}^{\infty} g_{N,i} e^{-\beta E_V(N,i)} \tag{24.14}$$

The free energy density then follows from equation (8.23):

$$\tilde{f}(\beta, \rho) = \sup_{\mu} \left[\rho \mu + \omega(\beta, \mu) \right] \tag{24.15}$$

and the particle number density follows from the supremum condition

$$\rho(\beta, \mu) = -\left(\frac{\partial \omega}{\partial \mu} \right)_{\beta}. \tag{24.16}$$

(As in the case of the canonical ensemble, we often consider $\tilde{f}$ as a function of $\beta = 1/k_B T$ and $\rho = 1/v$ instead of v and T.) The pressure $p(v, T)$ is given by equation (8.20): $p(v, T) = -\omega(\beta, \mu(\beta, v))$. Here $\mu(\beta, v)$ is obtained by inverting equation (24.16) at constant β which is possible by the fact that ω is a concave function of μ. (One often considers p to be a function of β and ρ instead.)

The entropy density and the internal energy density follow from equation (8.18):

$$\tilde{s}(\beta, \mu) = -\left(\frac{\partial \omega}{\partial T} \right)_{\mu} = k_B \beta^2 \left(\frac{\partial \omega}{\partial \beta} \right)_{\mu} \tag{24.17}$$

and

$$\tilde{u}(\beta, \mu) = \rho \mu + \omega + \beta \left(\frac{\partial \omega}{\partial \beta} \right)_{\mu}. \tag{24.18}$$

These formulas embody the **grand-canonical ensemble** in statistical mechanics.

CHAPTER 25

Existence of the Thermodynamic Limit

In this chapter, following the exposition by Hugenholtz (1982), we shall prove that the thermodynamic limit of the free energy density, that is the limit (24.8) exists in the case of a general class of models: **classical spin systems with finite-range pair interaction**. These are models of systems of localized particles as in chapter 21, where each particle can assume only a finite number of different states which we label by $s_x \in \{1, 2, \ldots, q\}$, where $x \in \mathbb{Z}^\nu$ is the position of the particle (spin). Here the particles are not assumed to be independent, however, but are allowed to have some mutual interaction which for simplicity we take to be given by a pair potential with finite range. This means that the energy levels of the system are given by

$$E_\Lambda\left(\{s_x\}_{x\in\Lambda}\right) = \sum_{x\in\Lambda}\left(\Psi(s_x) + \sum_{y:|x-y|\leqslant R}\Phi_{x-y}(s_x, s_y)\right). \qquad (25.1)$$

Here Λ is some finite region of ν-dimensional space and the sum is over all lattice points inside this region (the particles or **spins** are assumed to be localized on a lattice, $\mathbb{Z}^\nu$ for definiteness). R is the range of the interaction and Φ and Ψ are functions determining the interaction. The model is called **classical** because quantum spins are **operators** and if one diagonalizes an operator of the form (25.1), the energy eigenvalues are in general not of the form (25.1). One can nevertheless prove the existence of the thermodynamic limit (see for example the book by Ruelle (1969)) but we shall not go into that here. (The proof is very similar!) Specific models of the type (25.1) will be discussed in Part III. Note that we assume that the **interaction potential** Φ_{x-y} depends only on the difference vector $x - y$. This means that the interaction is **translation invariant.** We now want to prove the existence of the limit (24.8), where $N = |\Lambda|$ is the number of lattice sites in Λ. We therefore take a sequence Λ_l $(l = 1, 2, \ldots)$ of regions with volume tending to infinity and define the finite-volume free energy

$$F_{\Lambda_l}(\beta) = -\frac{1}{\beta} \ln Z_{\Lambda_l}(\beta), \tag{25.2}$$

where $Z_{\Lambda_l}(\beta)$ is the partition function

$$Z_{\Lambda_l}(\beta) = \sum_{\{s_x \mid x \in \Lambda_l\}} \exp\left[-\beta E_{\Lambda_l}(\{s_x\})\right]. \tag{25.3}$$

We want to show that the limit $\lim_{l \to \infty} F_{\Lambda_l}(\beta)/|\Lambda_l|$ exists and is independent of the sequence Λ_l. However, this can obviously not be true without restrictions on this sequence. We could for example take a sequence of volumes that grows only in one direction. In fact, we may expect that the regions must cover the whole ν-dimensional space in the limit and moreover, that they grow in each direction at a comparable rate. A sufficient condition was formulated by Van Hove.

Definition 25.1 *Consider the cubic lattice $\mathbb{Z}^\nu$ and divide this lattice into ν-dimensional cubes of side $a \in \{1, 2, \dots\}$. For any given finite region $\Lambda \subset \mathbb{Z}^\nu$, let Λ_a^- be the union of all such cubes contained in Λ and let Λ_a^+ be the union of all such cubes which contain at least one point of Λ. Then a sequence Λ_l of regions is said to tend to $\mathbb{Z}^\nu$ in the sense of Van Hove if $|\Lambda_l| \to \infty$ and for all $a \in \{1, 2, \dots\}$,*

$$\lim_{l \to \infty} \left(\frac{\left| (\Lambda_l)_a^+ \right|}{\left| (\Lambda_l)_a^- \right|} \right) = 1. \tag{25.4}$$

This means roughly that the boundary of Λ_l is of smaller order than its volume as $l \to \infty$ on *any scale*. It is a rather weak condition but sufficient for the limit of the free energy density to exist. To prove this, we first consider the case that Λ_l is a sequence of cubes.

Lemma 25.1 *Suppose that $K_i(a)$, $i = 1, 2, \dots, n$ are n non-over-lapping cubes in $\mathbb{Z}^\nu$ with side a. Then*

$$\lim_{a \to \infty} \left(\frac{1}{na^\nu} F_{\cup_i K_i(a)}(\beta) - \frac{1}{a^\nu} F_{K(a)} \right) = 0 \tag{25.5}$$

uniformly in n.

Proof. We can decompose the energy of $\cup_i K_i(a)$ as follows:

$$E_{\cup_i K_i(a)}(\{s_x\}) = \sum_{i=1}^n E_{K_i(a)}(\{s_x\}) + \sum_{i=1}^n \sum_{\substack{j=1 \\ j \neq i}}^n W_{i,j}(\{s_x\}), \tag{25.6}$$

where

$$W_{i,j}(\{s_x\}) = \sum_{x \in K_i(a)} \sum_{y \in K_j(a):|x-y| \leqslant R} \Phi_{x-y}(s_x, s_y). \tag{25.7}$$

The latter term is the interaction energy between two blocks. Analogous to equation (24.2), this term should be small compared with the first term on the right-hand side of (25.6). Indeed, each $E_{K_i(a)}$ will be of the order a^ν whereas $W_{i,j}$ will be of the order $a^{\nu-1}R$. To make this precise, let

$$M = \sum_{y:|x-y| \leqslant R} \max_{s_x, s_y} |\Phi_{x-y}(s_x, s_y)|. \tag{25.8}$$

Note that this is independent of x and that the sum and the maximum are over finite sets. It follows that

$$\left| \sum_{i=1}^{n} \sum_{j=1, j \neq i}^{n} W_{i,j}(\{s_x\}) \right| \leqslant n[a^\nu - (a - 2R)^\nu]M \tag{25.9}$$

because, for any $i = 1, \ldots, n$ only those terms in equation (25.7) contribute for which x is within R from the boundary of $K_i(a)$ and there are a total of $a^\nu - (a - 2R)^\nu$ points inside $K_i(a)$ within R from its boundary. This implies that

$$\exp\left[-|\beta|n(a^\nu - (a - 2R)^\nu)M\right] \leqslant \exp\left(-\beta \sum_{i \neq j} W_{i,j}(\{s_x\})\right)$$
$$\leqslant \exp\left[|\beta|n(a^\nu - (a - 2R)^\nu)M\right]$$

and after taking logarithms,

$$\left| F_{\cup_i K_i(a)}(\beta) - nF_{K(a)}(\beta) \right| \leqslant n[a^\nu - (a - 2R)^\nu]M. \tag{25.10}$$

Dividing by na^ν and using the simple inequality

$$a^\nu - (a - 2R)^\nu \leqslant 2\nu Ra^{\nu-1} \tag{25.11}$$

we see that this tends to zero as $a \to \infty$ uniformly in n. ∎

Corollary 25.1 *If $\Lambda_l = K(l)$ is a sequence of cubes then the thermodynamic limit of the free energy density*

$$\lim_{l \to \infty} l^{-\nu} F_{K(l)}(\beta) = f(\beta)$$

exists for all $\beta \neq 0$.

Proof. We have to show that the sequence $l^{-\nu}F_{K(l)}(\beta)$ is a Cauchy sequence. Now, by the lemma, given $\epsilon > 0$ there exists l_0 independent of n such that

$$\left|(nl)^{-\nu}F_{K(nl)} - l^{-\nu}F_{K(l)}\right| \leqslant \frac{\epsilon}{2}$$

for $l \geqslant l_0$. Hence, if $l_1, l_2 \geqslant l_0$ then

$$\left|l_1^{-\nu}F_{K(l_1)} - l_2^{-\nu}F_{K(l_2)}\right| \leqslant \left|(l_1 l_2)^{-\nu}F_{K(l_1 l_2)} - l_1^{-\nu}F_{K(l_1)}\right|$$
$$+ \left|(l_1 l_2)^{-\nu}F_{K(l_1 l_2)} - l_2^{-\nu}F_{K(l_2)}\right|$$
$$\leqslant \epsilon.$$

■

REMARK 25.1: *Negative temperatures again.*
Note that we allow also negative temperatures. This is due to the fact that the energy per particle is bounded above; compare remark 21.1. In fact it is better to consider $\beta f(\beta)$ which is also well-defined for $\beta = 0$. This is the rate function for the distribution of the energy density as discussed in chapter 26.

We can now generalize this to a general Van Hove sequence.

Theorem 25.1 *If Λ_l is a sequence of regions in $\mathbb{Z}^\nu$ tending to $\mathbb{Z}^\nu$ in the sense of Van Hove then the limit*

$$f(\beta) = \lim_{l \to \infty} \frac{1}{|\Lambda_l|}F_{\Lambda_l}(\beta) \tag{25.12}$$

exists for all $\beta \neq 0$.

Proof. We shall write $\Lambda_l^{\pm}$ instead of $(\Lambda_l)_a^{\pm}$. Analogous to equations (25.6) and (25.7) we can split the energy into three terms:

$$E_{\Lambda_l} = E_{\Lambda_l^-} + E_{\Lambda_l \setminus \Lambda_l^-} + W$$

where

$$W = \sum_{x \in \Lambda_l^-} \sum_{y \in \Lambda_l \setminus \Lambda_l^- : |x-y| \leqslant R} \Phi_{x-y}(s_x, s_y).$$

In the expression for the interaction energy W between the interior and the exterior of Λ_l^-, the points y must lie on the boundary of $\Lambda_l \setminus \Lambda_l^- \subset \Lambda_l^+ \setminus \Lambda_l^-$ and therefore certainly in the boundary of each of its constituent cells (cubes). The number of such points y is bounded by

$$\frac{a^\nu - (a - 2R)^\nu}{a^\nu}|\Lambda_l^+ \setminus \Lambda_l^-| \leqslant 2\nu\frac{R}{a}|\Lambda_l^+ \setminus \Lambda_l^-|.$$

It follows that

$$|W| \leqslant 2\nu\frac{R}{a}|\Lambda_l^+ \setminus \Lambda_l^-|M$$

where M is given by equation (25.8), and therefore,

$$\left| F_{\Lambda_l} - F_{\Lambda_l^-} - F_{\Lambda_l \setminus \Lambda_l^-} \right| \leqslant 2\nu \frac{R}{a} |\Lambda_l^+ \setminus \Lambda_l^-| M. \tag{25.13}$$

Now, if $A = \max_{s=1}^q |\Psi(s)|$ then $\left| E_{\Lambda_l \setminus \Lambda_l^-} \right| \leqslant (M + A)|\Lambda_l^+ \setminus \Lambda_l^-|$. This implies that

$$\left| F_{\Lambda_l \setminus \Lambda_l^-} \right| \leqslant c |\Lambda_l^+ \setminus \Lambda_l^-| \tag{25.14}$$

where the constant $c = |\beta|^{-1} \ln q + M + A$. By lemma 25.1 and the definition of Van Hove convergence, we now have that for any given $\epsilon > 0$,

$$\left| \frac{1}{|\Lambda_l|} F_{\Lambda_l^-}(\beta) - f(\beta) \right| \leqslant \frac{\epsilon}{2} \tag{25.15}$$

provided that a and l are large enough. Combining equations (25.13), (25.14), and (25.15) we obtain

$$\left| |\Lambda_l|^{-1} F_{\Lambda_l}(\beta) - f(\beta) \right| \leqslant \frac{\epsilon}{2} + \left(2\nu \frac{R}{a} M + c \right) \frac{|\Lambda_l^+ \setminus \Lambda_l^-|}{|\Lambda_l|}$$

where the last term can be made smaller than $\epsilon/2$ by taking l large enough. ∎

We finally prove that the function $-\beta f(\beta)$ is convex.

Theorem 25.2 *The function $-\beta f(\beta)$ defined in theorem 25.1 is convex.*

Proof. We can write $Z_{\Lambda_l}(\beta)$ as a multiple of the expectation of the function $\exp[-\beta X_l]$, where X_l is the random variable (the energy) with distribution function

$$F_l(x) = q^{-|\Lambda_l|} \# \left\{ \{s_x\}_{x \in \Lambda_l} \middle| E_{\Lambda_l}(\{s_x\}) \leqslant x \right\}. \tag{25.16}$$

It then follows from Hölder's inequality that for any $c \in [0, 1]$,

$$
\begin{aligned}
Z_{\Lambda_l}(c\beta_1 + (1 - c)\beta_2) &= q^{|\Lambda_l|} \mathbb{E}\left(\exp\left[-(c\beta_1 + (1 - c)\beta_2) X_l \right] \right) \\
&\leqslant \left(q^{|\Lambda_l|} \mathbb{E}[\exp(-\beta_1 X_l)] \right)^c \left(q^{|\Lambda_l|} \mathbb{E}[\exp(-\beta_2 X_l)] \right)^{1-c} \\
&= Z_{\Lambda_l}(\beta_1)^c Z_{\Lambda_l}(\beta_2)^{1-c}.
\end{aligned}
$$

Taking logarithms, dividing by $|\Lambda_l|$, and taking the limit $l \to \infty$ yields the convexity inequality. ∎

REMARK 25.2: *Cumulant generating function for the energy.*
The function

$$C(t) = -\ln q + t f(-t) = \lim_{l \to \infty} \frac{1}{|\Lambda_l|} \ln \mathbb{E}\left[\exp(t X_l) \right] \tag{25.17}$$

is the cumulant generating function for the energy density. We shall use this fact in the next chapter to prove the large-deviation property for the random variables $u_l = X_l / |\Lambda_l|$.

CHAPTER 26

Large Deviations for the Energy Density

In this chapter, we prove that the energy density of a classical spin system with pair interaction as introduced in chapter 25 has the large-deviation property. This enables us then as before to prove that the entropy density exists in the sense defined by equation (21.3). The proof of the large-deviation property is very similar to that of Cramér's theorem (theorem 20.2):

Theorem 26.1 *The energy density u_l of the classical spin system on a Van Hove sequence of finite regions $\Lambda_l \subset \mathbb{Z}^\nu$ given by the energy function $E_{\Lambda_l}\left(\{s_x\}_{x\in\Lambda_l}\right)$ is defined (in the micro-canonical ensemble) to be a random variable with the distribution function*

$$F_l(u) = q^{-|\Lambda_l|} \# \left\{ \{s_x\}_{x\in\Lambda_l} \middle| E_{\Lambda_l}\left(\{s_x\}_{x\in\Lambda_l}\right) \leqslant |\Lambda_l|\, u \right\}, \qquad (26.1)$$

where q is the number of values an individual spin s_x can take. The random variable u_l satisfies the large-deviation property with constants $N_l = |\Lambda_l|$ and rate function given by the Legendre transform of the function $C(t) = -\ln q + t f(-t)$, where f is the free energy density defined by theorem 25.1.

Proof. First note that

$$\left|E_{\Lambda_l}\left(\{s_x\}_{x\in\Lambda_l}\right)\right| \leqslant (M + A)\,|\Lambda_l| \qquad (26.2)$$

which implies that $|C(t)| \leqslant (M + A)|t|$ so that $I(x) = +\infty$ for $|x| > M + A$. The level sets of I are therefore bounded. They are also closed because if $x_n \to x$ then $t x_n - C(t) \to tx - C(t)$ for all t. I is therefore a valid candidate for the rate function.

The proof of the large-deviation upper bound is very similar to that in the proof of Cramér's theorem. Let $[x_-, x_+] = \{x \in \mathbb{R} \mid I(x) = 0\}$. If $[a, b] \cap [x_-, x_+] \neq \emptyset$ then $\inf_{x\in[a,b]} I(x) = 0$ and there is nothing to prove. Next suppose $b < x_-$. Then $\inf_{x \in [a,b]} I(x) = I(b)$ and we can use again the fact that $e^{N_l t(x-b)} \geqslant 1_{[a,b]}$ for $t \leqslant 0$ to conclude that

$$\mathbb{P}(u_l \in [a,b]) \leqslant \int_{-\infty}^{\infty} e^{N_l t(x-b)} \mathrm{d}F_l(x).$$

This implies

$$\limsup_{l \to \infty} \frac{1}{N_l} \ln \mathbb{P}(u_l \in [a,b]) \leqslant \lim_{l \to \infty} \frac{1}{N_l} \ln \int_{-\infty}^{\infty} e^{N_l t(x-b)} \mathrm{d}F_l(x)$$
$$= -tb + C(t).$$

Since this holds for arbitrary $t \leqslant 0$, we conclude that

$$\limsup_{l \to \infty} \frac{1}{N_l} \ln \mathbb{P}(u_l \in [a,b]) \leqslant -I(b) = -\inf_{x \in [a,b]} I(x).$$

(Note that the supremum of $tb - C(t)$ is attained for $t < 0$ because $I(x)$ is decreasing for $x < x_-$.) In the case $a > x_+$ we use the inequality $e^{N_l t(x-a)} \geqslant 1_{[a,b]}$ for $t \geqslant 0$.

In the proof of the lower bound, we cannot now use the central limit theorem because we do not have independent random variables. Instead, we use the fact that if the Laplace transform of a sequence of probability distributions F_l on $\mathbb{R}$ converges to the Laplace transform of a probability distribution F then F_l tends to F weakly (see appendix A). Choose $\epsilon > 0$ arbitrary and let $x_0 \in (a,b)$ be such that $I(x) > I(x_0) - \epsilon$ for all $x \in (a,b)$. We may assume that we can choose x_0 such that there exists τ for which $C'(\tau) = x_0$. (This follows from the fact that $C(t)$ is almost everywhere differentiable and $I(x)$ is continuous on its essential domain.) Again, we define shifted random variables z_l with distribution functions

$$G_l(z) = \int_{-\infty}^{z} e^{N_l(\tau x - C_l(\tau))} \mathrm{d}F_l(x) \tag{26.3}$$

where C_l is the finite-volume cumulant generating function:

$$C_l(t) = \frac{1}{N_l} \ln \int_{-\infty}^{\infty} e^{N_l t x} \mathrm{d}F_l(x). \tag{26.4}$$

Taking $\delta > 0$ so small that $(x_0 - \delta, x_0 + \delta) \subset (a,b)$, we have

$$\mathbb{P}(u_l \in (a,b)) \geqslant \int_{x_0-\delta}^{x_0+\delta} e^{N_l(C_l(\tau)-\tau z)} \mathrm{d}G_l(z)$$
$$\geqslant e^{N_l(C_l(\tau)-\tau x_0-|\tau|\delta)} \mathbb{P}(z_l \in (x_0 - \delta, x_0 + \delta)). \tag{26.5}$$

Using the convergence of Laplace transforms, we presently show that the probability in the right-hand side converges to 1. Given that, we have again

$$\liminf_{l\to\infty} \frac{1}{N_l} \ln \mathbb{P}(u_l \in (a,b)) \geqslant C(\tau) - \tau x_0 - \delta|\tau|$$

$$= -I(x_0) - \delta|\tau|$$

$$> - \inf_{x\in(a,b)} I(x) - \epsilon - \delta|\tau|. \qquad (26.6)$$

This proves the large deviation lower bound as ϵ and δ can both be taken arbitrarily small.

To prove that

$$\mathbb{P}\big(z_l \in (x_0 - \delta, x_0 + \delta)\big) \to 1 \qquad (26.7)$$

we compute the Laplace transform of the distribution of z_l:

$$\int_\infty^\infty e^{tz} dG_l(z) = e^{-N_l C_l(\tau)} \int_{-\infty}^\infty e^{N_l(\tau+t/N_l)x} dF_l(x)$$

$$= \exp\left\{ N_l \left[C_l\left(\tau + \frac{t}{N_l}\right) - C_l(\tau) \right] \right\}. \qquad (26.8)$$

We now claim that

$$N_l \left[C_l\left(\tau + \frac{t}{N_l}\right) - C_l(\tau) \right] \to tC'(\tau). \qquad (26.9)$$

This implies that

$$\int_{-\infty}^\infty e^{tz} dG_l(z) \to e^{tC'(\tau)} = e^{tx_0} = \int_{-\infty}^\infty e^{tz} dG(z), \qquad (26.10)$$

where

$$G(z) = \begin{cases} 0 & \text{if } z < x_0 \\ 1 & \text{if } z \geqslant x_0 \end{cases}$$

is the distribution function of the random variable which equals x_0 with probability 1. This proves equation (26.7). To prove (26.9), which is known as **Griffiths' lemma**, note that, for any given $\epsilon > 0$, we can choose l so large that $|t|/N_l < \epsilon$ in which case, by the convexity of C_l,

$$\frac{C_l(\tau) - C_l(\tau - \epsilon)}{\epsilon} \leqslant \frac{N_l}{t} \left[C_l\left(\tau + \frac{t}{N_l}\right) - C_l(\tau) \right] \leqslant \frac{C_l(\tau + \epsilon) - C_l(\tau)}{\epsilon}.$$

Taking $l \to \infty$ with fixed ϵ, we obtain

$$\frac{C(\tau) - C(\tau - \epsilon)}{\epsilon} \leqslant \liminf_{l\to\infty} \frac{N_l}{t} \left[C_l\left(\tau + \frac{t}{N_l}\right) - C_l(\tau) \right]$$

$$\leqslant \limsup_{l\to\infty} \frac{N_l}{t} \left[C_l\left(\tau + \frac{t}{N_l}\right) - C_l(\tau) \right] \leqslant \frac{C(\tau + \epsilon) - C(\tau)}{\epsilon}.$$

Now taking $\epsilon \to 0$ this proves equation (26.9). ∎

The large-deviation property implies as before the existence of the entropy density in the thermodynamic limit (see equations (21.13) and (22.14)-(22.16)):

Corollary 26.1 *Let*

$$\Omega_{\Delta,l}(u) = \# \left\{ \{s_x\}_{x\in\Lambda_l} \,\big|\, |\Lambda_l|(u-\Delta) < E_{\Lambda_l}(\{s_x\}) < |\Lambda_l|(u+\Delta) \right\} \quad (26.11)$$

be the number of microstates with energy per particle in the interval $(u-\Delta, u+\Delta)$. Then the thermodynamic limit of the entropy density

$$s(u) = \lim_{\Delta\to 0} \lim_{l\to\infty} \frac{k_B}{|\Lambda_l|} \ln \Omega_{\Delta,l}(u) \quad (26.12)$$

exists and is given by equations (24.10) and (8.4):

$$s(u) = k_B \inf_{\beta\in\mathbb{R}} [\beta u - \beta f(\beta)]. \quad (26.13)$$

Proof. Note that

$$\Omega_{\Delta,l}(u) = q^{N_l} \mathbb{P}(u_l \in (u-\Delta, u+\Delta)).$$

Therefore

$$\liminf_{l\to\infty} \frac{k_B}{N_l} \ln \Omega_{\delta,l}(u) \geqslant k_B \left(\ln q - \inf_{x\in(u-\Delta,u+\Delta)} I(x) \right). \quad (26.14)$$

On the other hand

$$\limsup_{l\to\infty} \frac{k_B}{N_l} \ln \Omega_{\delta,l}(u) \leqslant k_B \left(\ln q - \inf_{x\in[u-\Delta,u+\Delta]} I(x) \right). \quad (26.15)$$

The infima on the right-hand side of equations (26.14) and (26.15) tend to $I(u)$ as $\Delta \to 0$. Therefore

$$\begin{aligned}
s(u) &= k_B[\ln q - I(u)] \\
&= k_B \ln q - k_B \sup_{t\in\mathbb{R}} [tu - C(t)] \\
&= k_B \inf_{t\in\mathbb{R}} [-tu + tf(-t)].
\end{aligned}$$

∎

Problems to Part II

Note: More difficult problems are indicated by an asterisk.

II-1. Show that the hydrogen spectrum (18.11) also follows from Bohr's postulates as outlined in remark 18.2.

II-2. Consider a plane circular current loop in a homogeneous external magnetic field $\vec{H}$. Assume that the normal to the plane of the loop makes an angle θ with the direction of the field. Let the current in the loop be I and the radius of the loop r. Show that the torque on the loop due to the magnetic field can be written as $\vec{T} = \vec{\mu} \times \vec{B}_0$ and give an expression for the magnetic dipole moment $\vec{\mu}$ of the loop. (The Lorentz force on a short section $d\vec{r}$ of the loop is given by $d\vec{F} = I\,d\vec{r} \times \vec{B}_0$.) Assuming that the current in fact consists of a single (or several) charged particles with (total) charge $-e$, express $\vec{\mu}$ in terms of the angular momentum $L = m\omega r^2$ as follows:

$$\vec{\mu} = -\frac{e}{2m}\vec{L}.$$

Note that if we consider the electron as a rotating charge, then this formula is wrong by a factor $g = 2$. This factor is called the **gyromagnetic ratio** of the electron. It is due to a quantum effect.

II-3. Using the formula $W = \int T\,d\theta$ for the work done by a torque T during rotation, compute the energy needed to rotate the current loop of the previous question from the neutral position $\theta = \pi/2$ to a general position. Compare your answer with equation (18.14).

II-4. The Hamiltonian matrix of a spin-$\frac{1}{2}$ particle in a magnetic field $\vec{H}$ is given by $\mathcal{H} = \mu_0 \vec{\mu} \cdot \vec{H}$ where $\vec{\mu} = g(q/2m)\vec{S}$. Here q is the charge of the particle, m is its mass and $\vec{S}$ is the spin angular momentum. g is the **gyromagnetic ratio**. For the electron $q = -e$ and $g = 2$. If we take $|\vec{S}| = \hbar/2$ we obtain equation (18.13). In fact the components $S^{(x)}$, $S^{(y)}$, and $S^{(z)}$ of $\vec{S}$ are matrices. They are usually expressed as $\vec{S} = (\hbar/2)\vec{\sigma}$, where

$$\sigma^{(x)} = \begin{pmatrix} 0 & 1 \\ 1 & 0 \end{pmatrix}, \quad \sigma^{(y)} = \begin{pmatrix} 0 & -i \\ i & 0 \end{pmatrix}, \quad \sigma^{(z)} = \begin{pmatrix} 1 & 0 \\ 0 & -1 \end{pmatrix}$$

are the **Pauli spin matrices**. Show that, no matter what the direction of $\vec{H}$, $\mathcal{H}$ always has two eigenvalues $\pm\epsilon$ where

$$\epsilon = g\mu_0|\vec{H}|\frac{q\hbar}{4m}.$$

Show also that $\vec{S}^2$ has a single eigenvalue equal to $\frac{3}{4}\hbar^2$.

II-5. Show that $\vec{F} = -\vec{\nabla}V$ is given by equation (18.8) if V is given by (18.10) and (18.9).

II-6. The Hamiltonian matrix of a particle on a line in a central force field $F = -kx$ as described in example 18.1 is given by $\mathcal{H} = \hbar\omega(A^*A + \frac{1}{2}I)$, where I is an infinitely large unit matrix and A is the infinite matrix with entries given by $A_{k,k+1} = \sqrt{k}$ for $k = 1, 2, \ldots$, and $A_{k,l} = 0$ if $l \neq k + 1$. Show that
(i) $A A^* - A^* A = I$; and
(ii) $\mathcal{H}$ is a diagonal matrix with eigenvalues (18.3).

II-7. Compute a numerical value for the so called **Bohr magneton** μ_B given by equation (18.13). Given $e = 1.6 \times 10^{-19}$ C and $m_e = 9.11 \times 10^{-31}$ kg, hence derive a rough estimate for the strength of the magnetic induction B_0 below which Curie's law would hold if the model of an electronic paramagnet were still valid (i.e. the interaction between spins negligible) at a temperature of 10 mK.

II-8. Draw graphs of the entropy as a function of the temperature $T > 0$ for two different values of the magnetic field H. (See figure 21.1.) **Adiabatic demagnetization** is a method for achieving low temperatures using a paramagnetic salt. It consists of two alternating steps. First the paramagnet is magnetized in a high field H_2 at a constant temperature T_2. It is then thermally isolated and the field is reduced slowly and adiabatically to a much lower value H_1. Show that in this second step the temperature decreases to a value $T_1 < T_2$ and compute T_1. These steps are then repeated to achieve lower and lower temperatures. Indicate this process in the diagram of s versus T.

II-9. Compute the free energy density $f(T)$ and the internal energy density $u(T)$ for the model of a paramagnetic salt described in chapters 19 and 21. Draw a graph of $u(T)$ and show that it is discontinuous at $T = 0$ but not at $T = \infty$ if negative temperatures are considered a continuation of large positive temperatures. Compute also the specific heat $c_V(T)$.

II-10. Prove that, if $\left(a_n^{(j)}\right)_{n=1}^{\infty}$ are sequences of real numbers $(j = 1, 2, \ldots, r)$, then

$$\limsup_{n\to\infty} \bigvee_{j=1}^{r} a_n^{(j)} = \bigvee_{j=1}^{r} \limsup_{n\to\infty} a_n^{(j)}.$$

Give also a counterexample to show that in general

$$\liminf_{n\to\infty} \bigvee_{j=1}^{r} a_n^{(j)} \neq \bigvee_{j=1}^{r} \liminf_{n\to\infty} a_n^{(j)}.$$

II-11. Suppose that $\left(X^{(k)}\right)_{k=1}^{\infty}$ is a sequence of independent identically distributed random variables with distribution function F satisfying

$$\ln \int e^{tx} dF(x) = \sqrt{1 + t^2} - 1 = C(t).$$

Show that $C(t)$ is a convex function and compute the rate function $I(x)$ for the random variables $X_N = N^{-1} \sum_{k=1}^{N} X^{(k)}$.

II-12. Suppose that the sequence of random variables $(X_N)_{N=1}^{\infty}$ satisfies the LDP with rate function $I : \mathbb{R} \to [0, +\infty]$. Prove that the random variables Y_N $(N = 1, 2, \ldots)$ defined by $Y_N = |X_N|$ satisfy the LDP with rate function J given by

$$J(x) = \begin{cases} I(x) \wedge I(-x) & \text{if } x \geqslant 0, \\ +\infty & \text{if } x < 0, \end{cases}$$

where $\wedge$ denotes 'the minimum of'.

II-13*. Prove that if a sequence of random variables $(X_N)_{N=1}^{\infty}$ satisfies the LDP then the corresponding rate function I is unique.
(Hint: Assume that there are two rate functions I_1 and I_2 which differ in a point x_0. Then use the compactness of the level sets and the LDP for $(X_N)_{N=1}^{\infty}$.)

II-14. Prove the validity of remark 20.1.

II-15. Compute

$$\lim_{N \to \infty} \frac{1}{N} \ln \left(\sum_{M=0}^{N} e^{\beta M^2 / N} \right)$$

for all real values of β.

II-16. Complete the proof of the LDP in example 20.5. Next show that the function $G(x) = -|x - 1|$ satisfies the conditions for Varadhan's theorem and hence compute

$$\lim_{N \to \infty} \frac{1}{N} \ln \int_{-\infty}^{\infty} e^{N G(x)} dF_N(x).$$

II-17. Let the random variables X_N $(N = 1, 2, \ldots)$ have the distribution functions F_N defined by

$$F_N(x) = \frac{1}{Z_N} \sum_{k=-\infty}^{[x]} e^{-N k^4},$$

where Z_N is a normalization constant and $[x]$ denotes the integer part of x. Show that the sequence $(X_N)_{N=1}^{\infty}$ satisfies the LDP with rate function

$$I(x) = \begin{cases} +\infty & \text{if } x \notin \mathbb{Z}, \\ x^4 & \text{if } x \in \mathbb{Z}. \end{cases}$$

Use this result to compute

$$\lim_{N \to \infty} \frac{1}{N} \ln \sum_{k=-\infty}^{\infty} e^{N(3k + 2k^2 - 2k^4)}.$$

II-18. The canonical partition function of a classical hard-core gas in one dimension is given by

$$Z(N, L) = \frac{1}{N!}(L - (N-1)b)^N.$$

Here N is the number of particles ('molecules'), b is the linear size of a molecule and L is the length of the one-dimensional container. Use Stirling's formula to compute the free energy density $f(\beta, \rho)$ in the thermodynamic limit as a function of $\beta = 1/k_\mathrm{B}T$ and the particle number density ρ. Hence obtain the pressure $p(v, T)$ for this model.

II-19. Let $\{X_n\}_{n=1}^\infty$ be a sequence of independent and identically distributed random variables taking values in a finite set $\{a_1, \ldots, a_r\}$ with probabilities $p_1, \ldots, p_r$. Define the random variables

$$N_{n,k} = \frac{1}{n} \#\{i \leqslant n : X_i = a_k\} \qquad (k = 1, \ldots, r).$$

Using Cramér's theorem, show that the r-dimensional random variables $\vec{N}_n = (N_{n,1}, \ldots, N_{n,r})$ satisfy the LDP and compute the corresponding rate function $I(\vec{x})$. (This result is a special case of **Sanov's theorem**.)

II-20. Large-deviation theory has many applications other than in statistical mechanics. In this problem, we consider a simple application in information theory. See the books by McEliece (1977) and Roman (1992) for an introduction to information theory. If a variable can have two possible values then giving its value provides one **bit** of information. If it can have four possible values then we would get two bits of information because two yes/no questions are sufficient to determine its value. In general, the knowledge of a variable with N possible values contains $\log_2 N$ bits of information. Now compute the average gain in information if we learn the value of a random variable taking r values with probabilities $p_1, \ldots, p_r$ such that $p_1 + \cdots + p_r = 1$ in a large number N of trials. It is given by $H(p_1, \ldots, p_r) = \lim_{N \to \infty} \frac{1}{N} \log_2 \Omega_N(p_1, \ldots, p_r)$, where Ω_N is the number of different combinations of outcomes, i.e.

$$\Omega_N(p_1, \ldots, p_r) = \frac{N!}{(p_1 N)! \ldots (p_r N)!}$$

since for large N the number of times the random variable takes the k-th value is approximately $(p_k N)!$. In fact, of course, these numbers fluctuate so a more accurate definition is required. In light of the definition (21.3), give a more precise definition of the average information gain H and use the result of the previous problem to rederive the expression for $H(p_1, \ldots, p_r)$.

II-21. Equation (20.19) contains more information than is used in the proof of theorem 20.2. Consider for example a sequence of independent random variables X_k $(k = 1, 2, \ldots)$ with mean zero and $|X_k| \leqslant M$ for some constant

M. Show that there is a constant K such that $C(t) \leqslant \frac{1}{2}t^2\sigma^2 + Kt^3$ for $t < 1$, where $\sigma^2 = \mathbb{E}(X_k^2)$. Insert this into equation (20.19) and take $t = a/\sigma^2$ and $a = c\sqrt{(\ln N)/N}$ to conclude that for large N,

$$\mathbb{P}(S_N \geqslant c\sqrt{N \ln N}) \leqslant \exp\left[-\left(\frac{c^2}{2\sigma^2} - \epsilon\right)\ln N\right]$$

for any $\epsilon > 0$ if $S_N = \sum_{k=1}^N X_k$. Finally use the Borel-Cantelli lemma (lemma A.1 in appendix A) to show that for large N, $|S_N| \leqslant 2\sigma\sqrt{N \ln N}$ with probability 1. This result is due to Hardy and Littlewood (1914). It has been strengthened by Khintchine (1924), who showed that in fact $|S_N| = \mathcal{O}(\sqrt{N \ln \ln N})$ with probability 1.

II-22. Suppose that $C : \mathbb{R} \to \mathbb{R}$ is differentiable and convex. Show that if I is the Legendre transform, $I = C^*$ then

$$\inf_{x>0} \frac{I(x)}{x} = \sup\{t : C'(t) < 0\}.$$

II-23*. Let $\{X_n\}_{n=1}^\infty$ be a sequence of independent random variables with identical distribution functions $F(x)$. Suppose that the cumulant generating function $C(t)$ exists for all $t \in \mathbb{R}$ and let $S_N = \sum_{n=1}^N X_n$. Prove that

$$\lim_{b\to+\infty} \frac{1}{b} \ln \mathbb{P}(\sup_N S_N > b) = -\sup\{t : C'(t) < 0\}.$$

This result is important in queueing theory. If X_n is the difference between the incoming workload and the amount of work that can be dealt with at (discrete) times n then the event $\sup_N S_N > b$ represents a **buffer overflow** so that the above result means that the buffer overflow probability is exponentially small. For an introduction to queueing theory see e.g. the books by Conolly (1975) and Asmussen (1987). For generalizations of the above result see Glynn and Whitt (1993), and also Duffield and O'Connell (1995). [Hints: Prove upper and lower bounds for lim sup and lim inf. For the lower bound, use the simple fact that $\mathbb{P}(\sup_N S_N > b) \geqslant \sup_N \mathbb{P}(S_N > b)$ together with the LDP and the result of problem II-22. For the upper bound, use the estimate $\mathbb{P}(\sup_N S_N > b) \leqslant \sum_{N=1}^\infty \mathbb{P}(S_N > b)$ and split the sum into two parts: one with $N \leqslant Kb$ and one with $N \geqslant Kb$, taking K sufficiently large.]

II-24*. Some nuclei have spin 1. Consider a (classical) model of a spin-1 paramagnet where localized independent spins s_i ($i = 1, 2, \ldots, N$) can take the values 0 and $\pm\hbar$. The corresponding energy levels in a magnetic field H are 0 and $\pm\epsilon$ where $\epsilon = \mu_0\mu H$. Find an expression for the number of allowed microstates $\Omega(E)$ at energy $E = m\epsilon$ where $m = n_+ - n_-$ is the number of spins in the $+\epsilon$ state minus that in the $-\epsilon$ state. Use the theory of large deviations to compute the entropy density $s(u)$. Hence derive the free energy density $f(T)$.

Note how much easier it is to compute $f(T)$ directly: see the next problem. Compute the susceptibility for small fields and show that Curie's law holds.

II-25. Consider a model for a spin-J paramagnet consisting of independent localized spins s_i ($i = 1, 2, \ldots, N$) taking the values $S = j\hbar$, with $j \in \{-J, -J+1, \ldots, J\}$. In an external magnetic field H each spin has $2J+1$ possible energy levels $j\epsilon$, where $\epsilon = \mu_0 \mu H$ and where $\mu = g\frac{q}{2m}S$ is the magnetic dipole moment (see problems II-3 and II-4). Compute the free energy density $f(T, H)$ for this model. Use equation (10.9) to compute the magnetization $m(T, H)$ and show that it can be expressed in terms of the **Brillioun function**

$$B_J(z) = \frac{2J+1}{2J} \coth\left(\frac{(2J+1)z}{2J}\right) - \frac{1}{2J} \coth\left(\frac{z}{2J}\right).$$

Finally, compute the susceptibility $\chi(T, H)$ and show that Curie's law holds for small fields.

II-26. A diatomic molecule may to good approximation be considered to vibrate in one-dimensional harmonic motion. It thus has an infinite set of vibrational energy levels given by equation (18.3). Assuming that the other energy levels can be ignored, write down an expression for the canonical partition function Z_N for an assembly of N such molecules assuming that they are distinguishable. Hence derive an expression for the vibrational free energy density $f(T)$. Next obtain an expression for the specific heat of such a system. What are the high- and low-temperature asymptotics of this specific heat? Obtain the molar vibrational specific heat of nitrogen (N_2) at 1000 K given that the level spacing is $\hbar\omega = 0.3$eV. (1 eV= 1.6×10^{-19}J.)

II-27. Rederive equations (23.31) and (23.30) for the pressure and the density of a free fermion gas using the general grand-canonical formalism of chapter 24. Then prove the inequality $\ln(1 + x) > x/(1 + x)$ for $x > 0$ and use this to prove that $p_{\text{fermion gas}}(v, T) > p_{\text{ideal gas}}(v, T)$ for all positive values of v an T.

II-28. Express the density ρ of the free quantum gases as integrals $\int_0^\infty \tilde{\rho}(\epsilon)d\epsilon$ over the energy. Derive also equations (23.36), (23.37), and (23.38). Use integration by parts to prove that $u = \frac{3}{2}pv$. Hence derive a formula for the entropy density per unit volume $\tilde{s}$ of a free quantum gas (fermion or boson) in terms of β, μ, $\omega(\beta, \mu)$, and ρ.

II-29. Show that in the classical limit the density ρ of the free quantum gases tends to $\lambda_T^{-3} e^{\beta\mu} \int_0^\infty \phi_{\text{MB}}(\epsilon) \, d\epsilon$, where the **Maxwell-Boltzmann distribution**

$$\phi_{\text{MB}}(\epsilon) = \frac{2}{\sqrt{\pi}} \beta^{3/2} \epsilon^{1/2} e^{-\beta\epsilon}$$

is a probability distribution. Rewrite the Maxwell-Boltzmann distribution as a distribution over the molecular speeds v using $\epsilon = \frac{1}{2}mv^2$.

II-30. Consider a boson gas of diatomic molecules vibrating in one-dimensional harmonic motion as in problem II-26. The particle energies are given by $\epsilon_{i_1,i_2,i_3,r} = \epsilon^0_{i_1,i_2,i_3} + \hbar\omega(r+\frac{1}{2})$. The first term represents the kinetic energy (18.7) of the molecules and the second term is the energy due to vibration as in (18.3). Derive an expression for the grand potential. Show that in the classical limit it splits into a product of two factors: one due to the kinetic energy and the other due to the vibration. Deduce that the specific heat is the sum of two terms in this limit: the specific heat of the ideal gas and the vibrational specific heat computed in problem II-26.

III

Models in Statistical Mechanics

As mentioned in the introduction to this course, **models** are caricatures of physical systems designed to isolate and understand the principal reasons for physical phenomena. In statistical mechanics, phase transitions in particular are phenomena that are still poorly understood from the microscopic point of view. In this final part of the course we consider some simple models and determine their thermodynamic behaviour from the rules of statistical mechanics developed in Part II. These will also illustrate how phase transitions can come about.

In chapter 27 we consider a very simple model for rubber elasticity. Rubber consists of long chain molecules which are linked together at a few points by cross links. We model each section of a chain by a random walk. In chapter 28 we consider the Ising model for a ferromagnet. We show that it can be solved exactly in one dimension by means of a trick: the so-called **transfer matrix**. In two dimensions it can still be solved by essentially the same trick but the transfer matrix is then infinite-dimensional and one has to determine its maximum eigenvalue. This is possible but very difficult. We therefore only give the final result. At the end of this chapter there is a proof of the existence of a phase transition for the two-dimensional Ising model: the celebrated Peierls argument. In chapter 29 we discuss a simplification of the Ising model: the Weiss-Ising model. It can be solved in any dimension but is less realistic in that the solution is essentially dimension-independent. We show that the solution can be obtained using Varadhan's theorem. In chapter 30 we show that a reformulation of the Weiss-Ising model leads to a lattice model of a gas. In chapter 31 it is argued that the free fermion gas is a reasonable model for the conduction electrons in a metal. In chapter 32 it is explained that the free fermion gas is also a good model for so-called white-dwarf stars. These are very compact stars in which the atoms have mostly been stripped of their electrons and the pressure of which is determined by the resulting electron plasma. This chapter depends on chapter 17 about the stability of a normal star. In chapter 33 we show that a modification of the free boson gas yields a model for the vibrations of the atoms of a crystal. Chapters 33, 34, and 35 are more advanced and require a knowledge of the Hilbert space formalism of quantum mechanics (see appendix B). In chapter 34 we consider two simple models for helium. They are both inadequate for describing the interesting phenomenon of superfluidity in liquid helium but nevertheless give an idea of the mechanism behind it. In chapter 35 we discuss a simplified version of the BCS model for superconductivity.

In the new chapters 36 and 37 we introduce **cluster expansions**. First, in chapter 36 we consider the Mayer expansion which is a series expansion for the logarithm of the grand partition function of a classical gas in powers of the so-called fugacity $z = e^{\beta\mu}$. We prove that it converges absolutely for small z, i.e. low densities. This implies that there is no phase transition for low densities. Finally, in chapter 37 we extend this method to lattice models. This leads to a representation in terms of a polymer system and a convergent expansion valid for high temperatures. The proof of convergence is complicated and relies on the theory of chapter 25.

A Simple Model for Rubber

Rubber consists of very long flexible molecules. We shall model them as little sticks linked in a chain. The chains are cross-linked at random but relatively sparse positions. We model here a single chain segment (figure 27.1).

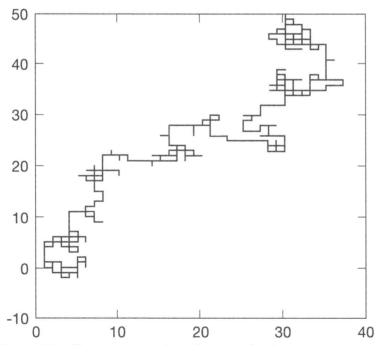

Figure 27.1 Chain segment of a polymer molecule as a random walk.

For simplicity, we assume that each stick can have only six positions: forward and backward in each of the coordinate directions. We also assume that each stick has length ℓ. If each stick is completely free to bend in any of these six directions, the energy of the chain is independent of the configuration of the chain but highly degenerate. The square of the end-to-end length of the chain L is then the sum of the squares of random variables X, Y, and Z. The vector (X, Y, Z) is a sum of independent random variables: $(X^{(k)}, Y^{(k)}, Z^{(k)})$ $(k = 1, 2, \ldots, N)$ with joint distribution given by

$$\mathbb{P}[(X^{(k)}, Y^{(k)}, Z^{(k)}) = (\pm\ell, 0, 0)] = 1/6,$$
$$\mathbb{P}[(X^{(k)}, Y^{(k)}, Z^{(k)}) = (0, \pm\ell, 0)] = 1/6,$$
$$\mathbb{P}[(X^{(k)}, Y^{(k)}, Z^{(k)}) = (0, 0, \pm\ell)] = 1/6. \tag{27.1}$$

The corresponding cumulant generating function is

$$C(t_1, t_2, t_3) = \ln \int e^{t_1 x + t_2 y + t_3 z} \mathrm{d}F(x, y, z)$$
$$= \ln[\cosh(t_1\ell) + \cosh(t_2\ell) + \cosh(t_3\ell)] - \ln 3. \tag{27.2}$$

Let (X_N, Y_N, Z_N) be the three-dimensional random variable defined by

$$X_N = \frac{1}{N} \sum_{k=1}^{N} X^{(k)} \qquad Y_N = \frac{1}{N} \sum_{k=1}^{N} Y^{(k)} \qquad Z_N = \frac{1}{N} \sum_{k=1}^{N} Z^{(k)}. \tag{27.3}$$

Then by Cramér's theorem (X_N, Y_N, Z_N) satisfies the LDP with rate function

$$I(x, y, z) = \sup_{t_1, t_2, t_3} \left[t_1 x + t_2 y + t_3 z - \ln \sum_{i=1}^{3} \cosh(t_i \ell) \right] + \ln 3. \tag{27.4}$$

What does this mean for the rubber chain? If we assume that the energy of the chain is independent of the configuration, i.e. the sticks are free to rotate, then we have

$$f(x, y, z; \beta) = u_0 - T s(x, y, z), \tag{27.5}$$

where u_0 is the energy of a single stick and $s(x, y, z)$ is given by (cf. equation (21.13))

$$s(x, y, z) = k_\mathrm{B} \ln 6 - k_\mathrm{B} I(x, y, z). \tag{27.6}$$

Normally, the chain is highly contracted, that is, $L = N r_N \ll N\ell$. In that case we can approximate the maximizers $t_1, t_2,$ and t_3 in equation (27.4) by $t_1 \approx 3x/\ell^2$, $t_2 \approx 3y/\ell^2$ and $t_3 \approx 3z/\ell^2$. Thus

$$I(x, y, z) \approx \frac{3(x^2 + y^2 + z^2)}{2\ell^2} \tag{27.7}$$

and

$$s(x, y, z) \approx k_\mathrm{B} \left(\ln 6 - \frac{3}{2\ell^2}(x^2 + y^2 + z^2) \right). \tag{27.8}$$

The important observable of rubber as regards its elasticity is the **tension** $\mathcal{F}$. To compute this quantity assume that the rubber chain is stretched in one direction and at the same time contracts in the others so as to keep its volume constant. Then we can write $x = \alpha_x x_0$, $y = \alpha_y y_0$, and $z = \alpha_z z_0$ where $\alpha_x = \alpha$ say, and $\alpha_y = \alpha_z = 1/\sqrt{\alpha}$. The tension of a string is a one-dimensional analogue of minus the pressure. It is given by the analogue of equation (8.8):

$$\mathcal{F} = \frac{\partial f}{\partial \lambda} \tag{27.9}$$

where $\lambda = x$ is the length of the chain per number of sticks in the direction which is stretched. Clearly, $\alpha = x/x_0 = \lambda/\lambda_0$, so

$$\mathcal{F} = \frac{1}{\lambda_0}\frac{\partial f}{\partial \alpha} \approx \frac{3k_{\mathrm{B}}T}{2\lambda_0 \ell^2}\left(2\alpha x_0^2 - \frac{1}{\alpha^2}(y_0^2 + z_0^2)\right). \tag{27.10}$$

If the chain is originally at rest, we expect its relative dimension in each direction to be the same:

$$x_0^2 = y_0^2 = z_0^2 = \mathbb{E}(r^2)/3 \tag{27.11}$$

and we get

$$\mathcal{F} \approx \frac{k_{\mathrm{B}}T}{\ell^2 \lambda_0}\mathbb{E}(r^2)\left(\alpha - \frac{1}{\alpha^2}\right). \tag{27.12}$$

Note in particular that the tension is directly proportional to the temperature. This means that, if the temperature is increased at constant tension, α and therefore the length (per stick) λ decreases: *rubber contracts when it is heated*. This is caused by the fact that the free energy (27.5) is independent of energy, i.e. completely entropic. This means that there is complete degeneracy. This is of course an oversimplification. The sticks (elementary monomers in the chain) are not completely free to rotate. Although this model is highly simplistic, it brings out the essential property of rubber; the entropy term dominates the internal energy: $T s \gg u$ for normal temperatures. At low temperatures this is no longer the case: u then dominates $T s$ and the rubber becomes rigid.

REMARK 27.1: *The central limit theorem.*

The large deviation result for (X_N, Y_N, Z_N) means roughly that

$$\mathbb{P}[(X_N, Y_N, Z_N) \in (x, x + \mathrm{d}x) \times (y, y + \mathrm{d}y) \times (z, z + \mathrm{d}z)]$$
$$\propto \exp\left[-NI(x, y, z)\right]\mathrm{d}x\,\mathrm{d}y\,\mathrm{d}z$$
$$\approx \exp\left(-\frac{3N(x^2 + y^2 + z^2)}{2\ell^2}\right)\mathrm{d}x\,\mathrm{d}y\,\mathrm{d}z. \tag{27.13}$$

This is a Gaussian (normal) distribution. This statement is a weak form of the central limit theorem. Using this distribution, we have

$$\mathbb{E}(x^2) = \frac{\int_{-\infty}^{\infty} x^2 \exp\left[-3Nx^2/2\ell^2\right]\mathrm{d}x}{\int_{-\infty}^{\infty} \exp\left[-3Nx^2/2\ell^2\right]\mathrm{d}x} = \frac{\ell^2}{3N} \tag{27.14}$$

and similarly for $\mathbb{E}(y^2)$ and $\mathbb{E}(z^2)$. We can thus write equation (27.12) in the form

$$\mathcal{F} \approx \frac{k_{\mathrm{B}}T}{L_0}\left(\alpha - \frac{1}{\alpha^2}\right) \qquad (27.15)$$

where $L_0 = N\lambda_0$ is the equilibrium length of the chain.

CHAPTER 28

The Ising Model

In chapters 19 and 21, we have discussed a simple model for a paramagnetic salt. Paramagnetism could be explained by assuming that all spins are independent. However, as can be seen from the graphs in figure 19.1, this model does not explain permanent magnetization. Indeed, if the field strength H is reduced to zero the magnetization disappears. To explain the phenomenon of permanent magnetization or **ferromagnetism** therefore, we must take into account the *interaction* between spins. Unfortunately, this is not an easy matter. Understanding magnetism from a microscopic point of view is a delicate problem and has by no means been solved completely.

Table 28.1. The Periodic table.

Element	1s	2s	2p	3s	3p	3d	4s
Cr	2	2	6	2	6	5	1
Mn	2	2	6	2	6	5	2
Fe	2	2	6	2	6	6	2
Co	2	2	6	2	6	7	2
Ni	2	2	6	2	6	8	2
Cu	2	2	6	2	6	10	1

For example, it is still not feasible to predict on theoretical grounds alone (*ab initio*) whether a given substance is ferromagnetic or paramagnetic (or diamagnetic). Only the following few elements are ferromagnetic: iron (Fe), cobalt (Co), nickel (Ni), gadolinium (Ga), and dysprosium (Dy). If we look at the electronic configuration of the elements near iron in the periodic table, we can see how delicate the situation actually is (table 28.1).

Apparently the 3d electrons play a crucial role. These are the outer electrons or **conduction electrons**. (The 3d band and the 4s band overlap and are both partly filled. The 3d electrons are more important for ferromagnetism, however.) We have seen in chapter 18 that the electron has a spin angular momentum and an associated magnetic moment. If we consider it as a little bar magnet then we see that two neighbouring electrons have a tendency to *anti*-align (figure 28.1).

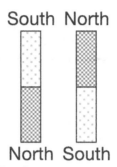

South North

North South

Figure 28.1 Two bar magnets tend to anti-align.

This largely cancels the total magnetization! In ferromagnets there must therefore be *another force* that tends to align the elementary magnets or spins. The mechanism for this alignment force was discovered by Heisenberg in 1928. It is quantum mechanical in origin and is called the **exchange force**. It is a subtle interplay between the Pauli principle and the Coulomb (electric) potential. We outline Heisenberg's derivation here. It requires a knowledge of quantum mechanics beyond that of chapter 18 but it is not essential for the remainder of this chapter and can be omitted. The result is given by equation (28.1).

Consider first two electrons moving in the individual potentials of two atoms but without interaction between them. Then the eigenfunctions for the system of two electrons are given by the singlet (B.38d) and the triplet (B.38a)–(B.38c) in appendix B. These four states are degenerate with energy $E_1 + E_2$ if E_1 and E_2 are the energies of the individual electrons. If we now want to take into account the interaction H_{int} between the electrons then the change in energy is, to first order in perturbation theory, given by

$$\Delta E = \int \int \overline{\psi(\vec{x}_1, \vec{x}_2)} H_{\text{int}} \psi(\vec{x}_1, \vec{x}_2) \, \mathrm{d}^3 x_1 \, \mathrm{d}^3 x_2$$

where ψ is one of the singlet or triplet eigenfunctions. A straightforward calculation shows that $\Delta E = K + J$ for the singlet state and $\Delta E = K - J$ for the triplet states. Here K is given by

$$K = \int \int \overline{\psi_1(\vec{x}_1)\psi_2(\vec{x}_2)} H_{\text{int}} \psi_1(\vec{x}_1)\psi_2(\vec{x}_2) \, \mathrm{d}^3 x_1 \, \mathrm{d}^3 x_2$$

and J is given by

$$J = \int \int \overline{\psi_1(\vec{x}_2)\psi_2(\vec{x}_1)} H_{\text{int}} \psi_1(\vec{x}_1)\psi_2(\vec{x}_2) \, \mathrm{d}^3 x_1 \, \mathrm{d}^3 x_2.$$

J is called the **exchange integral**; when it is positive then the triplet has lower energy, which means that parallel spins are favoured since two-thirds of these states have parallel spins. In appendix B we compute that $(\vec{S}_1 + \vec{S}_2)^2 = 2\hbar^2$ for triplet states and $(\vec{S}_1 + \vec{S}_2)^2 = 0$ for the singlet state, but $\vec{S}_1^2 = \vec{S}_2^2 = \frac{3}{4}\hbar^2$ so $2\vec{S}_1 \cdot \vec{S}_2 = \frac{1}{2}\hbar^2$ and $-\frac{3}{2}\hbar^2$ respectively. Therefore we can write

$$\Delta E = K - \frac{1}{2}J - 2\hbar^{-2}J\vec{S}_1 \cdot \vec{S}_2.$$

The first two terms are spin-independent; the last term gives rise to ferromagnetism if $J > 0$. This is the case for the 3d electrons of iron and nickel.

The effective interaction energy between the 3d-electrons of different atoms is of the form

$$E_{\text{int}} = -2\hbar^{-2}J\,\vec{S}_1 \cdot \vec{S}_2, \tag{28.1}$$

where $\vec{S}_1$ is the spin of the electron of atom 1 and $\vec{S}_2$ is the spin of the electron of atom 2. For ferromagnets, the **exchange integral** J is positive and E_{int} dominates the small magnetic dipole interaction energy (which, incidentally, is of the same form with $J < 0$). Let us now make another drastic simplification here. For electrons the spin $\vec{S}$ can have only two directions, one diametrically opposite the other. Let us assume that this direction is the same for all electrons and coincides with that of the external field $\vec{H}$ which we can take to be the z-direction. Then we can replace $\vec{S}$ by $S^{(z)} = \frac{1}{2}\hbar s$ for each electron, where $s = \pm 1$. Let us also assume that the atoms are arranged (figure 28.2) in a cubic lattice (in most cases the lattice is of a more complicated type but that is not essential).

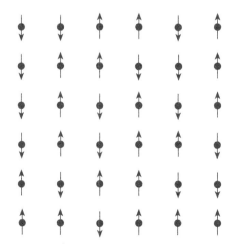

Figure 28.2 A cubic lattice of elementary spins.

We finally assume that their 3d electrons are localized near the corresponding atom. This latter assumption was originally also made by Heisenberg, but is contrary to the fact that the 3d electrons are conduction electrons and can move easily between atom cores. The assumption is valid for non-metallic ferromagnets (salts), however.

Given both these assumptions, we can write the total energy of the system of electrons as

$$E = -\frac{1}{4} \sum_{x \in \mathcal{L}} \sum_{y \in \mathcal{L}: y \neq x} J_{xy} \, s_x \, s_y + \mu_0 \mu H \sum_{x \in \mathcal{L}} s_x. \tag{28.2}$$

Here $\mathcal{L}$ denotes the lattice, H is the external field and $s_x = \pm 1$ for each $x \in \mathcal{L}$. (Note that every pair of sites (x, y) is counted twice in the double sum.) This model for ferromagnetism is called the **Ising model.** The exchange integrals J_{xy} normally decay rapidly with the distance $|x - y|$ so it is a good approximation to assume that $J_{xy} = 0$ unless x and y are nearest neighbours. Assuming also that the interaction is isotropic, and absorbing the constants $\frac{1}{2}$ into J and $-\mu_0 \mu$ into H we obtain the simple **nearest neighbour Ising model:**

$$\boxed{E = -\frac{J}{2} \sum_{x} \sum_{y: \, |y-x|=1} s_x \, s_y - H \sum_{x} s_x} \tag{28.3}$$

(We retain a factor $\frac{1}{2}$ in the exchange term because the double sum counts every nearest-neighbour pair x, y twice. Also, by changing the sign of H, the field term now tends to align the spin, not the magnetic moment, with the field.) This model has been analysed in great detail in the literature. Although it looks very simple, it turns out that it is still by no means easy to analyse. We shall compute the free energy density here for the one-dimensional case only. In two dimensions it is still possible to compute $f(\beta, J, H)$ in case $H = 0$ but if $H \neq 0$ or the number of dimensions is three (which, of course, is the most interesting case) then no explicit solution is known for $f(\beta, J, H)$ and one has to resort to approximations.

In one dimension the lattice is a linear array of spins $s_x = \pm 1$, which we may label by the integers between 1 and N: $x = 1, 2, \ldots, N-1, N$, (figure 28.3).

1   2   . . .   . . .   N-1   N

Figure 28.3 A one-dimensional spin chain.

Every spin except s_1 and s_N interacts with two neighbours. One speaks of **free boundary conditions** in this case. To simplify the analysis we shall restore the symmetry by introducing one extra interaction term in the expression (28.3) for the energy namely $-J s_1 s_N$. In that case one speaks of **periodic boundary conditions.** One can show that this extra term does not affect the result for the free energy density in the thermodynamic limit

$(N \to \infty)$ (see chapter 25). By equation (24.8) the free energy density is given by

$$f(\beta, J, H) = -\frac{1}{\beta} \lim_{N \to \infty} \frac{1}{N} \ln Z_N(\beta, J, H). \qquad (28.4)$$

For a given **configuration** of spins $\{s_x\} = \{s_1, s_2, \ldots, s_N\}$ the energy is given by

$$E_{\text{pbc}}(\{s_x\}) = -J s_1 s_N - J \sum_{x=2}^{N} s_{x-1} s_x - H \sum_{x=1}^{N} s_x. \qquad (28.5)$$

(Here pbc indicates the periodic boundary conditions.) This has to be inserted in the expression for the (canonical) partition function:

$$Z_N(\beta, J, H) = \sum_{\{s_x\}} \exp\left[-\beta E_{\text{pbc}}(\{s_x\})\right]. \qquad (28.6)$$

The limit (28.4) can be computed using the following clever but standard trick. We define a 2×2 matrix A, the so-called **transfer matrix**, by

$$A = \begin{pmatrix} e^{\beta(J+H)} & e^{-\beta J} \\ e^{-\beta J} & e^{\beta(J-H)} \end{pmatrix}. \qquad (28.7)$$

This can also be written as follows:

$$A_{i,j} = \exp\left(\beta J s s' + \beta H \frac{s + s'}{2}\right), \qquad (28.8)$$

where

$$s = \begin{cases} +1 & \text{if } i = 1 \\ -1 & \text{if } i = 2 \end{cases} \quad \text{and} \quad s' = \begin{cases} +1 & \text{if } j = 1 \\ -1 & \text{if } j = 2. \end{cases}$$

Using this latter expression we can write

$$Z_N(\beta, J, H) = \sum_{i_1=1,2} \cdots \sum_{i_N=1,2} A_{i_N, i_1} A_{i_1, i_2} \cdots A_{i_{N-1}, i_N}$$

$$= \sum_{i=1,2} \left(A^N\right)_{i,i} = \text{Tr}\left(A^N\right). \qquad (28.9)$$

Thus we obtain:

$$\boxed{f(\beta, J, H) = -\frac{1}{\beta} \lim_{N \to \infty} \frac{1}{N} \ln\left[\text{Tr}\left(A^N\right)\right]} \qquad (28.10)$$

Now assume that λ_1 and $\lambda_2 < \lambda_1$ are the two eigenvalues of A. Then

$$\text{Tr}\left(A^N\right) = \lambda_1^N + \lambda_2^N \qquad (28.11)$$

and λ_2^N is negligible with respect to λ_1^N if N is large. More precisely,

$$f(\beta, J, H) = -\frac{1}{\beta} \lim_{N \to \infty} \left[\ln \lambda_1 + \frac{1}{N} \ln\left(1 + \frac{\lambda_2^N}{\lambda_1^N}\right)\right],$$

which yields

$$f(\beta, J, H) = -\frac{1}{\beta} \ln \lambda_1 \qquad (28.12)$$

It remains to determine the eigenvalues $\lambda_{1,2}$ of A:

$$\lambda_{1,2} = e^{\beta J} \cosh(\beta H) \pm \sqrt{e^{2\beta J} \sinh^2(\beta H) + e^{-2\beta J}}. \qquad (28.13)$$

Clearly, the plus sign gives the maximum value so that

$$f(\beta, J, H) = -\frac{1}{\beta} \ln \left[e^{\beta J} \cosh(\beta H) + \left(e^{2\beta J} \sinh^2(\beta H) + e^{-2\beta J} \right)^{1/2} \right]. \qquad (28.14)$$

This expression for the one-dimensional model is already quite complicated. In the case $H = 0$ it simplifies considerably, however:

$$f_{H=0}(\beta, J) = -\frac{1}{\beta} \ln 2 \cosh \beta J. \qquad (28.15)$$

We can now define the **magnetization** m as the average of $\frac{1}{N} \sum_x s_x$ for the canonical distribution (24.3):

$$m = \lim_{N \to \infty} \frac{1}{N} \frac{1}{Z_N(\beta, J, H)} \sum_{\{s_x\}} \left(\sum_x s_x \right) e^{-\beta E_{\text{pbc}}(\{s_x\})} \qquad (28.16)$$

which we can write as

$$m = \lim_{N \to \infty} \frac{1}{N} \frac{\partial}{\partial H} \left(\frac{1}{\beta} \ln Z_N(\beta, J, H) \right). \qquad (28.17)$$

Interchanging limit and derivative we obtain

$$m = -\left(\frac{\partial f}{\partial H} \right)_\beta \qquad (28.18)$$

(If we restore the constant $-\mu_0 \mu$ in front of H in the expression for the energy we must define m as

$$m = -\frac{\mu}{v} \lim_{N \to \infty} \left(\frac{1}{N} \sum_x s_x \right)$$

as in equation (19.13). The above derivation then yields that

$$m = -\frac{1}{\mu_0 v} \frac{\partial f}{\partial H}$$

in accordance with equation (10.9).) Performing the derivative we find

$$m = \frac{\sinh(\beta H)}{\sqrt{\sinh^2(\beta H) + e^{-4\beta J}}} \qquad (28.19)$$

The general shape of the graph of m versus H is the same as in figure 19.1. The susceptibility is given by

$$\chi(\beta, J, H) = \frac{\beta \cosh(\beta H) e^{-4\beta J}}{\left(\sinh^2(\beta H) + e^{-4\beta J}\right)^{3/2}}. \tag{28.20}$$

In particular, for $H = 0$ we get

$$\chi(\beta, J, 0) = \beta e^{2\beta J} \approx \frac{1}{k_B T} \text{ if } \beta J \ll 1. \tag{28.21}$$

The one-dimensional Ising model behaves therefore as a paramagnet! When Ising first derived this result, he conjectured that the higher-dimensional analogues would probably behave in the same way. This has turned out to be a faulty conclusion. The free energy per spin of the 2-dimensional Ising model with zero external field was computed by Onsager (1944). This is a real *tour de force* and we shall not give this derivation here but simply state the result:

$$-\beta f(\beta) = \ln 2$$
$$+ \frac{1}{2\pi^2} \int_0^\pi d\theta_1 \int_0^\pi d\theta_2 \ln[\cosh^2(2\beta J) - \sinh(2\beta J)(\cos\theta_1 + \cos\theta_2)]. \tag{28.22}$$

Figure 28.4 shows the graph of $f(\beta)$ together with a horizontal line at $-2J$ and the graph of $-(\ln 2)/\beta$. These are the asymptotic limits of $f(\beta)$ for $\beta \to \infty$ and $\beta \to 0$, respectively, as can be easily deduced from equation (28.22). It is obvious from the figure that they are in fact very good approximations. It is important to understand why these are the correct asymptotics. In the limit $\beta \to 0$ the temperature is high. The entropy should then dominate as is also clear from the formula for the partition function (cf. equation (28.6)). The spins have essentially the same probability of being $+1$ or -1 and $Z_N \approx 2^N$. Taking the logarithm, dividing by N and multiplying by $-1/\beta$ gives $-(\ln 2)/\beta$. Note that this argument holds in any dimension and for any lattice, so this limiting behaviour always applies. On the other hand, if $\beta \to \infty$ the temperature is low and we expect that the energy term dominates. The spins should then align and the energy becomes $-2JN$ ($-J$ times the number of nearest-neighbour pairs). The partition function is thus $Z_N \approx 2e^{2\beta JN}$ (a factor 2 for the two alignment directions) and $f(\beta) \approx -2J$. This argument assumes that the energy term does indeed dominate, which means that there is a phase transition: see below.

Note that the non-analyticity of the free energy is not apparent in figure 28.4. This is because the derivative is in fact continuous as we shall see shortly. The phase transition at the point $\beta = \beta_c$ given below is a second-order transition. From equation (28.22), we can derive the internal energy per spin

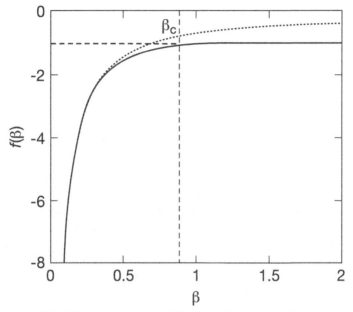

Figure 28.4 The free energy of the two-dimensional Ising model.

using (24.11):

$$u(\beta) = -J \coth(2\beta J)$$
$$\times \left[1 + \frac{\sinh^2(2\beta J) - 1}{\pi^2} \int_0^\pi \int_0^\pi \frac{d\theta_1 \, d\theta_2}{\cosh^2(2\beta J) - \sinh(2\beta J)(\cos\theta_1 + \cos\theta_2)} \right].$$
$$(28.23)$$

The integral in this expression diverges logarithmically at the origin $\theta_1 = \theta_2 = 0$ when

$$\delta := \cosh^2(2\beta J) - 2\sinh(2\beta J) = 0. \tag{28.24}$$

To see this, write $\cos\theta_1 + \cos\theta_2 \sim 2 - \frac{1}{2}(\theta_1^2 + \theta_2^2)$ for small θ_1 and θ_2. For small δ we then have

$$\frac{1}{\pi^2} \int_0^\pi \int_0^\pi \frac{d\theta_1 \, d\theta_2}{\cosh^2(2\beta J) - \sinh(2\beta J)(\cos\theta_1 + \cos\theta_2)}$$
$$\sim \frac{1}{\pi^2} \int_0^\pi \int_0^\pi \frac{d\theta_1 \, d\theta_2}{\delta + \frac{1}{2}\sinh(2\beta J)(\theta_1^2 + \theta_2^2)}$$
$$\sim \frac{1}{2\pi} \int_0^\pi \frac{r \, dr}{\delta + \frac{1}{2}\sinh(2\beta J)\, r^2}$$
$$\sim -\frac{1}{2\pi\sinh(2\beta J)} \ln|\delta|.$$

This singularity means that there must be a **phase transition** at the value $\beta = \beta_c$ for which equation (28.24) holds, that is,

$$\sinh(2\beta_c J) = 1 \text{ or } \beta_c J \approx 0.44 \tag{28.25}$$

Note that the divergent integral in equation (28.23) is multiplied by the factor $\sinh^2(2\beta J) - 1$, which is zero for $\beta = \beta_c$, so that the internal energy is in fact continuous at $\beta = \beta_c$.

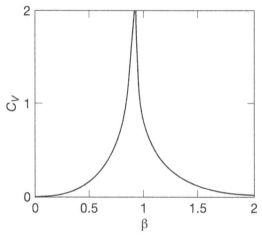

Figure 28.5 The specific heat of the two-dimensional Ising model.

For $\beta \sim \beta_c$,

$$u(\beta) \sim -\sqrt{2}J\left[1 - AJ(\beta - \beta_c)\ln|J(\beta - \beta_c)|\right], \tag{28.26}$$

where $A = 16\sqrt{2}/\pi$ is a constant. However, the specific heat $c_V = -k_B\beta^2(\partial u/\partial \beta)$ diverges logarithmically (figure 28.5)

$$c_V \sim -B\ln|J(\beta - \beta_c)|, \tag{28.27}$$

which means that the phase transition is **second-order** as mentioned above.

Unfortunately, no solution is known for the two-dimensional Ising model with external field $H \neq 0$. The magnetization $m(\beta, J, H)$ can therefore not be determined. Nevertheless there is a way of computing exactly the permanent magnetization $m_0(\beta, H)$ in two dimensions at zero field. This was first done by Yang (1952) (see also Schultz *et al.* (1964) and Baxter (1982)). The result is:

$$m_0(\beta) = \{1 - [\sinh(2\beta J)]^{-4}\}^{1/8} \quad (\beta > \beta_c). \tag{28.28}$$

As a function of T this has the general shape of figure 10.1. In any dimension $\geqslant 2$ it is possible to prove the existence of permanent magnetization m_0 for $H = 0$ when β is sufficiently large, i.e. at low temperatures. That means that the graph of $H(m)$ has a horizontal piece between $-m_0(\beta)$ and $m_0(\beta)$ as in figure 10.2. This proof is called the **Peierls argument** and will be presented below. The important lesson to be learnt from this model is that *the existence of a phase transition depends critically on the dimension.* Indeed, one can prove that *no one-dimensional spin model with short-range interaction can exhibit spontaneous magnetization.*

Note that the existence of permanent magnetization is analogous to the existence of two phases of matter for low temperatures, for instance liquid and gas. For a ferromagnet the two coexisting phases are those with magnetization $m = m_0(\beta)$ and $m = -m_0(\beta)$. The variable m that distinguishes between the two phases is called an **order parameter.** For the liquid-gas transition the order parameter is the density ρ. This analogy will be made more explicit in chapter 30.

The Peierls argument

We now present a proof, due essentially to Peierls (1936) but corrected by Griffiths (1964) and Dobrushin (1965), that the nearest-neighbour Ising model in dimensions greater than 1 has a phase transition. This is a very powerful argument and has been extended to a variety of situations, including the case of more general types of interactions which do not have a spin-up vs. spin-down symmetry. (This is called **Pirogov-Sinai theory** (Sinai (1982).) For simplicity we consider the two-dimensional case. The main idea is to separate regions of +-spins from regions of −-spins by lines between points of the dual lattice. For the square lattice, the dual lattice is also a square lattice consisting of points at the centre of elementary squares (plaquettes) of the original lattice (figure 28-6).

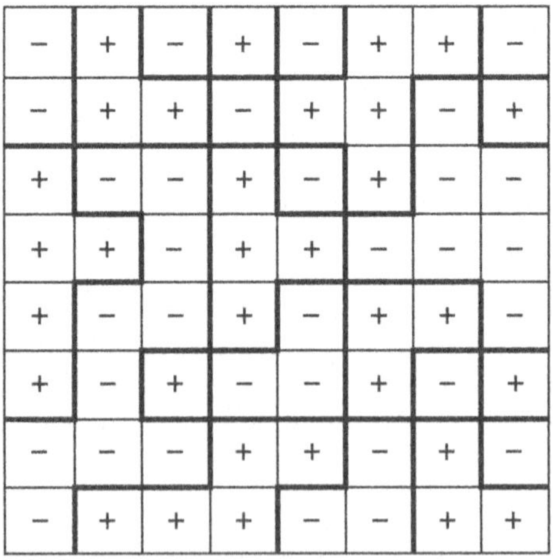

−	+	−	+	−	+	+	−
−	+	+	−	+	+	−	+
+	−	−	+	−	+	−	−
+	+	−	+	+	−	−	−
+	−	−	+	−	+	+	−
+	−	+	−	−	+	−	+
−	−	−	+	+	−	+	−
−	+	+	+	−	−	+	+

Figure 28.6 Contours on the dual lattice.

In three dimensions the + and − regions are separated by surfaces consisting of elementary squares of the dual lattice. Note that, for a given configuration, the lines on the dual lattice separate into connected pieces which are either closed curves or curves which begin and end at the boundary of the lattice. These curves are called **contours**. Any given consistent set of contours defines a unique configuration. (There is a possible ambiguity when a contour self-intersects. This can be resolved in various ways. One can impose the condition that the number of contours must be maximal or one can decide to always introduce a thin channel in the direction southwest to northeast as in figure 28.7.)

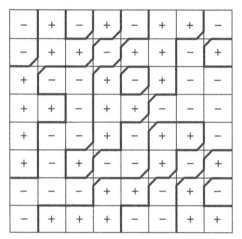

Figure 28.7 Modified contours.

The Ising model thus becomes equivalent to a **contour model**. This leads to a **contour expansion** for the free energy which one can prove converges for low temperatures. We shall not enter into this here, however.

Peierls' argument now proceeds as follows. Suppose that $H = 0$ and that the temperature is low, that is β large. Then we expect that most spins align either in the +-direction or in the −-direction resulting in a positive respectively negative magnetization. To distinguish these two possibilities, we consider a finite but large region Λ of the lattice and assume first that the boundary spins of Λ are all positive (positive boundary conditions). This should tip the balance towards a positive magnetization. There are now *only closed contours* and any −-spin must be surrounded by at least one contour. Note that the magnetization in Λ is given by

$$m_\Lambda = \frac{1}{N}\mathbb{E}_\beta(N_+ - N_-),\qquad(28.29)$$

where $N_\pm$ is the number of $\pm$-spins. The expectation of N_- at a given temperature $T = 1/k_\mathrm{B}\beta$ is given by the canonical probability distribution (24.3):

$$\mathbb{E}_\beta(N_-) = \frac{1}{Z_\beta}\sum_{\{s_x\}_{x\in\Lambda}} N_-(\{s_x\})\exp\left[-\beta E(\{s_x\})\right],\qquad(28.30)$$

where the sum is restricted to configurations with $+$ boundary conditions. Note that this implies that all contours must be closed loops. We can classify the contours according to their length l. It is easily seen that in two dimensions, the number of different contours with a given length l is bounded by

$$n(l) \leqslant \frac{4N3^{l-1}}{l} \tag{28.31}$$

for there are N possible starting points of the contour and 4 possible directions in which to start moving. At every subsequent point there are only 3 directions left which gives rise to the factor 3^{l-1} and we are overcounting by a factor l because we obtain the same contour starting at any of its (dual) lattice points. Let $X_l^{(i)}$ be the random variable which equals 1 if the i-th contour of length l occurs in a given configuration and 0 otherwise. Its probability distribution is given by the canonical distribution of the configurations:

$$\mathbb{E}_\beta(X_l^{(i)}) = \frac{1}{Z_\beta} \sideset{}{'}\sum_{\{s_x\}} \exp\left[-\beta E(\{s_x\})\right], \tag{28.32}$$

where the primed sum is over all configurations containing the i-th contour of length l. Let γ denote the i-th contour of length l. If in any configuration $\{s_x\}$ containing the contour γ, all spins inside γ are reversed (flipped), we obtain another valid configuration in which the contour γ is now absent. Denoting the resulting configuration $\{s_x\}^*$ its energy is lower by an amount $2Jl$:

$$E(\{s_x\}) = E(\{s_x\}^*) + 2Jl. \tag{28.33}$$

In the defining expression for Z_β we can restrict the sum to all configurations which can be obtained from a configuration containing the i-th contour of length l:

$$Z_\beta = \sum_{\{s_x\}} \exp\left[-\beta E(\{s_x\})\right] \geqslant \sideset{}{'}\sum_{\{s_x\}} \exp\left[-\beta E(\{s_x\}^*)\right]. \tag{28.34}$$

Inserting equations (28.33) and (28.34) into (28.32) we get a simple upper bound for $X_l^{(i)}$:

$$\mathbb{E}_\beta\left(X_l^{(i)}\right) \leqslant e^{-2J\beta l}. \tag{28.35}$$

Now, the number N_- of minus spins $s_x = -1$ is bounded by the total number of spins inside all the contours and since a contour of length l encloses at most $(l/4)^2$ lattice sites,

$$\mathbb{E}_\beta\left(N_-\right) \leqslant \sum_{l=4,6,8,\ldots} \frac{l^2}{16} \sum_{i=1}^{n(l)} \mathbb{E}_\beta\left(X_l^{(i)}\right)$$

$$\leqslant \frac{N}{12} \sum_{l=4,6,8,\ldots} l3^l e^{-2\beta Jl}$$

$$= \frac{N\kappa^4}{6} \frac{2 - \kappa^2}{(1 - \kappa^2)^2}, \tag{28.36}$$

provided

$$\kappa = 3e^{-2\beta J} < 1. \tag{28.37}$$

It is easy to see that the function of κ on the right-hand side of equation (28.36) is increasing and equals $0.396N$ for $\kappa = 3/4$. It follows, using equation (28.29), that for $\kappa < 3/4$, $m_\Lambda(\beta) > 0.2$. Note that this bound is independent of the size of Λ. In exactly the same way we find that for negative boundary conditions $m_\Lambda(\beta) < -0.2$ if $\beta J > \ln 2$.

How does this prove the existence of a phase transition? In general, one says that a system exhibits a phase transition if the free energy is not analytic at a point in the state space. If the first derivative of f has a discontinuity the phase transition is called a **first-order phase transition**. If the first derivative exists but the second derivative has a jump or does not exist it is called a **second-order phase transition**. These are the main two types of phase transitions. We shall now show that, for low temperatures the derivative of $f(\beta, J, H)$ with respect to H has a discontinuity at $H = 0$. Recall (chapter 25) that $f(\beta, J, H) = \lim_{\Lambda \to \mathbb{Z}^2} |\Lambda|^{-1} F_\Lambda^b(\beta, J, H)$ is independent of the boundary conditions b. Now, we have seen (see the derivation of equation (28.18) but without taking the limit $N \to \infty$) that

$$M_\Lambda^b(\beta, J, H) = \mathbb{E}_\beta \left(\sum_{x \in \Lambda} s_x \right) = -\frac{\partial}{\partial H} F_\Lambda^b(\beta, J, H) \tag{28.38}$$

and since F_Λ^b is concave in H, we have that

$$-\frac{F_\Lambda^b(\beta, J, 0) - F_\Lambda^b(\beta, J, -H)}{H} \leqslant M_\Lambda^b(\beta, J, 0)$$
$$\leqslant -\frac{F_\Lambda^b(\beta, J, H) - F_\Lambda^b(\beta, J, 0)}{H} \tag{28.39}$$

for $H > 0$. Dividing by $N = |\Lambda|$ and taking the limit $\Lambda \to \mathbb{Z}^2$ (in the sense of Van Hove) we obtain

$$-\frac{f(\beta, J, 0) - f(\beta, J, -H)}{H} \leqslant \liminf_{\Lambda \to \mathbb{Z}^2} \left(\frac{M_\Lambda^b(\beta, J, 0)}{N} \right)$$
$$\leqslant \limsup_{\Lambda \to \mathbb{Z}^2} \left(\frac{M_\Lambda^b(\beta, J, 0)}{N} \right) \leqslant -\frac{f(\beta, J, H) - f(\beta, J, 0)}{H}$$

and, taking the limit $H \to 0$,

$$-\frac{\partial f}{\partial H}(\beta, J, 0^-) \leqslant \liminf_{\Lambda \to \mathbb{Z}^2} \left(\frac{M_\Lambda^b(\beta, J, 0)}{N} \right)$$
$$\leqslant \limsup_{\Lambda \to \mathbb{Z}^2} \left(\frac{M_\Lambda^b(\beta, J, 0)}{N} \right) \leqslant -\frac{\partial f}{\partial H}(\beta, J, 0^+). \tag{28.40}$$

This is a more precise formulation of equation (28.18); if the derivative of $f(H)$ exists at $H = 0$ then we obtain (28.18). These inequalities hold for any arbitrary boundary condition b. Taking $+$ and $-$ boundary conditions we have seen that for $\beta J > \ln 2$,

$$\frac{M_\Lambda^{(+)}(\beta, J, 0)}{N} > 0.2 \text{ and } \frac{M_\Lambda^{(-)}(\beta, J, 0)}{N} < -0.2. \tag{28.41}$$

Thus

$$-\frac{\partial f}{\partial H}(\beta, J, 0^-) < -0.2 < 0.2 < -\frac{\partial f}{\partial H}(\beta, J, 0^+) \tag{28.42}$$

for $\beta J > \ln 2$, which implies that the derivative of f with respect to H is discontinuous at $H = 0$.

CHAPTER 29

The Weiss-Ising Model

Since the Ising model is so difficult to solve one may try to find variations of this model that are easier to treat mathematically. After all, the Ising model itself is also an idealization of the true physical situation. In particular, we have neglected in equation (28.3) all interaction terms of equation (28.2) other than the nearest-neighbour terms. We can also go to the other extreme and assume that each spin interacts equally with every other spin. Alternatively, one can say that each spin s_x interacts with the **mean field** $\frac{1}{N-1} \sum_{y \neq x} s_y$ due to the other spins. The resulting model is the **Weiss-Ising model.** Its energy levels are given by

$$E_N\left(\{s_x\}\right) = -\frac{J}{N-1} \sum_x \sum_{y \neq x} s_x \, s_y - H \sum_x s_x. \tag{29.1}$$

Note that in this expression the structure of the lattice is irrelevant. This means that *the free energy of this model is dimension-independent.* To compute the partition function for this model, we use the theory of large deviations and define random variables

$$X_N = \frac{1}{N} \sum_x s_x \tag{29.2}$$

with distribution function

$$F_N(s) = \frac{1}{2^N} \# \left\{ \{s_x\} \mid \sum_x s_x \leqslant sN \right\}. \tag{29.3}$$

Clearly, X_N is the mean of independent random variables $s_x = \pm 1$ with probability $\frac{1}{2}$, i.e. with distribution function

$$F(s) = \begin{cases} 0 & \text{if } s < -1 \\ \frac{1}{2} & \text{if } -1 \leqslant s < 1 \\ 1 & \text{if } s \geqslant 1. \end{cases}$$

By Cramér's theorem, this means that the large deviation principle holds with rate function given by

$$I(s) = \sup_{t \in \mathbb{R}}[st - C(t)] \tag{29.4}$$

where

$$C(t) = \ln \int_{-\infty}^{\infty} e^{st}\, dF(s) = \ln(\cosh t). \tag{29.5}$$

Hence

$$I(s) = \frac{1}{2}[(1+s)\ln(1+s) + (1-s)\ln(1-s)] \tag{29.6}$$

for $-1 \leqslant s \leqslant 1$ and $I(s) = +\infty$ for $|s| > 1$ (cf. equations (21.8), (21.9), and (21.12)). We now want to use Varadhan's theorem to compute the free energy density. To that end we must write the partition function $Z_N(\beta)$ as an integral with respect to F_N. Using the fact that $s_x^2 = 1$ we can write E_N in terms of X_N:

$$E_N = -\frac{N^2 J}{N-1}X_N^2 + \frac{NJ}{N-1} - NHX_N. \tag{29.7}$$

We therefore have

$$Z_N(\beta) = \sum_{\{s_x\}} e^{-\beta E_N(\{s_x\})}$$

$$= 2^N \int_{-\infty}^{\infty} e^{-\beta N \mathcal{E}_N(s)} dF_N(s) \tag{29.8}$$

where $\mathcal{E}_N(s)$ is $\frac{1}{N}E_N$ with X_N replaced by s:

$$\mathcal{E}_N(s) = -\frac{NJ}{N-1}s^2 + \frac{1}{N-1}J - Hs. \tag{29.9}$$

The free energy per spin therefore becomes

$$f(\beta) = -\frac{1}{\beta} \lim_{N \to \infty} \frac{1}{N} \ln Z_N(\beta)$$

$$= -\frac{1}{\beta} \lim_{N \to \infty} \frac{1}{N} \ln \int_{-\infty}^{\infty} e^{-\beta N \mathcal{E}_N(s)} dF_N(s) - \frac{1}{\beta}\ln 2. \tag{29.10}$$

Note that this is *almost* of the form (20.7) with the function in the exponent replaced by $-\beta \mathcal{E}_N(s)$. The only problem is that $\mathcal{E}_N(s)$ depends on N. However, as $N \to \infty$, $-\beta \mathcal{E}_N(s) \to G(s) = \beta J s^2 + \beta H s$ so that we can replace $-\beta \mathcal{E}_N$ by G and apply Varadhan's theorem to conclude that

$$f(\beta) = -\frac{1}{\beta}\ln 2 - \frac{1}{\beta} \sup_{s \in [-1,1]} [G(s) - I(s)]$$

$$= -\frac{1}{\beta}\ln 2 - \frac{1}{\beta} \sup_{s \in [-1,1]} \{\beta J s^2 + \beta H s$$

$$- \frac{1}{2}[(1+s)\ln(1+s) + (1-s)\ln(1-s)]\}. \tag{29.11}$$

Differentiating, we find the following equation for the value of s where the maximum is attained:

$$s = \tanh\left(2\beta Js + \beta H\right) \qquad (29.12)$$

The case $H = 0$ is especially interesting. Figure 29.1 illustrates the graphical solution of (29.12) in this case.

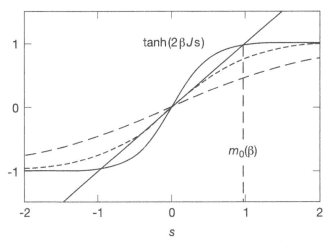

Figure 29.1 Solving the implicit equation for $s = m_0(\beta)$ in case $H = 0$.

If $2\beta J \leqslant 1$ then there is only one value of s satisfying equation (29.12), namely $s = 0$, and therefore

$$f(\beta) = -\frac{1}{\beta}\ln 2 \qquad (2\beta J \leqslant 1). \qquad (29.13)$$

However, if $2\beta J > 1$ then there are three solutions: $s = 0$ and $s = \pm m_0(\beta)$, and it is easy to see that the maximum is attained for $s = \pm m_0(\beta)$ (see figure 29.1). Now, it follows immediately from equation (28.18) that at any value of H, the maximizing value for s is just the magnetization $m(\beta, J, H)$. Taking $H = 0$ we conclude that there is spontaneous magnetization for temperatures $T < T_c$, where the **critical temperature** is given by

$$k_{\mathrm{B}}T_c = 2J. \qquad (29.14)$$

The spontaneous magnetization satisfies (29.12):

$$m_0(\beta) = \tanh\left[2\beta Jm_0(\beta)\right]. \qquad (29.15)$$

Figure 29.2 shows f as a function of the temperature at $H = 0$ (upper graph). It is linear for $T > T_c$.

When $H > 0$ but small then there are still three solutions of equation (29.12) if $T < T_c$ but the supremum in (29.11) is now attained at a unique

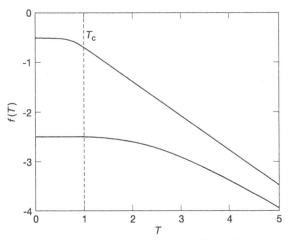

Figure 29.2 The free energy of the Weiss-Ising model.

value of s namely the positive solution of (29.12). This means that $m(\beta)$ is the positive solution of:

$$m(\beta) = \tanh[2\beta J m(\beta) + \beta H]. \tag{29.16}$$

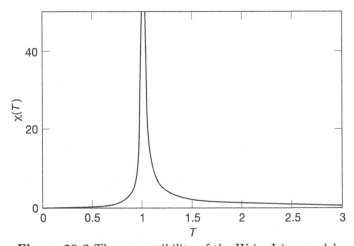

Figure 29.3 The susceptibility of the Weiss-Ising model.

Differentiating we find the following expression for the susceptibility:

$$\chi(\beta) = \beta \left(\frac{1}{1 - m(\beta)^2} - 2\beta J \right)^{-1}. \tag{29.17}$$

It follows from this that $\chi(\beta)$ diverges for $\beta = \beta_c = 1/k_B T_c$ and $H = 0$ since $m_0(\beta_c) = 0$ and $2\beta_c J = 1$. Figure 29.3 depicts the susceptibility as a function of T. Figure 29.2 is a graph of the free energy $f(T)$ for two values of the magnetic field: $H = 0$ and $H > 0$. Note that since $m > 0$ for $H > 0$, the free energy is decreasing in H.

CHAPTER 30

The Mean-Field Lattice Gas

We have considered models of free particles (fermions or bosons) in Part II. In order to understand the liquid-gas phase transition, however, we need to take into account the interaction between the particles just as we need to take into account the interaction between spins in order to explain the ferromagnetic phase transition (Curie transition). A rough way of doing this was discussed in chapter 9. We shall now discuss a microscopic model with very similar results. This highly simplified model is obtained by dividing the volume V into microscopic cells which are assumed to be so small that they can contain at most one gas molecule (figure 30.1).

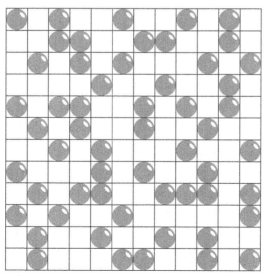

Figure 30.1 Dividing the volume into cells.

If there is a molecule in cell x, we put $n_x = 1$, otherwise $n_x = 0$. Let $\mathcal{V}(x - y)$ be the interaction potential (see figure 13.1). Then we can write the total energy of this **lattice gas** in the form

$$E_V\left(\{n_x\}\right) = \sum_x \sum_{y \neq x} \mathcal{V}(x - y)\, n_x\, n_y. \tag{30.1}$$

Note that $\mathcal{V}$ should not include the repulsive part of the potential for small r because this part of the potential has already been taken into account by the assumption that there can be no more than one particle in each cell. (This is called a **hard-core interaction.**)

As in the case of spins we can now make either of two approximations. If we assume that $\mathcal{V}$ has a very short range then we can put

$$E_V\left(\{n_x\}\right) = -\lambda \sum_{x,y:\, |x-y|=1} n_x\, n_y. \tag{30.2}$$

(We have included a minus sign so that $\lambda > 0$ for attractive interaction.) This is the **nearest-neighbour lattice gas** introduced by Lee and Yang (1952). Alternatively, we can assume that $\mathcal{V}$ is weak but has very long range. Analogous to equation (29.1) we can then put

$$E_V\left(\{n_x\}\right) = -\frac{\lambda}{V-1} \sum_x \sum_{y \neq x} n_x\, n_y. \tag{30.3}$$

(We identify V with the number of cells.) This is the **mean-field lattice gas.** We consider the latter model. To determine the thermodynamics of this model, we must compute the grand-canonical partition function $\mathcal{Z}_V(\beta, \mu)$ given by equation (24.14). The grand potential is then given by equation (24.13). We could compute this limit in the same way as in equation (29.10) of the previous chapter. Instead, we follow Lee and Yang and remark that *there is a direct relation between the grand-canonical ensemble of the lattice gas and the canonical ensemble of an Ising model.* To see this let us define spin variables s_x by

$$n_x = \frac{1}{2}(1 + s_x) \tag{30.4}$$

These variables then take the values ± 1 and we can write

$$\mathcal{Z}_V(\beta, \mu) = \sum_{\{s_x\}} \exp\left[-\beta \tilde{E}_V\left(\{s_x\}\right)\right] \tag{30.5}$$

where

$$\tilde{E}_V\left(\{s_x\}\right) = -\frac{\lambda}{4(V-1)} \sum_x \sum_{y \neq x} s_x\, s_y - \frac{1}{2}(\lambda + \mu) \sum_x s_x - \frac{1}{4}(\lambda + 2\mu)V. \tag{30.6}$$

We therefore conclude that

$$\omega(\beta, \mu) = f_{\mathrm{WI}}\left(\beta, \frac{1}{4}\lambda, \frac{1}{2}(\lambda + \mu)\right) - \frac{1}{4}\lambda - \frac{1}{2}\mu, \tag{30.7}$$

where $f_{\mathrm{WI}}(\beta, J, H)$ is the free energy density of the Weiss-Ising model given by equation (29.11). Transforming back to the more natural variable $\rho = \frac{1}{2}(1+s)$ we obtain the elegant formula

$$\boxed{\omega(\beta, \mu) = -\frac{1}{\beta} \sup_{\rho \in [0,1]} \left[\beta\lambda\rho^2 + \beta\mu\rho - \rho\ln\rho - (1 - \rho)\ln(1 - \rho)\right].} \tag{30.8}$$

Note that, by equation (24.16), the maximizing ρ in this formula is exactly the equilibrium density. It is tempting to conclude from equations (30.8) and (24.15) that the free energy density of the lattice gas is given by

$$\tilde{f}(\beta, \rho) = -\lambda\rho^2 + \frac{1}{\beta}[\rho\ln\rho + (1 - \rho)\ln(1 - \rho)]. \tag{30.9}$$

This is, however, not entirely correct. Indeed, we have seen that $\tilde{f}$ must be a convex function of ρ and the function (30.9) is not convex: see figure 30.2. In fact, $\tilde{f}(\beta, \rho)$ is given by the **convex hull** of the function (30.9), i.e. the largest convex function bounded above by (30.9) (cf. theorem 7.2). To see this consider the maximizer ρ in equation (30.8). It must satisfy

$$2\beta\lambda\rho + \beta\mu = \ln\left(\frac{\rho}{1 - \rho}\right)$$

or, equivalently,

$$\rho = \left(e^{-\beta(2\lambda\rho + \mu)} + 1\right)^{-1}. \tag{30.10}$$

If $\beta\lambda \leqslant 2$ then the derivative of the right-hand side of equation (30.10) is bounded above by 1 so that there is a unique solution. If $\beta\lambda > 2$ then the equation has three solutions for values of μ in the neighbourhood of $\mu = -\lambda$. The largest and smallest solutions correspond to local maxima, and the middle solution to a local minimum.

If $\mu = -\lambda$ we denote the maximizers by $\rho_-(\beta)$ and $\rho_+(\beta)$. They satisfy $\rho_+ = 1 - \rho_-$ and the corresponding maxima are equal. As μ decreases both maxima decrease but the high-density maximum decreases faster so that the supremum in equation (30.8) is given by $\rho = \rho_-(\beta, \mu)$ if $\mu < -\lambda$, where $\rho_-(\beta, \mu)$ is the minimal solution of (30.10):

$$\omega(\beta, \mu) = -\lambda\rho_-(\beta, \mu)^2 - \mu\rho_-(\beta, \mu)$$
$$+ \frac{1}{\beta}\{\rho_-(\beta, \mu)\ln\rho_-(\beta, \mu) + (1 - \rho_-(\beta, \mu))\ln(1 - \rho_-(\beta, \mu))\}, \tag{30.11}$$

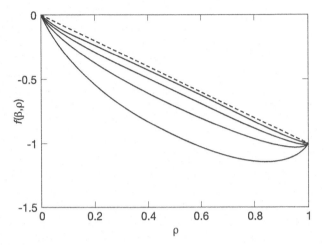

Figure 30.2 The free energy density of the mean-field lattice gas.
(The broken line is a guide to the eye.)

Similarly, if $\mu > -\lambda$,

$$\omega(\beta, \mu) = -\lambda \rho_+(\beta, \mu)^2 - \mu \rho_+(\beta, \mu)$$
$$+ \frac{1}{\beta} \left\{ \rho_+(\beta, \mu) \ln \rho_+(\beta, \mu) + (1 - \rho_+(\beta, \mu)) \ln(1 - \rho_+(\beta, \mu)) \right\},$$
$$(30.12)$$

The maximizer $\rho(\beta, \mu)$ therefore has a **jump** at $\mu = -\lambda$ (figure 30.3).

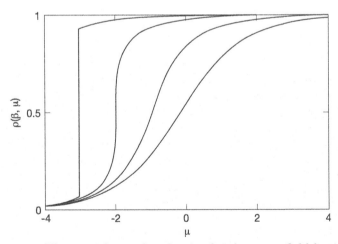

Figure 30.3 The particle number density for the mean-field lattice gas.

It follows that the grand potential has a **kink** at the value $\mu = -\lambda$ since, by equation (24.16),

$$\rho(\beta, \mu) = -\frac{\partial \omega}{\partial \mu}(\beta, \mu).$$

(30.13)

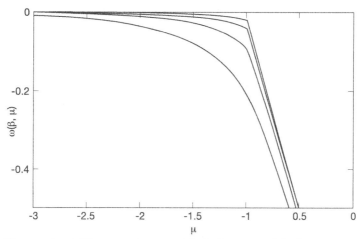

Figure 30.4 The grand potential of the mean-field lattice gas.

Now let us consider again the variational formula (24.15) for the free energy density. This formula says that $\tilde{f}(\beta, \rho)$ is minus the vertical distance between the graph of $-\omega(\beta, \mu)$ and the straight line $\mu \mapsto \rho\mu$. This distance can be found by shifting the line until it touches the graph of $\mu \mapsto -\omega(\beta, \mu)$. This means that the maximizer μ is determined by $\rho = \rho(\beta, \mu)$ if $\rho \leqslant \rho_-(\beta)$ or $\rho \geqslant \rho_+(\beta)$, but $\mu = -\lambda$ if $\rho_-(\beta) < \rho < \rho_+(\beta)$! In the former case equation (30.10) yields

$$\mu = -2\lambda\rho + \frac{1}{\beta} \ln\left(\frac{\rho}{1-\rho}\right)$$

(30.14)

and inserting this into equation (24.15) we find that $\tilde{f}(\beta, \rho)$ is indeed given by equation (30.9). However, if $\rho_-(\beta) < \rho < \rho_+(\beta)$ then

$$\tilde{f}(\beta, \rho) = \tilde{f}(\beta, \rho_\pm(\beta)) - \lambda(\rho - \rho_\pm(\beta))$$

(30.15)

where we can take either ρ_+ or ρ_- because they both correspond to the supremum in equation (30.8).

Finally, we can determine the pressure $p(\beta, \rho)$, that is the equation of state. For $\rho \leqslant \rho_-(\beta)$ and $\rho \geqslant \rho_+(\beta)$ we can simply insert equation (30.14) into (30.8), obtaining

$$\boxed{p(\beta, \rho) = -\lambda\rho^2 - \frac{1}{\beta} \ln(1 - \rho).}$$

(30.16)

This equation of state is very similar to the van de Waals' equation. Note in particular that for $\rho \ll 1$, $p(\beta, \rho) \approx \rho/\beta = \rho k_B T$, the ideal gas law. For $\rho_-(\beta) \leqslant \rho \leqslant \rho_+(\beta)$, $p(\beta, \rho) = -\omega(\beta, -\lambda)$ is independent of ρ, that is, the pressure is constant! Let us check by direct computation that this constant is given by **Maxwell's equal-area construction.**

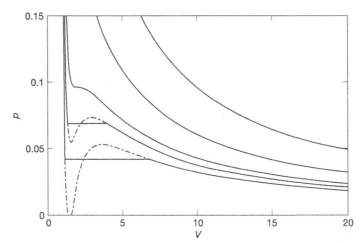

Figure 30.5 The pressure of the mean-field lattice gas.

If we denote the constant by $p_0(\beta)$ then

$$p_0(\beta) = -\lambda \rho_+(\beta)^2 - (1/\beta) \ln \rho_-(\beta). \tag{30.17}$$

We must prove that

$$\int_{v_-}^{v_+} \tilde{p}(\beta, v^{-1}) \, dv = (v_+ - v_-) p_0(\beta) \tag{30.18}$$

where $\tilde{p}(\beta, \rho)$ is the function (30.16) and $v = 1/\rho$ denotes the specific volume. (Of course, $v_+ = 1/\rho_-$ is the specific volume of the gas phase and $v_- = 1/\rho_+$ is the specific volume of the liquid phase.) A somewhat tedious computation confirms equation (30.18):

$$\int_{v_-}^{v_+} \tilde{p}(\beta, v^{-1}) dv = -\lambda(\rho_+ - \rho_-) - \frac{1}{\beta}[(v_+ - 1)\ln(v_+ - 1)$$
$$- (v_- - 1)\ln(v_- - 1) - v_+ \ln v_+ + v_- \ln v_-]$$
$$= -\lambda(v_+ - v_-)(\rho_+ - \rho_+ \rho_-) - (v_+ - v_-)(1/\beta)\ln \rho_-$$
$$= (v_+ - v_-)p_0(\beta).$$

Figure 30.5 shows the function $\tilde{p}(\beta, \rho)$ and the pressure p, both as a function of $v = 1/\rho$ for different values of the temperature. Note the vertical incline at $v = 1$, which is due to the hard-core potential which prevents more than one particle from occupying the same cell, so that $v \geqslant 1$.

CHAPTER 31

Metals and the Electron Gas

Metals have the property that the outermost atomic energy level or **shell** contains few electrons. These so-called **conduction electrons** are therefore very weakly bound to the core of the atom. In a macroscopic piece of metal, the conduction electrons are not bound to one particular atom but can move about quite freely inside the metal. It is the sharing of conduction electrons that causes the binding between the metal atoms. Many of the properties of a metal are due to the conduction electrons, and in order to understand these properties, we must therefore study the gas of conduction electrons. Since the electrons are so highly non-localized and so weakly bound to the atom cores we can, as a first approximation, assume that they are completely free within the confines of the metal. As electrons are fermions this leads us to believe that the free fermion gas considered in chapter 23 is a reasonable model for the conduction electrons in a metal. There is, however, one modification that we have to make. In chapter 23 we did not allow for the **spin degeneracy** of the energy states. Since electrons have spin $\frac{1}{2}$ each one-particle energy level can contain *two* electrons (of opposite spin). This means that the degeneracy factor g_i in equation (23.7) acquires an extra factor 2 and hence also $\gamma(\lambda, \beta)$ in equation (23.12). It follows that the pressure is now given by

$$p(v, T) = \frac{2k_{\mathrm{B}}T}{(2\pi)^3} \int_{\mathbb{R}^3} \ln(1 + e^{-(\epsilon(k) - \mu(v, T))/k_{\mathrm{B}}T}) \mathrm{d}^3k. \qquad (31.1)$$

REMARK 31.1
To derive this result directly from equations (24.13) and (23.14), it is important to realize that electrons with opposite spin are distinguishable! Writing $n_{j\downarrow}$ and $n_{j\uparrow}$ for the number of electrons with spin down and up respectively

in the jth single-particle energy level $\epsilon_V(j)$ we can write

$$E_V(N,j) = \sum_{j=0}^{\infty} (n_{j\uparrow} + n_{j\downarrow})\epsilon_V(j)$$

and the grand partition function becomes

$$\begin{aligned}
\mathcal{Z}_V(\beta,\mu) &= \sum_{\{n_{j\uparrow}\}} \sum_{\{n_{j\downarrow}\}} \exp\left(\beta \sum_{j=0}^{\infty}(n_{j\uparrow} + n_{j\downarrow})(\mu - \epsilon_V(j))\right) \\
&= \prod_{j=0}^{\infty} \left(1 + \exp\left[\beta(\mu - \epsilon_V(j))\right]\right)^2 .
\end{aligned}$$

The density of electrons given by equation (23.26) (or (24.16)) also gets an additional factor 2:

$$\rho(\beta,\mu) = \frac{2}{(2\pi)^3} \int_{\mathbb{R}^3} \frac{d^3k}{e^{\beta(\epsilon(k)-\mu)} + 1}. \tag{31.2}$$

Changing variables to $\epsilon = \hbar^2 k^2/2m$ we obtain

$$\rho = \frac{1}{2\pi^2}\left(\frac{2m}{\hbar^2}\right)^{3/2} \int_0^{\infty} \frac{\epsilon^{1/2}d\epsilon}{e^{\beta(\epsilon-\mu)} + 1}, \tag{31.3}$$

which we shall write as follows:

$$\rho = \int_0^{\infty} f_F(\epsilon)\, g(\epsilon)\, d\epsilon. \tag{31.4}$$

The function

$$g(\epsilon) = \frac{1}{2\pi^2}\left(\frac{2m}{\hbar^2}\right)^{3/2} \epsilon^{1/2} \tag{31.5}$$

represents the **electron density of states**. Its graph is depicted in figure 31.1.

The function

$$\boxed{f_F(\epsilon) = \frac{1}{e^{\beta(\epsilon-\mu)} + 1}} \tag{31.6}$$

is the Fermi-Dirac expression for the **expectation of occupancy** of a state with energy ϵ. (Do not confuse it with the free energy density $\tilde{f}(\beta,\rho)$!) Its graph is sketched in figure 31.2.

The function $f_F(\epsilon)$ is very important in the theory of the degenerate electron gas. (The electron gas is called **degenerate** if the temperature is so low and the density so high that quantum effects, in particular the Pauli exclusion

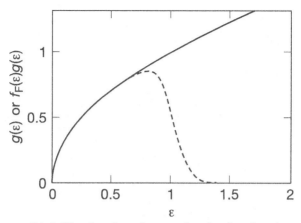

Figure 31.1 The density of states for the free fermion gas.

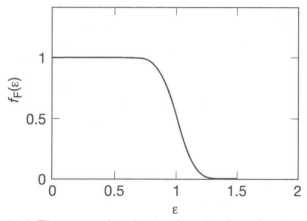

Figure 31.2 The energy distribution function for a free fermion gas.

principle, become important.) The chemical potential μ is called the **Fermi energy** or **Fermi level** and is usually denoted ϵ_F. It is the function $f_F(\epsilon)$ that makes the electron gas behave differently from an ideal gas. Consider the low-temperature limit. If $\beta \to \infty$ then $f_F(\epsilon) \to 1$ for $\epsilon < \epsilon_{F0}$ and $f_F(\epsilon) \to 0$ for $\epsilon > \epsilon_{F0}$, where ϵ_{F0} is the Fermi energy at zero temperature. This means that all energy levels below ϵ_{F0} are occupied and all levels above ϵ_{F0} are empty. We can compute ϵ_{F0} from equation (31.3):

$$\rho = \frac{1}{2\pi^2}\left(\frac{2m}{\hbar^2}\right)^{3/2}\int_0^{\epsilon_{F0}}\epsilon^{1/2}d\epsilon = \frac{1}{3\pi^2}\left(\frac{2m\epsilon_{F0}}{\hbar^2}\right)^{3/2}. \tag{31.7}$$

Now, metals usually contain more than 10^{28} conduction electrons per cubic meter. Inserting this into equation (31.7) gives a value of several electron-volts (eV) for ϵ_{F0}. (1 eV $= 1.6 \times 10^{-19}$ J.) On the other hand, at room temperature, $k_B T \approx 0.025$ eV. This means that the width of the region over which $f_F(\epsilon)$ decreases from 1 to 0 is very small compared to ϵ_{F0}. (The slope of the curve in figure 31.2 should in fact be much steeper.) At fixed ρ the Fermi level ϵ_F depends on the temperature. Integrating equation (31.3) by parts we get

$$\rho = \frac{\beta}{3\pi^2} \left(\frac{2m}{\hbar^2}\right)^{3/2} \int_0^\infty \frac{\epsilon^{3/2} e^{\beta(\epsilon-\mu)}}{(e^{\beta(\epsilon-\mu)}+1)^2} d\epsilon.$$

Then $\epsilon^{3/2}$ can be expanded around $\epsilon = \mu = \epsilon_F$:

$$\epsilon^{3/2} = \mu^{3/2} \left(1 + \frac{\epsilon-\mu}{\mu}\right)^{3/2} \approx \mu^{3/2} \left(1 + \frac{3x}{2\beta\mu} + \frac{3x^2}{8\beta^2\mu^2} + \cdots \right)$$

where $x = \beta(\epsilon - \mu)$. If we transform the integral over ϵ to an integral over x we can replace the lower bound $-\beta\epsilon_F$ by $-\infty$ and obtain

$$\rho \approx \frac{1}{3\pi^2} \left(\frac{2m}{\hbar^2}\right)^{3/2} \epsilon_F^{3/2} \left(1 + \frac{3}{8\beta^2\epsilon_F^2} \int_{-\infty}^\infty \frac{x^2 e^x \, dx}{(e^x+1)^2}\right).$$

Now,

$$\int_{-\infty}^\infty \frac{x^2 e^x \, dx}{(e^x+1)^2} = \frac{\pi^2}{3}$$

so we conclude that

$$\epsilon_{F0}^{3/2} \approx \epsilon_F^{3/2} \left(1 + \frac{\pi^2}{8\beta^2\epsilon_F^2}\right).$$

Inverting this relation we find:

$$\boxed{\epsilon_F \approx \epsilon_{F0} \left(1 - \frac{\pi^2}{12\beta^2\epsilon_{F0}^2}\right) \qquad (\beta\epsilon_{F0} \gg 1)} \qquad (31.8)$$

It is not possible to measure the pressure of the electron gas. However one can measure the heat capacity. (The heat capacity is a so-called **response function**; response functions are usually the easiest quantities to measure.) We first compute the internal energy using equation (24.18). This gives:

$$\tilde{u}(\beta, \mu) = \int_0^\infty \epsilon \, f_F(\epsilon) \, g(\epsilon) \, d\epsilon$$

$$= \frac{1}{2\pi^2} \left(\frac{2m}{\hbar^2}\right)^{3/2} \int_0^\infty \frac{\epsilon^{3/2} d\epsilon}{e^{\beta(\epsilon-\epsilon_F)}+1}. \qquad (31.9)$$

Again, we can integrate by parts and expand to estimate this integral:

$$\int_0^\infty \frac{\epsilon^{3/2}d\epsilon}{e^{\beta(\epsilon-\mu)}+1} \approx \frac{2}{5}\epsilon_F^{5/2}\left(1 + \frac{5}{8}\frac{\pi^2}{\beta^2\epsilon_F^2}\right). \tag{31.10}$$

Inserting equation (31.8) this becomes

$$\tilde{u}(\beta,\mu) \approx \frac{1}{5\pi^2}\left(\frac{2m}{\hbar^2}\right)^{3/2}\epsilon_{F0}^{5/2}\left(1 + \frac{5}{12}\frac{\pi^2}{\beta^2\epsilon_{F0}^2}\right). \tag{31.11}$$

Differentiating w.r.t. T we obtain

$$c_e \approx \frac{\pi^2\rho k_B^2 T}{2\epsilon_F}. \tag{31.12}$$

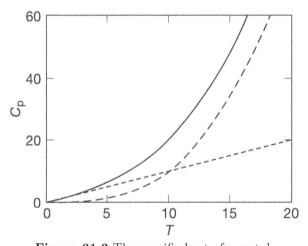

Figure 31.3 The specific heat of a metal.

This formula shows that the electronic specific heat c_e is *linear* in the temperature. c_e is not the only contribution to the specific heat of a metal. The other main contribution is due to lattice vibrations, or **phonons** (see chapter 33). It behaves like T^3 for low temperatures. The total specific heat is therefore of the form

$$c_p = AT + BT^3. \tag{31.13}$$

It is clear that for very low temperatures the electronic specific heat dominates (see figure 31.3).

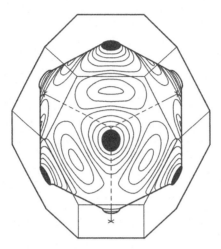

Figure 31.4 The Fermi surface of copper.

Measuring c_e gives an estimate for ϵ_F. Note that in this free-electron model the occupied states form a ball in k-space (given by $\epsilon(k) = \epsilon_F$). In more sophisticated models which take into account the attraction by the atom cores arranged in a regular lattice, the shape of the **Fermi sphere** changes. The atoms can be assumed to be arranged in a periodic lattice. As a result the k-space is also periodic. The unit cell of this space is called a **Brillouin zone**. Figure 31.4 shows the Fermi surface inside a Brillouin zone for copper.

CHAPTER 32

White-Dwarf Stars

In chapter 17 we considered a simple model for the equilibrium of a star. This model is valid for ordinary, so-called **main sequence stars.** The luminosity of main sequence stars is approximately proportional to their colour. They therefore lie on a straight line in a diagram of brightness versus colour, the so called **Herzsprung-Russell diagram**. When the hydrogen fuel of a main-sequence star runs out however, its luminosity is greatly diminished and it moves off the main sequence. The star collapses because the gas pressure is no longer sufficient to support the inward pull of gravity, and it becomes a **white dwarf.** The star consists mainly of helium but the helium is completely ionized and the pressure resisting gravity is now due to the electrons. As in the previous chapter, the electrons form a fermion gas which is highly degenerate. Indeed, the density of these stars is of the order of 10^7 g cm^{-3}. Given that a helium atom weighs 6.64×10^{-24} g, this implies that there are about $n \approx 3 \times 10^{30}$ electrons per cm^3. (We write n instead of ρ in this chapter for the number density to distinguish this quantity from the mass density, which we denote by ρ as in chapter 17.) This corresponds to a Fermi energy of (cf. equation (31.7))

$$\epsilon_F \approx \frac{\hbar^2}{2m}(3\pi^2 n)^{2/3} \approx 0.75 \text{ MeV}. \tag{32.1}$$

On the other hand, the temperature of a white dwarf is of the same order of magnitude as that of the sun: $T \approx 10^7$ K, so that

$$k_B T \approx 1000 \text{ eV} \ll \epsilon_F. \tag{32.2}$$

This means that the electron gas is indeed almost completely degenerate. Assuming first of all that the electron speed is not relativistic we can take the extreme degenerate limit of the pressure (31.1):

$$p(v, T) \approx \frac{2}{(2\pi)^3} 4\pi \int_0^{k_F} [\epsilon_F - \epsilon(k)] \, k^2 \, \mathrm{d}k$$

$$= \frac{\hbar^2 k_F^5}{15\pi^2 m} = \frac{1}{5}(3\pi^2)^{2/3}\frac{\hbar^2}{m} n^{5/3} \tag{32.3}$$

where k_F is defined by $\epsilon_F = \hbar^2 k_F^2/2m$ and we have used equation (31.7) and approximated ϵ_F by ϵ_{F0}.

REMARK 32.1: *Alternative derivation of the pressure* The same formula can also be obtained by differentiating $u(\beta, v) = v\tilde{u}(\beta, \mu)$ with respect to v:

$$p(v, T) \approx -\frac{\partial u}{\partial v} \tag{32.4}$$

for low temperatures, and using the lowest-order approximation to u:

$$u(\beta, v) \approx \frac{1}{5\pi^2}\left(\frac{2m}{\hbar^2}\right)^{3/2} v\epsilon_F^{5/2} \approx \frac{3\hbar^2}{10m}(3\pi^2 n)^{2/3}. \tag{32.5}$$

Since the white dwarf star consists predominantly of helium, which has a nucleus of two protons and two neutrons and liberates two electrons when completely ionized, we have the following relation between the mass density and the electron number density:

$$\rho = (m_{\mathrm{n}} + m_{\mathrm{p}})n \approx 2m_{\mathrm{p}}n, \tag{32.6}$$

where m_{p} and m_{n} are the mass of the proton and the neutron respectively. In terms of the mass density ρ the pressure is therefore given by

$$p = \kappa\rho^{5/3} \text{ with } \kappa = \frac{1}{5}(3\pi^2)^{2/3}\frac{\hbar^2}{m_e}\left(\frac{1}{2m_{\mathrm{p}}}\right)^{5/3}. \tag{32.7}$$

This is again a polytropic gas (cf. equation (17.5)). Analogous reasoning as in chapter 17 now yields the relation

$$\rho = \left(\frac{2\phi}{5\kappa}\right)^{3/2} \tag{32.8}$$

which inserted in the differential equation (17.6) gives

$$\frac{\mathrm{d}^2\phi}{\mathrm{d}r^2} + \frac{2}{r}\frac{\mathrm{d}\phi}{\mathrm{d}r} + \alpha^2\phi^{3/2} = 0 \tag{32.9}$$

where

$$\alpha^2 = 4\pi G\left(\frac{2}{5\kappa}\right)^{3/2}. \tag{32.10}$$

As in chapter 17 we transform to dimensionless variables:

$$u = \phi/\phi_0 \text{ and } z = \alpha\phi_0^{1/4}r. \tag{32.11}$$

The equation for $u(z)$ then reads

$$\frac{\mathrm{d}^2u}{\mathrm{d}z^2} + \frac{2}{z}\frac{\mathrm{d}u}{\mathrm{d}z} + u^{3/2} = 0. \tag{32.12}$$

This equation, the **Lane-Emden equation** with index $\frac{3}{2}$, can be solved numerically just like equation (17.19). The results are:

$$z|_{u=0} = 3.65 \text{ and } -z^2 \frac{du}{dz}\bigg|_{u=0} = 2.71 \tag{32.13}$$

The analogues of equations (17.21) and (17.22) are

$$M_{\text{WD}} = -\frac{1}{G\alpha} \phi_0^{3/4} \left(z^2 \frac{du}{dz}\right)_{u=0} \tag{32.14}$$

and

$$R_{\text{WD}} = \alpha^{-1} \phi_0^{-1/4} z|_{u=0}. \tag{32.15}$$

Inserting equation (32.15) into (32.14) yields the interesting conclusion that

$$M_{\text{WD}} R_{\text{WD}}^3 = \text{constant}. \tag{32.16}$$

The radius of a white dwarf therefore *decreases* when its mass increases! The above derivation assumes that the electron gas is non-relativistic. However, as the mass increases the density increases and hence the Fermi energy increases. The non-relativistic relation $\epsilon(k) = \hbar^2 k^2 / 2m$ is then no longer valid over a large part of the range $[0, \epsilon_F]$ but has to be replaced by the relativistic equation

$$\epsilon(k) = \sqrt{(\hbar k c)^2 + (mc^2)^2}. \tag{32.17}$$

($\hbar k$ is the momentum of the electron.) The pressure in the completely degenerate limit is then given by

$$\begin{aligned} p &= \frac{2}{(2\pi)^3} \int_0^{k_F} [\epsilon_F - \epsilon(k)] \, \mathrm{d}^3 k \\ &= \frac{1}{\pi^2} \left(\frac{1}{3} \epsilon_F k_F^3 - \int_0^{k_F} k^2 \epsilon(k) \, \mathrm{d}k\right). \end{aligned} \tag{32.18}$$

Introducing the dimensionless variable

$$x_F = \frac{\hbar k_F}{mc} = \frac{\hbar}{mc} (3\pi^2 n)^{1/3} \tag{32.19}$$

we can write this as

$$p = \frac{m^4 c^5}{\pi^2 \hbar^3} \left(\frac{1}{3} x_F^3 \sqrt{1 + x_F^2} - \int_0^{x_F} x^2 \sqrt{1 + x^2} \, \mathrm{d}x\right) = \frac{m^4 c^5}{24\pi^2 \hbar^3} f(x_F), \tag{32.20}$$

where the function $f(x_F)$ is given by

$$f(x_F) = x_F (2x_F^2 - 3) \sqrt{1 + x_F^2} + 3 \sinh^{-1}(x_F). \tag{32.21}$$

Inserting this expression into the differential equation (17.6) we have (using also equations (17.9), (32.6), and (32.19))

$$\frac{mc^2}{16m_p} \frac{1}{r^2} \frac{\mathrm{d}}{\mathrm{d}r} \left(\frac{r^2}{x_F^3} \frac{\mathrm{d}f(x_F)}{\mathrm{d}r} \right) = -\frac{8Gm^3 m_p c^3}{3\pi \hbar^3} x_F^3. \tag{32.22}$$

Since $f'(x) = 8x^4(1+x^2)^{-1/2}$, we have $x_F^{-3}(\mathrm{d}/\mathrm{d}r)[f(x_F)] = 8(\mathrm{d}/\mathrm{d}r)\sqrt{1+x_F^2}$ so that we can write equation (32.22) as

$$\frac{1}{r^2} \frac{\mathrm{d}}{\mathrm{d}r} \left(r^2 \frac{\mathrm{d}}{\mathrm{d}r} \sqrt{1+x_F^2} \right) = -\frac{16Gm^2 m_p^2 c}{3\pi \hbar^3} x_F^3. \tag{32.23}$$

Again, we transform to new variables. First we put y_0 equal to the value of $\sqrt{1+x_F^2}$ at the centre of the white dwarf star and then we define

$$u = \frac{1}{y_0} \sqrt{1+x_F^2} \quad \text{and} \quad z = \frac{y_0 r}{a} \tag{32.24}$$

where

$$a = \frac{1}{4mm_p} \sqrt{\frac{3\pi \hbar^3}{Gc}} \approx 3.86 \times 10^6 \text{ m} \approx 0.6 \, R_{\text{Earth}}. \tag{32.25}$$

Thus (32.23) transforms to

$$\frac{1}{z^2} \frac{\mathrm{d}}{\mathrm{d}z} \left(z^2 \frac{\mathrm{d}u}{\mathrm{d}z} \right) = -\left(u^2 - \frac{1}{y_0^2} \right)^{3/2}. \tag{32.26}$$

The initial conditions are as before: $u(0) = 1$ and $u'(0) = 0$. The radius of the star is now determined by the point where the density becomes zero, that is where $u = 1/y_0$. If we call this point z_1 then we have by equation (32.24),

$$R_{\text{WD}} = \frac{a z_1}{y_0}. \tag{32.27}$$

The mass is given by a formula analogous to equation (17.20):

$$\begin{aligned}
\frac{GM_{\text{WD}}}{R_{\text{WD}}^2} &= -\frac{\mathrm{d}\phi}{\mathrm{d}r} = -\frac{1}{\rho} \frac{\mathrm{d}p}{\mathrm{d}r} \\
&= -\frac{1}{\rho} \frac{m^4 c^5}{24\pi^2 \hbar^3} \frac{\mathrm{d}f(x_F)}{\mathrm{d}r} \\
&= -\frac{1}{\rho} \frac{m^4 c^5}{24\pi^2 \hbar^3} 8 x_F^3 y_0 \left(\frac{\mathrm{d}u}{\mathrm{d}r} \right)_{z=z_1} \\
&= -\frac{mc^2}{2m_p} \frac{y_0^2}{a} \frac{\mathrm{d}u}{\mathrm{d}z}(z_1)
\end{aligned}$$

Inserting equation (32.27),

$$M_{\text{WD}} = -\frac{mc^2}{2m_p} \frac{a}{G} \left(z^2 \frac{\mathrm{d}u}{\mathrm{d}z} \right)_{z=z_1}. \tag{32.28}$$

By computing these quantities for a series of values of y_0, one can obtain a graph of the radius versus the mass of the star. This graph is plotted in figure 32.1. It is apparent from this plot that the radius collapses to zero for a finite value of the mass. This remarkable result is due to Chandrasekhar (1935). Note that it is entirely due to the relativistic correction. (The figure also shows the non-relativistic equation (32.16) with the correct constant.)

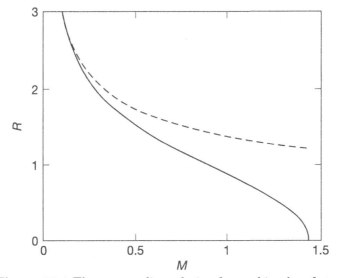

Figure 32.1 The mass-radius relation for a white-dwarf star.

The dimensions in the figure are as follows: the mass is expressed in units of the solar mass and the radius is expressed in units of the radius of the Earth. The limiting mass is called the **Chandrasekhar limit**. It is seen to be approximately 1.44 solar masses. This limiting value can also be computed as follows. As the density increases indefinitely, $y_0 \to \infty$ in equation (32.26). This equation then goes over into the Lane-Emden equation with index 3 considered in chapter 17! We can therefore use the results of Table 17.1 and compute the limiting mass according to

$$M_{\text{Chandrasekhar}} = 2.018 \frac{mc^2}{2m_p} \frac{a}{G}. \tag{32.29}$$

(The value is somewhat higher if the star is not completely ionized.) A star heavier than 1.5 solar masses can therefore not be supported by its electron degeneracy pressure and must collapse further. In fact, these stars become **neutron stars**; they are supported by the degeneracy pressure of neutrons. However, there is also a limiting mass for neutron stars. This limit is unfortunately not as accurately known. Stars heavier than this limit are thought to become black holes.

CHAPTER 33

Phonons

In chapter 13 we discussed the equation of state of a crystalline solid. The free energy included a term due to lattice vibrations. These vibrations are high-frequency sound waves which, like electromagnetic waves, have to be quantized. The quanta of these sound waves are called **phonons**. Here, we introduce a simple model due to Debye which describes these phonons.

We consider the solid to be made up of atoms of a single kind for simplicity arranged in a regular array or lattice. The simplest possibility is a simple cubic lattice but other arrangements are more common (see figure 12.3). If we assume that the forces between the atoms are approximately linear in the distance between atoms (Hooke's law; see example 18.1) the Hamiltonian is a sum of independent harmonic oscillator terms called **normal modes** with energies of the form (18.3), where the constant ω_i depends on the normal mode i. If the solid has N atoms the total number of normal modes equals $3N$ (each atom has 3 degrees of freedom). The total energy is therefore

$$E\{n_i\} = \sum_{i=1}^{3N} (n_i + \frac{1}{2})\hbar\omega_i \tag{33.1}$$

and the partition function is

$$Z_N(\beta) = \sum_{\{n_i\}_{i=1}^{3N}} e^{-\beta E(\{n_i\})} = \exp\left(-\frac{1}{2}\beta\hbar\sum_{i=1}^{3N}\omega_i\right)\prod_{i=1}^{3N}\frac{1}{1-e^{-\beta\hbar\omega_i}}. \tag{33.2}$$

To compute the thermodynamic limit we need to know the distribution of the frequencies ω_i. In 1907, Einstein proposed a model for the heat capacity of a solid in which he assumed all frequencies to be equal. This led to an explanation of the decrease of the specific heat at low temperatures observed experimentally and which could not be explained by classical mechanics. However, it is still a rather poor fit with experimental data and decreases too fast for low temperatures.

In 1912, Debye introduced an improvement on this assumption. He assumed, for the purpose of finding the angular frequencies ω_i, that the solid behaves approximately as an elastic continuum with **dispersion relation** $\omega = c_s k$, where c_s is the speed of sound in the solid. (This relation is simply the relation between the wavelength λ and the frequency ν, namely $\lambda\nu = c_s$. The angular frequency is given by $\omega = 2\pi\nu$ and the **wavenumber** k is defined by $k = 2\pi/\lambda$.) The **wave numbers** k_x, k_y, and k_z have the values given by standing waves in the crystal (cf. (23.11) and (18.7)):

$$\omega_i = c_s|\vec{k}| \tag{33.3}$$

with

$$k_{x,i} = \frac{\pi}{L}n_{x,i} \qquad k_{y,i} = \frac{\pi}{L}n_{y,i} \qquad k_{z,i} = \frac{\pi}{L}n_{z,i}.$$

Here, we have assumed a cubic lattice in a cube of side L. The integers $n_{x,i}$, $n_{y,i}$, and $n_{z,i}$ denote the particular energy level of the i-th oscillator and must not be confused with the **occupation numbers** n_i in equation (33.1). For every value of ω_i there are in fact 3 different modes: two for transversal waves and one for longitudinal waves. Debye also assumed that the allowed frequencies are given by those sets of non-negative integers $(n_{x,i}, n_{y,i}, n_{z,i})$ which are the coordinates of points lying inside a ball of radius such that the total number of such points equals N. (Each point contributes 3 wave modes.) (Because each coordinate is non-negative these points occupy only one octant of the ball.) That is, $\omega_i \leqslant \omega_D$, where

$$\#\left\{(n_x, n_y, n_z) : n_x^2 + n_y^2 + n_z^2 \leqslant \frac{L^2\omega_D^2}{\pi^2 c_s^2}\right\} = N. \tag{33.4}$$

This assumption has two aspects. Firstly, the fact that there is a maximum value for the frequency is due to fact that the atoms are arranged in a lattice so that the wavelength cannot be smaller than the distance between two neighbouring atoms. Secondly, by taking a ball in $\vec{k}$-space one essentially assumes that the elastic medium is isotropic.

With these assumptions the free energy becomes

$$f(\beta) = \lim_{N\to\infty} \frac{1}{N}\left\{\frac{\hbar}{2}\sum_{i=1}^{3N}\omega_i + \frac{1}{\beta}\sum_{i=1}^{3N}\ln(1 - e^{-\beta\hbar\omega_i})\right\}. \tag{33.5}$$

We can now argue as in the case of the free gas (see equations (23.7) and (23.8)). The sum over ω_i can be written as a sum over allowed $\vec{k}$-values and these become a Riemann integral in the limit $L \to \infty$. Since there are $(\pi/L)^3$ allowed $\vec{k}$-values per unit volume we get

$$f(\beta) = \frac{3\hbar v c_s}{2\pi^3}\int |\vec{k}|\,\mathrm{d}^3k + \frac{3v}{\pi^3\beta}\int \ln(1 - e^{-\beta\hbar c_s|\vec{k}|})\,\mathrm{d}^3k$$

$$= \frac{3\hbar v}{4\pi^2 c_s^3}\int_0^{\omega_D}\omega^3\,\mathrm{d}\omega + \frac{3v}{2\pi^2 c_s^3\beta}\int_0^{\omega_D}\ln(1 - e^{-\beta\hbar\omega})\omega^2\,\mathrm{d}\omega, \tag{33.6}$$

where the integrals in the first expression are over the first octant of a ball with radius equal to the maximum value of $|\vec{k}|$ allowed by equation (33.4). This gives rise to a factor $\frac{1}{8}$ in the second expression. The factor 3 is due to the three independent wave modes for each frequency mentioned above. The specific volume v satisfies (taking the limit $N \to \infty$ in equation (33.4))

$$\frac{1}{2}\pi\left(\frac{L}{\pi c_s}\right)^3\int_0^{\omega_D}\omega^2\,\mathrm{d}\omega = N \implies v = \frac{6\pi^2 c_s^3}{\omega_D^3}. \tag{33.7}$$

Inserting this we can write equation (33.6) in the form

$$f(\beta) = \frac{9\hbar}{2\omega_D^3}\int_0^{\omega_D}\omega^3\,\mathrm{d}\omega + \frac{9}{\omega_D^3\beta}\int_0^{\omega_D}\ln(1-e^{-\beta\hbar\omega})\omega^2\,\mathrm{d}\omega. \tag{33.8}$$

The first term in this expression is due to the zero-point energy of the harmonic oscillators. It is independent of the temperature and can be included in the term u_c of equation (13.1). The second term can be written in the form (13.3):

$$f_{\mathrm{ph}}(\beta) = T\,f_D\left(\frac{\Theta_D}{T}\right) \tag{33.9}$$

with

$$\Theta_D = \frac{\hbar\omega_D}{k_B} \tag{33.10}$$

and

$$f_D\left(\frac{\Theta_D}{T}\right) = 9k_B\left(\frac{T}{\Theta_D}\right)^3\int_0^{\Theta_D/T}\ln(1-e^{-x})x^2\,\mathrm{d}x. \tag{33.11}$$

Differentiating twice we get the specific heat

$$c_V = -T\left(\frac{\partial^2 f}{\partial T^2}\right)_V = 9k_B\frac{T^3}{\Theta_D^3}\int_0^{\Theta_D/T}\frac{x^4 e^x}{(e^x-1)^2}\,\mathrm{d}x. \tag{33.12}$$

Figure 33.1 shows the behaviour of this function. (The dashed curve is Einstein's original approximation.) Note that for low temperatures

$$c_V \approx \frac{9k_B T^3}{\Theta_D^3}\int_0^\infty\frac{x^4 e^x}{(e^x-1)^2}\,\mathrm{d}x = \text{constant} \times T^3 \qquad (T \ll \Theta_D) \tag{33.13}$$

in accordance with the assertion at the end of chapter 31. On the other hand, for high temperatures $c_v \approx 3k_B$ which is the law of Dulong and Petit (see equation (2.9)).

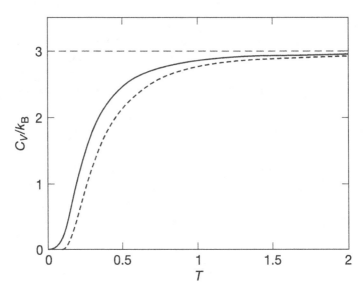

Figure 33.1 The specific heat of crystalline solids according to Debye.

CHAPTER 34

Boson Condensation and Liquid Helium

In chapter 23, we computed the pressure and the density of the free boson gas (see equations (23.25) and (23.26)). We then remarked that in fact equation (23.26) only holds if the density is less than the critical density ρ_c given by equation (23.27):

$$\rho_c = \frac{1}{(2\pi)^3} \int \frac{\mathrm{d}^3 k}{e^{\beta\epsilon(\vec{k})} - 1}. \tag{34.1}$$

If the density is higher than that value then $\mu = 0$ and the remaining particles *condense* into the zero-energy state (the ground state). Before we analyse this further let us write the grand potential in a different form (we write $\omega(\beta, \mu)$ instead of $\omega(\mu, T)$):

$$
\begin{aligned}
\omega(\beta, \mu) &= \frac{1}{(2\pi)^3 \beta} \int_{\mathbb{R}^3} \ln\left(1 - e^{\beta(\mu - \epsilon(\vec{k}))}\right) \mathrm{d}^3 k \\
&= \frac{1}{4\pi^2 \beta} \left(\frac{2m}{\hbar^2}\right)^{3/2} \int_0^\infty \epsilon^{1/2} \ln\left(1 - e^{\beta(\mu - \epsilon)}\right) \mathrm{d}\epsilon \\
&= -\frac{1}{4\pi^2 \beta} \left(\frac{2m}{\hbar^2}\right)^{3/2} \sum_{l=1}^\infty \frac{e^{\beta\mu l}}{l} \int_0^\infty \epsilon^{1/2} e^{-\beta\epsilon l} \mathrm{d}\epsilon \\
&= -\frac{1}{\beta\lambda_T^3} g_{5/2}(e^{\beta\mu})
\end{aligned}
\tag{34.2}
$$

where λ_T is the thermal wavelength given by equation (23.32):

$$\lambda_T = \left(\frac{2\pi\hbar^2}{mk_B T}\right)^{1/2} \tag{34.3}$$

and the function $g_\alpha(z)$ is defined, for general α by

$$g_\alpha(z) = \sum_{l=1}^\infty \frac{z^l}{l^\alpha}. \tag{34.4}$$

Note that $zg'_{5/2}(z) = g_{3/2}(z)$ and hence

$$\rho = \frac{1}{\lambda_T^3} g_{3/2}(e^{\beta\mu}) \tag{34.5}$$

provided $\rho \leqslant \rho_c$. In particular,

$$\rho_c = \frac{1}{\lambda_T^3} g_{3/2}(1). \tag{34.6}$$

Let us now come back to the question of Bose-Einstein condensation. To understand that the condensation is indeed into the ground state we need to consider the distribution of the particles over the allowed energy levels. At large but finite volume V the density is given by

$$\rho_V(\beta, \mu) = \frac{1}{V} \sum_{\vec{k}} \frac{1}{\exp\left[\beta(\epsilon(\vec{k}) - \mu)\right] - 1}. \tag{34.7}$$

Figure 34.1 shows the dependence of ρ_V on μ for an increasing set of volumes V.

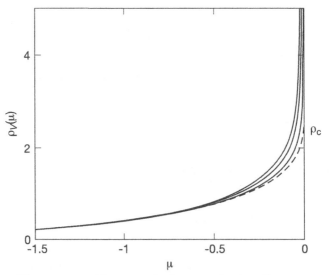

Figure 34.1 The particle density of a free Bose gas
for an increasing sequence of volumes.

It is seen that, at finite volume ρ_V becomes arbitrarily large as $\mu \uparrow 0$ but, as $V \to \infty$ the graphs move towards the axis above $\rho = \rho_c$ so that the limiting function $\rho(\mu)$ has a maximum at ρ_c. If we fix $\rho > \rho_c$ then at finite volume there is a strictly negative value of μ such that $\rho_V(\beta, \mu) = \rho$ but as V increases μ tends to 0. The individual terms in equation (34.7) correspond to the particle number density in each state given by $\vec{k}$. Let us single out the ground state term $\vec{k}_0 = \frac{\pi}{L}(1,1,1)$ with energy $\epsilon_0 = 3\pi^2\hbar^2/2mL^2$:

$$\rho_V(\beta,\mu) = \frac{1}{V}\frac{1}{e^{\beta(\epsilon_0-\mu)}-1} + \frac{1}{V}\sum_{\vec{k}\neq\vec{k}_0}\frac{1}{\exp[\beta(\epsilon(\vec{k})-\mu)]-1}.$$

The second term in the right-hand side tends to the integral (34.1) as $\mu \to 0$ and $V \to \infty$ but the first term is more delicate. We can let it tend to the value $\rho - \rho_c$ (the surplus density) if we take

$$\mu_V \sim \epsilon_0 - \frac{1}{\beta V(\rho-\rho_c)}. \tag{34.8}$$

Thus, the surplus particles all reside in the ground state $\vec{k} = \vec{k}_0$. Note that the next lowest eigenstates $\vec{k} = \frac{\pi}{L}(2,1,1)$, $\vec{k} = \frac{\pi}{L}(1,2,1)$, and $\vec{k} = \frac{\pi}{L}(1,1,2)$ do not contribute because the corresponding terms in equation (34.7) are of the order

$$\frac{1}{V}\left\{\exp\left[\beta\left(\frac{3\pi^2\hbar^2}{mL^2}-\mu_V\right)\right]-1\right\}^{-1} \sim \frac{1}{\beta V}\left(\frac{3\pi^2\hbar^2}{mL^2}-\mu_V\right)^{-1}$$

$$\sim \frac{2m}{3\beta\pi^2\hbar^2}L^{-1} \to 0 \tag{34.9}$$

as $L = V^{1/3} \to \infty$.

REMARK 34.1: *The complete particle distribution.*
One can derive a formula for the entropy and the free energy density in terms of the full distribution of particles over all the energy levels in the thermodynamic limit (cf. the remark at the end of chapter 22). The complete distribution is given by a *measure* m on $[0,\infty)$, so that $m([\epsilon,\epsilon'])$ is the number of particles with energies in the interval $[\epsilon,\epsilon']$ per unit volume. The singular part of the measure m does not contribute to the entropy in the thermodynamic limit. Writing the absolutely continuous part in the form

$$m_{\text{abs.cont.}}(\mathrm{d}\epsilon) = \frac{1}{4\pi^2}\left(\frac{2m}{\hbar^2}\right)^{3/2}\epsilon^{1/2}\rho(\epsilon)\,\mathrm{d}\epsilon, \tag{34.10}$$

where $\rho(\epsilon)$ is a density function, the entropy density with a given distribution measure m can be written as

$$\tilde{s}[m] = \frac{k_B}{4\pi^2}\left(\frac{2m}{\hbar^2}\right)^{3/2}\int_0^\infty s_0(\rho(\epsilon))\,\epsilon^{1/2}\mathrm{d}\epsilon, \tag{34.11}$$

in which the function $s_0(x)$ is given by

$$s_0(x) = (1+x)\ln(1+x) - x\ln x. \tag{34.12}$$

The free energy density for a given distribution is

$$\tilde{f}_F[m] = \int_0^\infty \epsilon\, m(\mathrm{d}\epsilon) - \frac{1}{k_B\beta}\tilde{s}[m] \tag{34.13}$$

and the free energy density of the free gas at fixed density ρ is then obtained by minimizing over the measures m with fixed density, i.e. $||m|| = \rho$:

$$f(\beta,\rho) = \inf_{m:\, ||m||=\rho} \tilde{f}_F[m]. \tag{34.14}$$

Since $\tilde{s}[m]$ is independent of the singular part of m, it follows immediately that the latter must be concentrated at $\epsilon = 0$. Moreover, minimization with respect to the absolutely continuous part yields easily that

$$\rho(\epsilon) = \frac{1}{e^{\beta(\epsilon-\mu)} - 1}.$$

If $\rho > \rho_c$ then the infimum is attained for $m(\{0\}) \neq 0$.

One may now ask: is the boson gas a model for a realistic physical system? And in particular, is there a system where Bose-Einstein condensation actually occurs? In fact the above analysis may give the impression that Bose-Einstein condensation is a peculiarity of the free gas. This is not the case as we demonstrate below. It was suggested in 1938 by Fritz London that the transition to superfluidity in helium might be an example of Bose-Einstein condensation. Helium is indeed a very remarkable substance. It has two isotopes: ^{3}He and ^{4}He. Both remain fluid down to the absolute zero of temperature, but otherwise they behave quite differently. Liquid ^{4}He has another phase transition, the so-called λ-**transition**, from a phase called He I to a phase He II. The latter is a superfluid phase, which means that the fluid flows with zero viscosity. The transition takes place at a temperature $T_\lambda = 2.18$ K along the vapour pressure curve and at specific volume $v_\lambda = 46.2 \times 10^{-24}$ cm^3 atom^{-1}. Figure 34.2 shows the phase diagram of ^{4}He (see also figure 12.6). At the superfluid transition the specific heat diverges (figure 34-3). The name λ-transition is derived from the shape of this curve.

It was discovered in 1979 by D. D. Osheroff, R. C. Richardson, and D. M. Lee that ^{3}He also has a superfluid phase but the transition takes place at a very much lower temperature. The difference in behaviour is clearly due to the fact that ^{4}He atoms are bosons and ^{3}He atoms are fermions because the former have an even number of spin-$\frac{1}{2}$ particles and the latter an odd number. (The total spin of an even number of spin-$\frac{1}{2}$ particles is an integer multiple of $\hbar$ (see appendix B.)) It is therefore reasonable to suspect that the transition in ^{4}He has something to do with the Bose-Einstein transition. This connection becomes even more likely if we compute the transition temperature for a boson gas with a specific volume given by v_λ. Taking the mass of a ^{4}He atom to be $m = 6.65 \times 10^{-24}$ g and using $g_{3/2}(1) = 2.612$, we get $T_c = 3.12$ K. Other than this remarkably good agreement of T_λ and T_c there are important

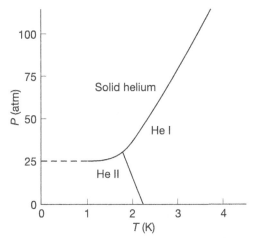

Figure 34.2. The phase diagram of ^{4}He.

differences, however. For example, the specific heat of a free boson gas does not diverge at the critical temperature but has a finite jump discontinuity. Also, the Bose-Einstein transition is a first-order transition whereas the λ-transition is second-order: there is no latent heat of transition. These differences must be due to the interactions (forces) between the helium atoms. These are very well known: the corresponding potential $\mathcal{V}(r)$ is given by a graph as in figure 13.1. Including the effect of this potential into the model makes it into a genuine **quantum many-body problem**, however, which is very difficult to analyse. We shall consider here only a rough approximation. The derivation of the Hamiltonian requires the theory of second quantization (see appendix B). The full quantum-mechanical N-body Hamiltonian is given by (cf. equation (B.47))

$$
\begin{aligned}
H_l = \sum_{\vec{k}} \epsilon_l(\vec{k}) a^*(\vec{k}) a(\vec{k}) \\
+ \frac{1}{2V_l} \sum_{\vec{k}_1} \sum_{\vec{k}_2} \sum_{\vec{q}} \widehat{\mathcal{V}}(\vec{q}) a^*(\vec{k}_1) a^*(\vec{k}_2) a(\vec{k}_2 - \vec{q}) a(\vec{k}_1 + \vec{q}),
\end{aligned}
\tag{34.15}
$$

where V_l $(l = 1, 2, \dots)$ is a sequence of volumes tending to infinity, $\widehat{\mathcal{V}}$ is the Fourier transform of $\mathcal{V}$,

$$
\widehat{\mathcal{V}}(\vec{q}) = \int_{V_l} \mathcal{V}(\vec{r}) e^{i\vec{q}\cdot\vec{r}} \, d\vec{r},
\tag{34.16}
$$

and the sum is over the allowed $\vec{k}$-values as in equation (33.9): $k_x = (\pi/L)i_x$, etc. for positive integers i_x, i_y, i_z. $a(\vec{k})$ and $a^*(\vec{k})$ are operators which annihilate repsectively create a particle with momentum $\vec{k}$. They satisfy respectively

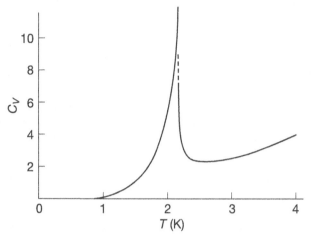

Figure 34.3. The specific heat of liquid ^{4}He
along the vapour pressure curve.

commutation relations

$$[a(\vec{k}), a(\vec{k}')] = 0, \qquad [a^*(\vec{k}), a^*(\vec{k}')] = 0 \qquad [a(\vec{k}), a^*(\vec{k}')] = \delta_{\vec{k},\vec{k}'}. \qquad (34.17)$$

In order to analyse the Hamiltonian (34.15), one has to make approximations. Here we shall begin by neglecting all terms which do not commute with the number operators $n(\vec{k}) = a^*(\vec{k})a(\vec{k})$. This means that we neglect all terms with $\vec{q} \neq 0$ and $q \neq \vec{k}_2 - \vec{k}_1$ because all number operators $n(\vec{k})$ commute with one another. The resulting interaction can be written in the form $U_{\text{diag}} = U_1 + U_2 + U_3$, where

$$U_1 = \frac{\widehat{\mathcal{V}}(0)}{2V_l} (N^2 - N) \qquad (34.18)$$

$$U_2 = \frac{\widehat{\mathcal{V}}(0)}{2V_l} \left(N^2 - \sum_{\vec{k}} n(\vec{k})^2 \right), \qquad (34.19)$$

and

$$U_3 = \frac{1}{2V_l} \sum_{\vec{k}_1} \sum_{\vec{k}_2} \left[\widehat{\mathcal{V}}(\vec{k}_1 - \vec{k}_2) - \widehat{\mathcal{V}}(0) \right] n(\vec{k}_1)n(\vec{k}_2). \qquad (34.20)$$

Here $N = \sum_{\vec{k}} n(\vec{k})$ is the total number operator. We now make a further approximation and assume that the potential $\mathcal{V}$ has very short range. In that case $\widehat{\mathcal{V}}(\vec{q})$, is almost independent of $\vec{q}$ and we can neglect the third term U_3 in the potential. In the thermodynamic limit N becomes very large so that U_1 can be simplified to $U_1 = (1/2V_l)\widehat{\mathcal{V}}(0) N^2$. The resulting interaction is called the **Huang-Yang-Luttinger** (HYL) (1957) model. (A rigorous analysis of the HYL model using large-deviation theory has been given by van den Berg *et al.* (1988, 1990); for a more general model, see Dorlas *et al.* (1993).) Writing $a = \widehat{V}(0)$ it is given by

$$U_{\text{HYL}} = \frac{a}{2V_l}\left(2N^2 - \sum_{\vec{k}} n(\vec{k})^2\right). \tag{34.21}$$

In order to compute the grand potential in the thermodynamic limit we want to use Varadhan's theorem. First we have to write it in terms of random variables, which we take to be the occupation numbers of the free boson gas. They have a probability distribution given by (cf. equation (23.4))

$$\mathbb{P}_l^{\beta,\mu_0}\left[\{n(\vec{k})\}\right] = \frac{1}{\mathcal{Z}_l(\beta,\mu_0)}\exp\left(\beta\sum_{\vec{k}} n(\vec{k})[\mu_0 - \epsilon(\vec{k})]\right). \tag{34.22}$$

Note that this distribution is only defined for $\mu_0 < 0$ whereas we shall find that the grand potential for the HYL model can be defined for all real values of μ. For that reason we introduce an artificial new parameter $\mu_0 < 0$: the chemical potential for the free gas. Obviously, this parameter should drop out of the final result. In terms of this probability distribution we have

$$\omega(\beta,\mu) = \omega_{\text{FG}}(\beta,\mu_0)$$
$$- \lim_{l\to\infty}\frac{1}{\beta V_l}\ln \mathbb{E}_l^{\beta,\mu_0}\left\{\exp\left[\beta(\mu - \mu_0)N - \frac{\beta a}{2V_l}\left(2N^2 - \sum_{\vec{k}} n(\vec{k})^2\right)\right]\right\} \tag{34.23}$$

where $\omega_{FG}(\beta,\mu_0)$ is the grand canonical potential of the free gas.

Consider the random variables $Y_l = V_l^{-1}N$ and $Z_l = V_l^{-1}n(0)$ with joint distribution function

$$F_l(y,z) = \sum_{\{n(\vec{k})\}:\, Y_l \leqslant y, Z_l \leqslant z} \mathbb{P}_l^{\beta,\mu_0}\left[\{n(\vec{k})\}\right]. \tag{34.24}$$

Computing the cumulant generating function (taking for convenience the scaling constants βV_l), it is straightforward to see that the sequence (Y_l, Z_l) satisfies the LDP with rate function given by

$$I^{\beta,\mu_0}(y,z) = -\omega_{\text{FG}}(\beta,\mu_0) + \tilde{f}_{\text{FG}}(\beta, y - z) - \mu_0 y \tag{34.25}$$

if $0 \leqslant z \leqslant y$ and infinite otherwise. Now note that since $\sum_{\vec{k}} n(\vec{k})^2 \geqslant n(0)^2$,

$$\omega(\beta,\mu) \leqslant \omega_{\text{FG}}(\beta,\mu_0)$$
$$- \lim_{l\to\infty}\frac{1}{\beta V_l}\ln \mathbb{E}_l^{\beta,\mu_0}\left\{\exp\left[\beta V_l\left((\mu - \mu_0)Y_l - \frac{a}{2}(2Y_l^2 - Z_l^2)\right)\right]\right\}.$$

Using Varadhan's theorem, we conclude that

$$\omega(\beta, \mu) \leqslant - \sup_{\rho \geqslant \rho_0 \geqslant 0} \left[\mu\rho - \tilde{f}_{FG}(\beta, \rho - \rho_0) - \frac{1}{2}a(2\rho^2 - \rho_0^2) \right].$$

Using a more sophisticated form of this argument one can show that the terms other than $n(0)^2$ in the sum $\sum_{\vec{k}} n(\vec{k})^2$ are in fact negligible so that $\omega(\beta, \mu)$ is actually equal to the right-hand side:

$$\boxed{\omega_{\mathrm{HYL}}(\beta, \mu) = - \sup_{\rho \geqslant \rho_0 \geqslant 0} \left\{ \mu\rho - \tilde{f}_{FG}(\beta, \rho - \rho_0) - \frac{1}{2}a(2\rho^2 - \rho_0^2) \right\}} \quad (34.26)$$

This variational problem can be analysed as follows. If we put $x = \rho - \rho_0$, $-\omega$ can be written as a supremum over x and ρ_0 and the supremum over the last variable can be easily performed. The result is:

$$\omega(\beta, \mu) = - \sup_{x \geqslant 0} \{g(x) - \tilde{f}_{FG}(x)\}, \quad (34.27)$$

where we have omitted the variable β in $\tilde{f}_{FG}$ and where the function $g(x)$ is given by

$$g(x) = \begin{cases} \frac{\mu^2}{2a} + ax^2 - \mu x, & \text{if } 0 \leqslant x \leqslant \frac{\mu}{2a}, \\ -ax^2 + \mu x, & \text{if } \frac{\mu}{2a} < x. \end{cases} \quad (34.28)$$

Now, $g(x)$ is decreasing for all $x \geqslant 0$ and $\tilde{f}_{FG}(x)$ is decreasing for $x \in [0, \rho_c)$ and constant for $x \geqslant \rho_c$. Moreover, $\tilde{f}'_{FG}(x) \to -\infty$ as $x \to 0$. The supremum is therefore attained for $x \in (0, \rho_c]$. This immediately implies that if $\rho > \rho_c$ then $\rho_0 > 0$, which is Bose-Einstein condensation. In fact, we shall see that the true critical value of ρ is slightly lower than ρ_c.

If $\mu > 2a\rho_c$ then the upper formula for g holds for all $x \in (0, \rho_c]$ and the maximum is attained at the point x where $\tilde{f}'_{FG}(x) = g'(x) = 2ax - \mu$. Now, $\tilde{f}'_{FG}(x) = \alpha(x) \leqslant 0$ where $\alpha(x)$ is the real root of $\rho_{FG}(\alpha) = -\omega'_{FG}(\alpha) = x$. We conclude that for $\mu > 2a\rho_c$,

$$\omega(\beta, \mu) = -\frac{\mu^2}{2a} - ax^2 + (\mu + \alpha)x + \omega_{FG}(\beta, \alpha) \quad (34.29)$$

where α is the unique root of

$$2a\rho_{FG}(\alpha) - \alpha = \mu, \quad (\alpha \leqslant 0) \quad (34.30)$$

and where $x = \rho_{FG}(\alpha)$. To find the pressure, note that by (34.27) and (34.28),

$$\rho = -\frac{\partial \omega}{\partial \mu} = \frac{\mu}{a} - x = x - \frac{\alpha}{a} = \rho_{FG}(\alpha) - \frac{\alpha}{a}. \quad (34.31)$$

Solving this equation for μ and α and setting $x = \rho_{FG}(\alpha)$, the pressure is given by

$$\boxed{p(v) = a\rho^2 - \frac{1}{2}a(\rho - x)^2 + p_{FG}(1/x)} \quad (34.32)$$

It remains to consider the case $\mu \leqslant 2a\rho_c$. The supremum in the range $x \in [0, \mu/(2a)]$ is again determined by the function

$$h(\alpha) = 2a\rho_{\mathrm{FG}}(\alpha) - \alpha \tag{34.33}$$

The graph of this function is plotted in figure 34.4.

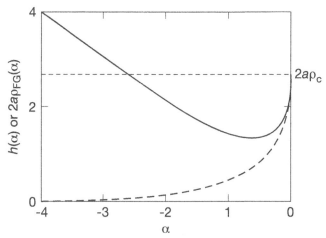

Figure 34.4 Graph of the function $h(\alpha) = 2a\rho_{\mathrm{FG}}(\alpha) - \alpha$ (full curve) and $2a\rho_{\mathrm{FG}}(\alpha)$ (broken curve).

If μ_{min} denotes the minimum value of this function then clearly, for $\mu < \mu_{\mathrm{min}}$, $h(\alpha) - \mu > 0$ and the supremum in equation (34.27) is attained for $x \in [\mu/(2a), \rho_c]$. In that case

$$\omega(\beta, \mu) = -a\rho^2 + \omega_{\mathrm{FG}}(\beta, \alpha') \tag{34.34}$$

where $\rho = \rho_{\mathrm{FG}}(\alpha')$ and α' is the solution of

$$2a\rho_{\mathrm{FG}}(\alpha') + \alpha' = \mu. \tag{34.35}$$

The situation is more complicated for $\mu_{\mathrm{min}} < \mu < 2a\rho_c$. A careful analysis shows that there is a critical value $\mu^* \in [\mu_{\mathrm{min}}, 2a\rho_c]$ such that for $\mu < \mu^*$ the supremum is attained for $x = \rho = \rho_{\mathrm{FG}}(\alpha')$, where α' is the unique solution of equation (34.35), whereas for $\mu > \mu^*$ the supremum is attained for $x = \rho_{\mathrm{FG}}(\alpha)$, where α is the most negative of the roots of equation (34.30). (Note that the other root is a local minimum of $g(x) - \tilde{f}_{\mathrm{FG}}(x)$.) Indeed, denote

$$\omega^{\pm}(\beta, \mu) = -\sup_{x \geqslant 0}[g^{\pm}(x) - \tilde{f}_{\mathrm{FG}}(x)],$$

where $g^+(x)$ is the upper function in (34.28), and $g^-(x)$ is the lower function. Then for $\mu \in [\mu_{\mathrm{min}}, 2a\rho_c]$, the difference $-\omega^+(\beta, \mu) + \omega^-(\beta, \mu)$ is increasing in μ, since $-\partial w^+/\partial \mu = \mu/a - \rho_{\mathrm{FG}}(\alpha)$ and $-\partial w^-/\partial \mu = \rho_{\mathrm{FG}}(\alpha')$. Inserting (34.30) and (34.35), we get

$$\frac{\partial}{\partial \mu}(-w^+(\beta, \mu) + w^-(\beta, \mu)) = \frac{-\alpha + \alpha'}{2a} > 0.$$

There is therefore a unique solution to $w^+(\beta, \mu) = w^-(\beta, \mu)$.

In case $\mu < \mu^*$, $\rho_0 = 0$ and the pressure is given by (34.34), i.e.

$$\boxed{p(v) = a\rho^2 + p_{\mathrm{FG}}(v)} \tag{34.36}$$

In the case $\mu > \mu^*$, there is Bose-Einstein condensation:

$$\rho_0 = -\frac{\alpha}{a} \tag{34.37}$$

(by equation (34.31)) and the pressure is given by equation (34.32).

The pressure for this interacting boson gas is shown in figure 34.5.

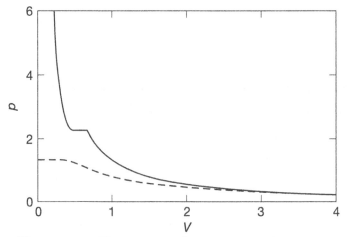

Figure 34.5. The pressure of an interacting boson gas.

REMARK 34.2: *Large deviations for the particle distribution.*
The grand-canonical ensemble (34.22) in volumes V_l determines a random distribution of the particle distribution measures m of remark 34.1. The random distribution of these measures also satisfies the LDP. The corresponding rate function $I^{\beta,\mu}$ is given by the analogue of equation (23.13) (but with constants βV_l)

$$I^{\beta,\mu}[m] = \sup_{\phi \in \mathcal{C}([0,+\infty)): \phi(\epsilon) < \epsilon - \mu} \left[\int \phi(\epsilon)\, m(\mathrm{d}\epsilon) - C^{\beta,\mu}[\phi] \right] \tag{34.38}$$

where, analogous to equation (23.8),

$$C^{\beta,\mu}(\phi) = -w[\beta, \mu + \phi(\cdot)] + w(\beta, \mu) \tag{34.39}$$

and the functional $\omega[\beta, \mu + \phi(\cdot)]$ is given by

$$\omega[\beta, \lambda(\cdot)] = \frac{1}{4\pi^2\beta} \left(\frac{2m}{\hbar^2}\right)^{3/2} \int_0^\infty \epsilon^{1/2} \ln(1 - e^{\beta(\lambda(\epsilon)-\epsilon)}) \, d\epsilon. \qquad (34.40)$$

The supremum can be evaluated, as a result of which $I^{\beta,\mu}$ can be written as (cf. equation (34.25))

$$I^{\beta,\mu}(m) = f_F[m] - \omega(\beta, \mu) - \mu \, ||m||. \qquad (34.41)$$

Here $f_F[m]$ is the free energy functional introduced in remark 34.1, equation (34.13). Using Varadhan's theorem, this infinite-dimensional large-deviation principle allows the evaluation of the grand potential for more general models (see Dorlas *et al.* (1993).)

The HYL model is obviously still unsatisfactory for explaining the properties of liquid ^{4}He. For example, figure 34.5 shows that the transition is still of first order. However, recently Bose-Einstein condensation has been observed in other, more artificial systems. These are produced by confining a very rare gas by means of laser beams. This technique was invented by S. Chu, C. Cohen-Tannoudji, and W. D. Phillips (see for example the review article by Phillips and Cohen-Tannoudji (1990)), who earned the 1997 Nobel prize for their contributions. The density in these systems is much lower than that of liquid ^{4}He and as a result the interaction between the atoms is much smaller. The HYL model might then be a reasonable approximation. The first experimental evidence for Bose-Einstein condensation in this kind of system was obtained in 1995 by M. H. Anderson *et al.* (1995).

CHAPTER 35

Superconductivity

Superconductivity is a remarkable phenomenon. It was discovered in 1911, by Heike Kamerlingh Onnes in Leiden. At the time Leiden had the best facilities for obtaining low temperatures and Kamerlingh Onnes set out to investigate systematically the properties of materials at very low temperatures.

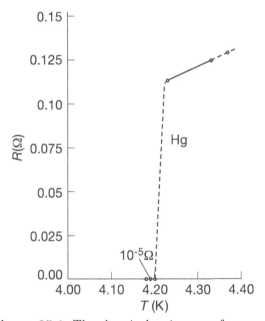

Figure 35.1. The electrical resistance of mercury.

He thus found quite unexpectedly that mercury becomes perfectly conducting at a temperature below 4.2 K. Other metals were then also found to have this property: tin below $T_c = 3.72$ K, lead below $T_c = 7.2$ K and aluminium below $T_c = 1.12$ K. The change in conductivity is quite abrupt as can be seen in Onnes' first measurements (figure 35.1).

This indicates that this phenomenon is a phase transition. Actually, this was not so obvious at the time because the absence of dissipation for electrical currents seemed to imply thermodynamic irreversibility.

To understand this, one first has to know that the superconducting state disappears when a sufficiently large magnetic field is applied. The critical field H_c depends on the temperature roughly as depicted in figure 35.2.

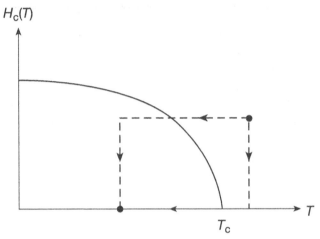

Figure 35.2. The critical magnetic field of a superconductor.

Now consider the following two experiments. Starting from a state at a temperature above the critical temperature and in the presence of a magnetic field below the maximal critical field, either first lower the temperature below T_c and then remove the external field or first remove the field and then lower the temperature. In the first experiment, the field is removed in the superconducting state and induces a circular current inside the superconductor which then gives rise to a remnant magnetization. In the second experiment, the field is removed in the normal state and the resulting current dies out very rapidly (dissipates). After cooling there is no remnant magnetization. The two experiments are therefore not equivalent although they have the same initial and final states in the $H - T$ diagram.

The above argument is in fact fallacious. This was discovered by Walther Meissner and Robert Ochsenfeld. They showed in 1933 that in fact magnetic fields are *expelled* from superconductors; superconductors are perfect diamagnets, that is $\vec{m} = -\vec{H}$ in equation (10.1). (In general, materials with $\chi < 0$ are called **diamagnets**.) This so called **Meissner effect** implies that in the first experiment above there is no induced current and therefore no remnant magnetization. It thus became clear that an explanation in terms of equilibrium thermodynamics might be possible.

Next came two important contributions by Fritz London. First he considered the solutions of Maxwell's equations for a superconductor and deduced that the magnetic field is not completely expelled from the interior of the superconductor after all, but penetrates the surface to a depth ξ of order 700Å. He then considered the quantum-mechanical electric current and reasoned that the Meissner effect can be explained if the wave function of the electrons does not change when a magnetic field is introduced. This 'rigidity'of the wave function could be explained if there were a gap in the spectrum. The assumption of a gap in the spectrum would also explain the observed behaviour of the specific heat which is exponential in $1/T$ for low temperatures instead of linear in T as for normal metals (see equation (31.13)). Figure 35.3 shows this change in behaviour.

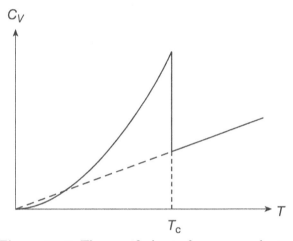

Figure 35.3. The specific heat of a superconductor.

The energy gap must be due to some interaction between the conduction electrons. One might think that this must be the Coulomb interaction (electrical repulsion) but that cannot be the case as it is much too large. One can estimate the energy gap to be of the order $\Delta \sim k_B T_c \sim 10^{-3}$ eV electron^{-1} whereas the Coulomb energy is of the order of 1 eV electron^{-1}. Moreover, it was discovered in 1950 that if one changes the isotope of the superconductor material the critical temperature changes. This suggests that the interaction has something to do with vibrations of the atomic lattice, i.e. phonons (see chapter 33). In 1950, Herbert Fröhlich therefore proposed a Hamiltonian of the form

$$H_{e-ph} = \sum_{\vec{k},\sigma} \epsilon(\vec{k})c^*(\vec{k},\sigma)c(\vec{k},\sigma) + \sum_{\vec{q}} \omega(\vec{q})a_{\vec{q}}^*a_{\vec{q}}$$

$$+ \sum_{\vec{k},\vec{k}',\sigma} M_{\vec{k},\vec{k}'}a_{\vec{k}'-\vec{k}}c^*(\vec{k}',\sigma)c(\vec{k},\sigma) + \sum_{\vec{k},\vec{k}',\sigma} M_{\vec{k},\vec{k}'}^*a_{\vec{k}-\vec{k}'}^*c^*(\vec{k}',\sigma)c(\vec{k},\sigma).$$

$$(35.1)$$

Here $c(\vec{k},\sigma)$ and $c^*(\vec{k},\sigma)$ are the annihilation and creation operators for electrons with momentum $\vec{k}$ and spin σ, and $a_{\vec{q}}$ and $a_{\vec{q}}^*$ are the annihilation and creation operators for phonons, and M^* is the adjoint matrix of M. The operators c and c^* are anti-commuting variables, a and a^* are commuting variables (for $\vec{k} \neq \vec{k}'$). $\epsilon(\vec{k})$ is the energy of an electron of momentum $\vec{k}$ and $\omega(\vec{q})$ is the energy of a free phonon. The last two terms represent the interaction between electrons and phonons. Note that it does not affect the spin of the electron.

This Hamiltonian can indeed give rise to an effective attraction between electrons. To see this we have to compute *second-order* terms in the perturbation series for the energy. The relevant terms are given by the Feynman diagrams of figure 35.4.

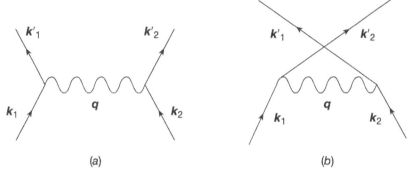

Figure 35.4. Second-order Feynman diagrams
for the interaction of electrons in a superconductor.

REMARK 35.1: *Normal resistance versus superconductivity.*
The interaction (35.1) is also responsible for normal resistance in metals. The electron loses energy by emitting a phonon. This is due to the *first-order* perturbation terms. These terms do not give rise to an effective interaction between electrons but describe the (inelastic) scattering of an electron off a phonon. High resistance at normal temperatures is caused by large scattering which means that the matrix elements of M are large. That in turn means that the metal is more likely to be superconducting at low temperatures!

Remember that in non-degenerate perturbation theory, the second-order energy shift in the i-th level is given by

$$\delta^2 E = \sum_{j \neq i} \frac{\langle \psi_i^0 \mid \mathcal{H} \mid \psi_j^0 \rangle \langle \psi_j^0 \mid \mathcal{H} \mid \psi_i^0 \rangle}{E_i^0 - E_j^0}. \tag{35.2}$$

The unperturbed energy eigenstates ψ_i^0 are given by the eigenstates of $c_{\vec{k}}^* c_{\vec{k}}$ and $a_{\vec{k}'}^*, a_{\vec{k}'}$. These are determined by the numbers of electrons, respectively phonons for each of the allowed momenta $\vec{k}$ and are all highly degenerate. We therefore need to use **degenerate perturbation theory**. In the case of degeneracy, the ith energy level splits due to the perturbation and the various energy shifts are given by the eigenvalues of a matrix with matrix elements given by a similar formula:

$$\langle \psi_{i'}^0 \mid H_{\text{eff}} \mid \psi_i^0 \rangle = \sum_{j:E_j^0 \neq E_i^0} \frac{\langle \psi_{i'}^0 \mid H \mid \psi_j^0 \rangle \, \langle \psi_j^0 \mid H \mid \psi_i^0 \rangle}{E_i^0 - E_j^0}. \tag{35.3}$$

As in equation (35.2), the intermediate states must be different, that is, they must be orthogonal to the eigenspace for the unperturbed energy E_i^0. The intermediate (or 'virtual') states ψ_j^0 are those where one of the electrons has created a phonon. In figure 35.4(a), the intermediate energy is given by $\epsilon(\vec{k}_1 - \vec{q}) + \epsilon(\vec{k}_2) + \hbar\omega(\vec{q})$ if it was emitted by electron # 1 and $\epsilon(\vec{k}_1) + \epsilon(\vec{k}_2 + \vec{q}) + \hbar\omega(\vec{q})$ if it was emitted by electron # 2. The terms corresponding to figure 35.4(a) are therefore given by

$$\langle \vec{k}_1', \sigma_1'; \vec{k}_2', \sigma_2', \mid H_{\text{eff}}^1 \mid \vec{k}_1, \sigma_1; \vec{k}_2, \sigma_2 \rangle =$$

$$M_{\vec{k}_2, \vec{k}_2'} \frac{1}{[\epsilon(\vec{k}_1) + \epsilon(\vec{k}_2)] - [\epsilon(\vec{k}_1') + \epsilon(\vec{k}_2) + \hbar\omega(\vec{q})]} M_{\vec{k}_1, \vec{k}_1'}^* \delta_{\sigma_1, \sigma_1'} \delta_{\sigma_2, \sigma_2'} \tag{35.4}$$

and

$$\langle \vec{k}_1', \sigma_1'; \vec{k}_2', \sigma_2' \mid H_{\text{eff}}^{1'} \mid \vec{k}_1, \sigma_1; \vec{k}_2, \sigma_2 \rangle =$$

$$M_{\vec{k}_1, \vec{k}_1'} \frac{1}{[\epsilon(\vec{k}_1) + \epsilon(\vec{k}_2)] - [\epsilon(\vec{k}_1) + \epsilon(\vec{k}_2') + \hbar\omega(\vec{q})]} M_{\vec{k}_2, \vec{k}_2'}^* \delta_{\sigma_1, \sigma_1'} \delta_{\sigma_2, \sigma_2'}. \tag{35.5}$$

Here $\vec{q} = \vec{k}_1 - \vec{k}_1' = \vec{k}_2' - \vec{k}_2$. Similarly, figure 35.4(b) contributes the **exchange terms**

$$\langle \vec{k}_1', \sigma_1'; \vec{k}_2', \sigma_2' \mid H_{\text{eff}}^2 \mid \vec{k}_1, \sigma_1; \vec{k}_2, \sigma_2 \rangle =$$

$$- M_{\vec{k}_2, \vec{k}_1'} \frac{1}{[\epsilon(\vec{k}_1) + \epsilon(\vec{k}_2)] - [\epsilon(\vec{k}_2) + \epsilon(\vec{k}_2') + \hbar\omega(\vec{q})]} M_{\vec{k}_1, \vec{k}_2'}^* \delta_{\sigma_1, \sigma_2'} \delta_{\sigma_2, \sigma_1'} \tag{35.6}$$

and

$$\langle \vec{k}_1', \sigma_1'; \vec{k}_2', \sigma_2' \mid H_{\text{eff}}^{2'} \mid \vec{k}_1, \sigma_1; \vec{k}_2, \sigma_2 \rangle =$$

$$- M_{\vec{k}_1, \vec{k}_2'} \frac{1}{[\epsilon(\vec{k}_1) + \epsilon(\vec{k}_2)] - [\epsilon(\vec{k}_1') + \epsilon(\vec{k}_1) + \hbar\omega(\vec{q})]} M_{\vec{k}_2, \vec{k}_1'}^* \delta_{\sigma_1, \sigma_2'} \delta_{\sigma_2, \sigma_1'}, \tag{35.7}$$

where $\vec{q} = \vec{k}_1 - \vec{k}_2' = \vec{k}_1' - \vec{k}_2$.

Note the important minus sign in front of these expressions which is due to the anti-commutation of c-operators. The effective interaction is the sum of these four terms. Because the matrix elements $M_{\vec{k}_1, \vec{k}_2}$ do not depend on the spin variables it is easy to diagonalize the effective Hamiltonian with respect to the spin variables. If the spins are equal, $\sigma_1 = \sigma_2$, then all terms are already diagonal with respect to the spin variables and the total contribution is small due to the minus signs in equations (35.6) and (35.7). If the spins are opposite then equations (35.4) and (35.5) are diagonal but (35.6) and (35.7) are off-diagonal terms. Diagonalizing this 2×2-matrix, we obtain one state in which there is again near-cancellation and another state where the energies add up. (This is analogous to the discussion leading up to equation (28.1): the states $|\vec{k}_1, \sigma_1; \vec{k}_2, \sigma_2\rangle$ are not the same as equation (B.38) if the spins are opposite and the Hamiltonian H_{int} in chapter 28 is not diagonal in terms of these states.) The dominant contribution thus comes from this last term corresponding to opposite spins.

The latter term is largest when the electrons have energies near the Fermi surface. We shall therefore make the approximation $\epsilon(\vec{k}) \approx \epsilon_F$. Assuming also that $M_{\vec{k}, \vec{k}'}$ only depends on $|\vec{k} - \vec{k}'|$ we obtain the following effective second-order Hamiltonian:

$$\langle \vec{k}'_1, \sigma'_1; \vec{k}'_2, \sigma'_2 | H_{eff} | \vec{k}_1, \sigma_1; \vec{k}_2, \sigma_2 \rangle =$$
$$- \frac{1}{\hbar \omega(\vec{q})} \left(|M_{\vec{k}_1 - \vec{k}'_1}|^2 + |M_{\vec{k}_1 - \vec{k}'_2}|^2 \right) \delta_{\sigma_1, -\sigma_2} \delta_{\sigma_1, \sigma'_1} \delta_{\sigma_2, \sigma'_2} \qquad (35.8)$$

assuming $\vec{k}_1 + \vec{k}_2 = \vec{k}'_1 + \vec{k}'_2$. This effective Hamiltonian clearly shows that electrons near the Fermi surface attract one another. The interaction between the electrons and the phonons thus gives rise to an effective attraction between the electrons. There is also a repulsive force between the electrons, of course: the Coulomb repulsion. If the net interaction is attractive, superconductivity results.

Including the Coulomb repulsion, we can now write an **effective Hamiltonian** for the conduction electrons:

$$H_{eff} = \sum_{\vec{k}, \sigma} \epsilon(\vec{k}) c^*(\vec{k}, \sigma) c(\vec{k}, \sigma)$$
$$+ \sum_{\sigma} \sum_{\vec{k}_1} \sum_{\vec{k}_2} \sum_{\vec{q}} U_{\vec{q}} c^*(\vec{k}_1 + \vec{q}, \sigma) c^*(\vec{k}_2 - \vec{q}, -\sigma) c(\vec{k}_2, -\sigma) c(\vec{k}_1, \sigma). \qquad (35.9)$$

To analyse this model, *Bardeen, Cooper, and Schrieffer (BCS) (1957)* introduced a very important simplification. They argued that the lowest energy states of equation (35.9) are given by configurations in which the $\vec{k}$ states are occupied in pairs, called **Cooper pairs**: whenever $(\vec{k}, \uparrow)$ is occupied then $(-\vec{k}, \downarrow)$ is also occupied and vice versa. One can then introduce creation and annihilation operators for pairs:

$$b_{\vec{k}} = c(-\vec{k}, \downarrow) c(\vec{k}, \uparrow) \text{ and } b_{\vec{k}}^* = c(\vec{k}, \uparrow)^* c(-\vec{k}, \downarrow)^*. \qquad (35.10)$$

They satisfy a mixed set of commutation relations:

$$\begin{cases} [b_{\vec{k}}, b_{\vec{k}'}^*] = [1 - n(\vec{k}, \uparrow) - n(-\vec{k}, \downarrow)] \delta_{\vec{k}, \vec{k}'}, \\ [b_{\vec{k}}, b_{\vec{k}'}] = 0, \\ \{b_{\vec{k}}, b_{\vec{k}'}\} = 2 b_{\vec{k}} b_{\vec{k}'} (1 - \delta_{\vec{k}, \vec{k}'}) \end{cases} \qquad (35.11)$$

(Here the curly brackets $\{,\}$ denote the **anti-commutator**: see equation (B.44)). In terms of these operators the **BCS Hamiltonian** becomes

$$H_{\text{BCS}} = 2 \sum_{\vec{k}} \epsilon(\vec{k}) b_{\vec{k}}^* b_{\vec{k}} - \sum_{\vec{k},\vec{k}'} U_{\vec{k},\vec{k}'} b_{\vec{k}'}^* b_{\vec{k}} \qquad (35.12)$$

where we have introduced a minus sign in front of the interaction potential to indicate that it is attractive for a superconductor. To analyse this, Hamiltonian we make another simplifying assumption: as the main contribution to the interaction comes from electrons near the Fermi surface one can replace U by a constant on a small energy interval around the Fermi energy: $\epsilon(\vec{k}) \in (\epsilon_F - \epsilon_0, \epsilon_F + \epsilon_0)$ and by zero outside this interval. In taking the thermodynamic limit it is, moreover, important to realize that this constant must be inversely proportional to the number of available states, that is, to the volume: $U_{\vec{k},\vec{k}'} = \lambda_0/V$.

To simplify the analysis even further, we shall make an even more drastic simplification and assume that $\epsilon(\vec{k})$ is also constant and equals ϵ_F. This is called the **strong-coupling approximation** (SCA). Note that $\sum_{\vec{k}} b_{\vec{k}}^* b_{\vec{k}}$ and $\left(\sum_{\vec{k}} b_{\vec{k}}^*\right)\left(\sum_{\vec{k}'} b_{\vec{k}'}\right)$ commute. Our simple model approximation thus becomes a sum of two commuting terms:

$$H_{\text{SCA}} = 2\epsilon_F \sum_{\vec{k}} b_{\vec{k}}^* b_{\vec{k}} - \frac{\lambda_0}{V} \sum_{\vec{k},\vec{k}'} b_{\vec{k}'}^* b_{\vec{k}} \qquad (35.13)$$

where the sums are restricted to the range of $\vec{k}$ such that $\epsilon_F - \epsilon_0 < \epsilon(\vec{k}) < \epsilon_F + \epsilon_0$. The remaining part of the Hilbert space can be disregarded. (To be precise, the Hilbert space is a tensor product $\mathcal{H} = \mathcal{H}_{\text{in shell}} \otimes \mathcal{H}_{\text{off shell}}$, where the subscripts in shell and off shell refer to states $\vec{k}$ inside and outside the range $[\epsilon_F - \epsilon_0, \epsilon_F + \epsilon_0]$; and the Hamiltonian splits as follows: $H = H_{\text{SCA}} \otimes 1 + 1 \otimes H_{\text{free}}$.)

Note that both terms in equation (35.13) do not change the number of unpaired states. This means that a general configuration of electrons can be described as follows. Let N denote the total number of pair states inside the shell. Thus $2N/V$ is the local density of states. Then there are a certain number q of these states singly occupied, whereas the remaining $n = N - q$ are either empty or doubly occupied. The interaction only operates on the latter. We denote the corresponding Hilbert space $\mathcal{H}_n$. There are $\binom{N}{n}$ ways of choosing the singly occupied pairs and 2^q ways of choosing which state in each pair is occupied. Within the complementary space $\mathcal{H}_n$, where each pair is either unoccupied or doubly occupied, the commutation relations for $b_{\vec{k}}^*$ and $b_{\vec{k}'}$ simplify:

$$\begin{cases} [b_{\vec{k}}, b_{\vec{k}'}] = [b_{\vec{k}}, b_{\vec{k}'}^*] = 0 & \text{if } \vec{k} \neq \vec{k}', \\ \{b_{\vec{k}}, b_{\vec{k}}\} = 0, \quad \{b_{\vec{k}}, b_{\vec{k}}^*\} = 1. \end{cases} \qquad (35.14)$$

Remarkably, these operators can be represented in terms of spin-$\frac{1}{2}$ operators:

$$b_{\vec{k}} = \frac{1}{2}(\sigma_{\vec{k}}^{(x)} - i\sigma_{\vec{k}}^{(y)}) = \sigma_{\vec{k}}^{(-)}. \tag{35.15}$$

A simple computation shows that $b_{\vec{k}}^* b_{\vec{k}} = \frac{1}{2}(1 + \sigma_{\vec{k}}^{(z)})$ so that the Hamiltonian restricted to the space of paired electrons can be written as

$$H_{\text{SCA}} = \epsilon_F \sum_{k=1}^{n}(1 + \sigma_k^{(z)}) - \frac{\lambda_0}{V} \sum_{k,k'=1}^{n} \sigma_{k'}^{(+)}\sigma_k^{(-)}. \tag{35.16}$$

Since $\sigma_{k'}^{(+)}\sigma_k^{(-)} + \sigma_k^{(+)}\sigma_{k'}^{(-)} = \frac{1}{2}\vec{\sigma}_{k'} \cdot \vec{\sigma}_k - \frac{1}{2}\sigma_{k'}^{(z)}\sigma_k^{(z)} + \sigma_k^{(z)}\delta_{k,k'}$ we can rewrite equation (35.16) in the form

$$H_{\text{SCA}} = \epsilon_F \sum_{k=1}^{n}(1 + \sigma_k^{(z)}) - \frac{\lambda_0}{2V} \sum_{k=1}^{n} \sigma_k^{(z)}$$
$$- \frac{\lambda_0}{4V}\left[\left(\sum_{k=1}^{n}\vec{\sigma}_k\right)^2 - \left(\sum_{k=1}^{n}\sigma_k^{(z)}\right)^2\right]. \tag{35.17}$$

The operators $\left(\sum_{k=1}^{n}\vec{\sigma}_k\right)^2$ and $\sum_{k=1}^{n}\sigma_k^{(z)}$ commute and their respective eigenvalues are given by $4J(J+1)$ and $2m$, where $J = \frac{1}{2}, \frac{3}{2}, \dots, \frac{n}{2}$ if the number n of summands is odd and $J = 0, 1, \dots, \frac{n}{2}$ if n is even; and $m = -J, -J+1, \dots, J$.

We now want to compute the grand potential $\omega(\beta, \mu)$ (equation (24.13)). To do this we must study the degeneracy $c(n, J)$ of the eigenvalues $4J(J+1)$ for a given number n of summands. Indeed, the grand partition function can be written as a sum over states which involves these degeneracies:

$$\mathcal{Z}_V(\beta, \mu) = \sum_{q=0}^{N}\binom{N}{q}2^q e^{\beta\mu q}e^{-\beta\epsilon_F q}\sum_{J=0 \text{ or } \frac{1}{2}}^{(N-q)/2}\sum_{m=-J}^{J}c(N-q, J)$$
$$\times \exp\left(\beta(\mu - \epsilon_F)(N - q + 2m) + \frac{\beta m\lambda_0}{V} + \frac{\beta\lambda_0}{4V}(4J(J+1) - 4m^2)\right). \tag{35.18}$$

Here q is the number of pair states that is singly occupied, so that $n = N - q$ is the number of pair states that is either doubly occupied or not occupied. Note that $n + 2m$ counts the number of paired electrons, i.e. twice the number of doubly occupied pair states, since $1 + \sigma_k^{(z)} = 2b_k^* b_k$. As usual, we shall rewrite equation (35.18) as an integral with respect to a distribution function. In this case, we need three random variables corresponding to the three variables n, J and m. We define these random variables as follows

$$X_N = \frac{n}{N}, \qquad Y_N = \frac{2J}{N}, \qquad Z_N = \frac{2m}{N} \tag{35.19}$$

with their joint distribution given by

$$F_N(x,y,z) = 2^{-N} \sum_{n=0}^{[Nx]} 2^{-n} \binom{N}{n} \sum_{J=0 \text{ or } \frac{1}{2}}^{\frac{1}{2}([y]\wedge n)} \sum_{m=-J}^{J\wedge\frac{1}{2}[Nz]} c(n,J), \qquad (35.20)$$

where the square brackets $[\ ]$ denote the integer part. Note that this is a genuine distribution function because $\sum_{J=0 \text{ or } \frac{1}{2}}^{n/2}(2J+1)c(n,J) = 2^n$, the dimension of the Hilbert space $\mathcal{H}_n$. In terms of this distribution function, we have (with $\rho_0 = N/V$)

$$\mathcal{Z}_{V_N}(\beta,\mu) = 2^{2N} \int\int\int \exp\Big\{\beta N(\mu - \epsilon_F)(1+z) + \frac{1}{2}\beta\rho_0 z\lambda_0$$
$$+\frac{1}{4}\beta\rho_0\lambda_0 N\Big[y\Big(y+\frac{2}{N}\Big) - z^2\Big]\Big\} d^3 F_N(x,y,z) \qquad (35.21)$$

We now claim that the triple (X_N, Y_N, Z_N) satisfies the LDP.

Theorem 35.1 *The 3-dimensional random variables (X_N, Y_N, Z_N) with distribution function given by equation (35.20) satisfy the LDP with rate function defined as follows. Let the set $\Gamma \subset \mathbb{R}^3$ be given by*

$$\Gamma = \{(x,y,z) \in [0,1]^2 \times [-1,1] : y \leqslant x \text{ and } |z| \leqslant y\}. \qquad (35.22)$$

Then, if $(x,y,z) \in \Gamma$,

$$I(x,y,z) = I_0(2x-1) + xI_0(y/x) \qquad (35.23)$$

where I_0 denotes the usual Ising rate function,

$$I_0(x) = \frac{1}{2}(1+x)\ln(1+x) + \frac{1}{2}(1-x)\ln(1-x), \qquad (35.24)$$

and $I(x,y,z) = +\infty$ for $(x,y,z) \notin \Gamma$.

Note that I is independent of z once $z \in [-y,y]$. As before we shall not actually prove this theorem but merely compute the cumulant generating function. It is given by

$$C_N(r,s,t) = \frac{1}{N}\ln \int\int\int e^{N(rx+sy+tz)} d^3 F_N(x,y,z)$$
$$= \frac{1}{N}\ln\Big(2^{-N}\sum_{n=0}^{N}\binom{N}{n}2^{-n}\sum_{J=0 \text{ or } \frac{1}{2}}^{n/2}\sum_{m=-J}^{J} c(n,J)e^{rn+2sJ+2tm}\Big). \qquad (35.25)$$

We can estimate this using the following identities:

$$\text{Tr}\exp\Big[s\sum_{k=1}^{n}\sigma_k^{(z)}\Big] = \prod_{k=1}^{n}\text{Tr}\,e^{s\sigma_k^{(z)}} = (e^s + e^{-s})^n$$
$$= \sum_{J=0 \text{ or } \frac{1}{2}}^{n/2} c(n,J)\sum_{m=-J}^{J} e^{2ms}. \qquad (35.26)$$

It follows that for $s \geqslant 0$,

$$
C_N(r, s, t) \geqslant \frac{1}{N} \ln \left[\sum_{n=0}^{N} 2^{-N-n} \binom{N}{n} \sum_{J=0 \text{ or } \frac{1}{2}}^{n/2} c(n, J) \sum_{m=-J}^{J} e^{rn + 2m(s+|t|)} \right]
$$

$$
= \ln \left\{ \frac{1}{2} \left[1 + e^r \cosh(s + |t|) \right] \right\} \tag{35.27}
$$

and on the other hand

$$
\sum_{m=-J}^{J} e^{2sJ} e^{2tm} \leqslant (2J+1) e^{2(s+|t|)J} \leqslant (2J+1) \sum_{m=-J}^{J} e^{2(s+|t|)m}
$$

and therefore

$$
C_N(r, s, t) \leqslant \frac{1}{N} \ln \left(\sum_{n=0}^{N} \binom{N}{n} 2^{-N-n} (n+1) e^{rn} \right.
$$

$$
\times \left. \sum_{J=0 \text{ or } \frac{1}{2}}^{n/2} c(n, J) \sum_{m=-J}^{J} e^{2m(s+|t|)} \right)
$$

$$
\leqslant \frac{1}{N} \ln(N+1) + \ln \left\{ \frac{1}{2} \left(1 + e^r \cosh(s + |t|) \right) . \right\}
$$

The upper and lower bound together imply that for $s \geqslant 0$,

$$
C(r, s, t) = \lim_{N \to \infty} C_N(r, s, t) = \ln \left\{ \frac{1}{2} \left[1 + e^r \cosh(s + |t|) \right] \right\}. \tag{35.28}
$$

For $s < 0$ we simply remark that the inequality (35.27) is reversed: $C(r, s, t) \leqslant \ln[\frac{1}{2}(1 + e^r \cosh(s + |t|)]$. This implies in particular, taking $s \to -\infty$ and $t = \pm s$, that $I(x, y, z) = +\infty$ if $y < |z|$. Similarly, $I(x, y, z) = +\infty$ if $x < 0$. The remaining analysis only involves $s \geqslant 0$ and differentiation shows that (35.23) holds.

Given that (X_N, Y_N, Z_N) satisfies the LDP we can write the grand potential as follows:

$$
\omega(\beta, \mu) = -\frac{\rho_0}{\beta} \sup_{(x,y,z) \in \Gamma} \left[\beta(\mu - \epsilon_F)(1 + z) + \frac{1}{4} \beta \rho_0 \lambda_0 (y^2 - z^2) \right.
$$

$$
\left. - I(x, y, z) + \ln 4 \right]. \tag{35.29}
$$

(The terms $\frac{1}{2} \beta \rho_0 \lambda_0 z$ and $\frac{1}{2} \beta \rho_0 \lambda_0 y$ are irrelevant because they are not proportional to N.) This variational problem can be solved as follows. We first introduce new variables

$$
\tilde{y} = \frac{y}{x} \text{ and } \tilde{z} = \frac{z}{y}
$$

so that we have to maximize the expression

$$G(x, \tilde{y}, \tilde{z}) = \beta \tilde{\mu} x \tilde{y} \tilde{z} + \frac{1}{4} \beta \lambda x^2 \tilde{y}^2 (1 - \tilde{z}^2) - I_0(2x - 1) - x I_0(\tilde{y}) \qquad (35.30)$$

where we have also denoted $\lambda = \lambda_0 \rho_0$ and $\tilde{\mu} = \mu - \epsilon_F$. Differentiating with respect to $\tilde{z}$ we get

$$\frac{\partial G}{\partial \tilde{z}} = \beta \tilde{\mu} x \tilde{y} - \frac{1}{2} \beta \lambda x^2 \tilde{y}^2 \tilde{z}. \qquad (35.31)$$

This has a zero at

$$\tilde{z} = z_0 = \frac{2\tilde{\mu}}{\lambda x \tilde{y}}, \qquad (35.32)$$

where it attains its maximum provided that $|z_0| \in [-1, 1]$. Otherwise, the maximum is attained at $\tilde{z} = \operatorname{sgn}(\tilde{\mu})$. In the latter case we get, upon differentiation with respect to $\tilde{y}$,

$$\frac{\partial G(x, \tilde{y}, \pm 1)}{\partial \tilde{y}} = \beta |\tilde{\mu}| x - x I_0'(\tilde{y}),$$

and hence

$$\tilde{y} = \tanh(\beta |\tilde{\mu}|). \qquad (35.33)$$

Differentiation with respect to x yields

$$\frac{\partial G(x, \tilde{y}, \pm 1)}{\partial x} = \beta |\tilde{\mu}| \tilde{y} - 2 I_0'(2x - 1) - I_0(\tilde{y}).$$

After inserting equation (35.33) we obtain

$$x = \frac{\cosh(\beta |\tilde{\mu}|)}{1 + \cosh(\beta |\tilde{\mu}|)}. \qquad (35.34)$$

It follows that $|z_0| \geqslant 1$ provided $\beta \leqslant \beta_c$, where

$$\boxed{\beta_c = \frac{2}{\tilde{\mu}} \tanh^{-1}\left(\frac{2\tilde{\mu}}{\lambda}\right)} \qquad (35.35)$$

(In the strong-coupling approximation, $\lambda \gg |\tilde{\mu}|$. In the limit $\tilde{\mu} \to 0$ this tends to $\beta_c \approx 4/\lambda = 4/\rho_0 \lambda_0$. The density of states per unit volume ρ_0 is proportional to $M^{-1/2}$ where M is the atomic mass. To see this, Bardeen, Cooper, and Schrieffer write $\rho_0 = 2\hbar\omega \mathcal{N}(0)/V$, where $\mathcal{N}(0)$ is the density of states per unit energy and $2\hbar\omega = 2\epsilon_0$ is the width of the shell around the Fermi sphere. Then ω is proportional to $M^{-1/2}$ as can be seen in example 18.1.)

If $\beta > \beta_c$ we are in the case $|z_0| < 1$ and we get

$$\frac{\partial G(x, \tilde{y}, z_0)}{\partial \tilde{y}} = \beta |\tilde{\mu}| x z_0 + \frac{1}{2} \beta \lambda x^2 \tilde{y} (1 - z_0^2) - x I_0'(\tilde{y})$$

$$= \frac{1}{2} \beta \lambda x^2 \tilde{y} - x \tanh^{-1}(\tilde{y}).$$

This yields the equation

$$\tilde{y} = \tanh\left(\frac{1}{2}\beta\lambda x\tilde{y}\right). \tag{35.36}$$

Also,

$$\begin{aligned}
\frac{\partial G(x, \tilde{y}, z_0)}{\partial x} &= \beta\tilde{\mu}\tilde{y}z_0 + \frac{1}{2}\beta\lambda x\tilde{y}^2(1 - z_0^2) - \ln\left(\frac{x}{1-x}\right) - I_0(\tilde{y}) \\
&= \frac{2\beta\tilde{\mu}^2}{\lambda x} + \frac{1}{2}\beta\lambda x\tilde{y}^2 - \frac{2\beta\tilde{\mu}^2}{\lambda x} - \ln\left(\frac{x}{1-x}\right) - I_0(\tilde{y}) \\
&= \ln\left(\frac{1-x}{x}\right) - \frac{1}{2}\ln(1 - \tilde{y}^2),
\end{aligned}$$

where we used equation (35.36), and hence

$$x = \frac{1}{1 + \sqrt{1 - \tilde{y}^2}}. \tag{35.37}$$

Inserting this into equation (35.36) we get the following implicit equation for $\tilde{y}$:

$$\tanh^{-1}(\tilde{y}) = \frac{1}{2}\beta\lambda\frac{\tilde{y}}{1 + \sqrt{1 - \tilde{y}^2}}. \tag{35.38}$$

It is not difficult to see (figure 35.5) that this equation has a unique positive solution if $\beta\lambda > 4$. Indeed, both sides are increasing functions of $\tilde{y}$. The derivative of the difference equals

$$\frac{1}{1 - \tilde{y}^2}\left(1 - \frac{\beta\lambda\sqrt{1 - \tilde{y}^2}}{2(1 + \sqrt{1 - \tilde{y}^2})}\right).$$

This is positive if $\beta\lambda < 4$, in which case the only solution of equation (35.38) is $\tilde{y} = 0$. If $\beta\lambda > 4$ the derivative is first negative then positive. This means that for small values of $\tilde{y}$, the left-hand side of (35.38) is less than the right-hand side and for larger values the left-hand side is larger than the right-hand side because the left-hand side tends to $+\infty$ as $\tilde{y} \to 1$. It also means that the solution for $\tilde{y}$ increases as $\beta\lambda$ increases. However, this solution only corresponds to the maximum of G if $|z_0| < 1$, i.e. $\lambda x\tilde{y} > 2|\tilde{\mu}|$. This obviously happens once $\beta > \beta_c$.

Inserting the maximizing values for x, $\tilde{y}$, and $\tilde{z}$ into equation (35.29) we find that

$$\omega(\beta, \mu) = \rho_0\left\{\epsilon_F - \mu - \frac{1}{\beta}\ln[2(1 + \cosh(\beta|\mu - \epsilon_F|))]\right\} \tag{35.39}$$

for $\beta \leqslant \beta_c$ and

$$\omega(\beta, \mu) = \rho_0\left\{\epsilon_F - \mu - \frac{(\mu - \epsilon_F)^2}{\lambda} + \frac{1}{4}\lambda(2x - 1) + \frac{1}{\beta}\ln\left(\frac{1-x}{2}\right)\right\} \tag{35.40}$$

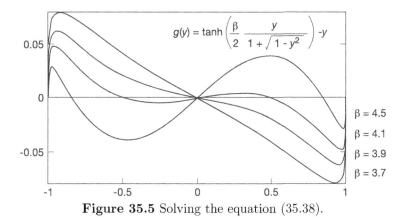

Figure 35.5 Solving the equation (35.38).

for $\beta > \beta_c$. Differentiating, we get the particle density and the entropy density (see equations (24.16) and (24.17)). The first is easy:

$$\rho(\beta, \mu) = -\frac{\partial \omega}{\partial \mu} = \begin{cases} \rho_0 \left(1 + \tanh\left[\frac{1}{2}\beta(\mu - \epsilon_F)\right]\right) & \text{if } \beta \leqslant \beta_c \\ \rho_0 \left(1 + \frac{2(\mu - \epsilon_F)}{\lambda}\right) & \text{if } \beta > \beta_c. \end{cases} \tag{35.41}$$

For the entropy density $\tilde{s}$, we have by equation (24.17),

$$\tilde{s}(\beta, \mu) = k_B \beta^2 \frac{\partial \omega}{\partial \beta} = k_B \beta \left(\frac{\partial}{\partial \beta}(\beta \omega(\beta, \mu)) - \omega(\beta, \mu)\right).$$

The first term is easily evaluated from equation (35.29) because the implicit derivatives cancel, so that

$$\tilde{s}(\beta, \mu) = k_B \rho_0 [\ln 4 - I(x, y, z)] \tag{35.42}$$

where (x, y, z) is the maximizer in equation (35.29). Hence we obtain

$$\tilde{s}(\beta, \mu) =$$
$$\begin{cases} k_B \rho_0 \left\{\ln\left[2(1 + \cosh[\beta(\mu - \epsilon_F)])\right] - \beta(\mu - \epsilon_F)\tanh[\frac{1}{2}\beta(\mu - \epsilon_F)]\right\} & \text{if } \beta \leqslant \beta_c \\ -\frac{1}{2}k_B \rho_0 \beta \lambda(2x - 1) - k_B \rho_0 \ln\left(\frac{1-x}{2}\right) & \text{if } \beta > \beta_c \end{cases}$$
$$\tag{35.43}$$

To derive the heat capacity we first need to differentiate x implicitly with respect to β. Note that by equation (35.37),

$$\frac{dx}{d\tilde{y}} = \frac{x^3 \tilde{y}}{1 - x}$$

and

$$\text{sech}^2(\frac{1}{2}\beta \lambda x \tilde{y}) = 1 - \tilde{y}^2.$$

This yields

$$\left[1 - \frac{1}{2}\beta\lambda(1 - x)\right]\frac{d\tilde{y}}{d\beta} = \frac{1}{2}\lambda x\tilde{y}(1 - \tilde{y}^2).$$

Using the fact that $x^2\tilde{y}^2 = 2x - 1$ it follows that

$$\left[1 - \frac{1}{2}\beta\lambda(1 - x)\right]\frac{dx}{d\beta} = \frac{1}{2}\lambda(1 - x)(2x - 1).$$

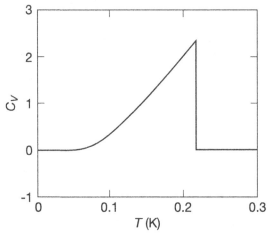

Figure 35.6 The specific heat of the BCS model in the SCA.

We can now obtain the heat capacity by means of the following formula:

$$c_V = T\left(\frac{\partial\tilde{s}}{\partial T}\right)_\rho = -\beta\left[\left(\frac{\partial\tilde{s}}{\partial\beta}\right)_\mu - k_B\beta^2\frac{(\partial\rho/\partial\beta)_\mu^2}{(\partial\rho/\partial\mu)_\beta}\right]. \tag{35.44}$$

This yields

$$c_V = \begin{cases} 0 & \text{if } \beta \leqslant \beta_c \\ \frac{1}{4}k_B\rho_0\beta^2\lambda^2\dfrac{(1 - x)(2x - 1)}{1 - \frac{1}{2}\beta\lambda(1 - x)} & \text{if } \beta > \beta_c. \end{cases} \tag{35.45}$$

Figure 35.6 shows a plot of c_V/k_B, where we arbitrarily put $\lambda = 1$ and $\tilde{\mu} = 0.3$. These have to be considered temperature units, so this corresponds to $\lambda \approx 8.6 \times 10^{-5}$ eV.

Note the similarity of figure 35.6 to figure 35.3. The linear part for $T > T_c$ is missing owing to the strong-coupling limit. This part is the normal metallic specific heat (31.12) (see figure 31.3).

The strong-coupling limit of the BCS model was first solved by Thouless (1960). A large-deviation treatment of the latter was given by Cegła *et al.* (1988) and was extended to the full BCS model by Duffield and Pulé (1988).

CHAPTER 36

Mayer Expansion for a Classical Gas

In chapter 28, we remarked that for high temperatures, the free energy density of an Ising model always behaves like $-k_B T \ln(2)$. Here $\ln(2)$ is the logarithm of the number of values the spin variable can take. In fact, we can develop the free energy density in terms of an expansion valid for high temperatures. In this chapter, we first consider a similar expansion for classical gases due to J. R. Mayer (1937). This expansion is particularly useful because it converges absolutely for low densities. The first proof of convergence is due to Groeneveld (1962) for non-negative potentials, and Penrose (1963) and Ruelle (1963) for more general potentials. See also the book of Ruelle (1969).

In chapter 30, we wrote the energy of a lattice gas in the form (30.1):

$$E_\Lambda(\{n_x\}_{x \in \Lambda}) = \sum_{\{x,y\} \subset \Lambda} \mathcal{V}(x - y) n_x n_y. \tag{36.1}$$

The corresponding grand-canonical partition function is

$$Z_\Lambda(\beta, \mu) = \sum_{\{n_x\}_{x \in \Lambda}} e^{\beta \mu \sum_{x \in \Lambda} n_x} e^{-\beta E_\Lambda(\{n_x\}_{x \in \Lambda})}. \tag{36.2}$$

In this chapter, we consider instead the analogous classical gas in the continuum, the energy of which is given by

$$E_N(x_1, \ldots, x_N) = \sum_{1 \leqslant i < j \leqslant N} \phi(x_i - x_j), \tag{36.3}$$

where N is the number of particles and the particle positions $x_1, \ldots, x_N$ are arbitrary points in a domain $\mathcal{R}$ in $\mathbb{R}^\nu$. ϕ is the interaction potential, which normally tends to infinity at short distances, and to 0 at long distances. (See figure 13.1.) The corresponding grand-canonical partition function in a bounded region $\mathcal{R}$ of $\mathbb{R}^\nu$ is given by the sum of multiple integrals

$$Z_\mathcal{R}(\beta, \mu) = \sum_{N=0}^{\infty} \frac{e^{\beta\mu N}}{N!} \int_\mathcal{R} \mathrm{d}^\nu x_1 \cdots \int_\mathcal{R} \mathrm{d}^\nu x_N e^{-\beta E_N(x_1,\dots,x_N)}, \qquad (36.4)$$

where there is a factor $1/N!$ because of equivalence of particles.

REMARK 36.1: *Classical Statistics*

As remarked in chapter 23, indistinguishability of microscopic particles is in fact a quantum-mechanical phenomenon. This means that the factor $1/N!$ in equation (36.4) cannot be justified classically. Before the advent of quantum mechanics it was simply put in by hand, which was a matter for considerable debate at the time.

Note also that we omitted the kinetic energy in equation (36.3). However, the resulting expression for the grand potential can be easily adjusted because the kinetic energy only depends on the particle momenta, whereas the potential energy E_N only depends on the particle positions, and in a classical partition function the momenta are integrated over independently of the positions. The result is simply an additional factor $1/\lambda_T^3$ as in equation (23.37).

The grand-canonical free energy is as usual defined by

$$\omega(\beta, \mu) = -\lim_{l\to\infty} \frac{1}{\beta V_l} \ln Z_{\mathcal{R}_l}(\beta, \mu). \qquad (36.5)$$

Here the limit is taken over a sequence of increasing regions $\mathcal{R}_l$ with volume V_l. We define the **fugacity** of the gas by $z = e^{\beta\mu}$. Mayer found a clever way to formally take the logarithm of $Z_\mathcal{R}$ in terms of a series expansion in powers of z. He defined

$$e^{-\beta\phi(x-y)} - 1 = f_\beta(x, y), \qquad (36.6)$$

and expanded the product

$$\prod_{1\leqslant i<j\leqslant N} e^{-\beta\phi(x_i-x_j)} = \prod_{1\leqslant i<j\leqslant N} (f_\beta(x_i, x_j) + 1)$$

$$= \sum_{p=1}^{N} \sum_{\{I_k\}_{k=1}^p \in \Pi_p(N)} \prod_{k=1}^{p} \left(\sum_{\Gamma\in G_c(I_k)} \prod_{(i,j)\in\Gamma} f_\beta(x_i, x_j) \right). \qquad (36.7)$$

Here, $\Pi_p(N)$ is the set of partitions of $\{1,\dots,N\}$ into p subsets $I_1,\dots,I_p$, and $G_c(I_k)$ is the set of connected graphs on I_k, where $(i,j)\in\Gamma$ denotes a line of the graph between vertices $i,j\in I_k$.

To understand this expression, consider the case $N=4$. Expanding the product $\prod_{1\leqslant i<j\leqslant N}(f_\beta(x_i, x_j) + 1)$, we then obtain $2^6 = 64$ terms. There is one term equal 1. This corresponds to the case $p=4$ where each $|I_k|=1$, the only graphs having no lines. Then there are 6 terms with a single $f_\beta(x_i, x_j)$.

This corresponds to taking $p = 3$ with one I_k having two points, the other two only one point. There are $\binom{6}{2} = 15$ terms with two factors f_β. These split into two categories: either they have a point in common, or they are disjoint. There are 3 disjoint terms. They correspond to the case $p = 2$ with $|I_1| = |I_2| = 2$. The terms with a point in common correspond to the case where $p = 2$ and $|I_1| = 3$ and $|I_2| = 1$, where the graph Γ on I_1 consists of two lines. There are 12 such terms: we can choose the single point in I_2 in 4 ways, and the double point in 3 ways. There are $\binom{6}{3} = 20$ terms with 3 factors f_β. Of these, 4 consist of factors where each point x_i occurs twice, whereas one point does not occur. These correspond to the complete graphs on 3 points, with one point unconnected, i.e. $p = 2$ with $|I_1| = 3$ and $|I_2| = 1$. The other 16 cases correspond to $p = 1$ with graphs connecting all 4 points by 3 lines. The terms with more than 3 factors contain factors f_β involving all 4 points and correspond to $p = 1$ with connected graphs on the 4 vertices of I_1 with 4, 5 or 6 lines.

In general, the subdivision $\{I_k\}_{k=1}^p$ corresponds to the connected components of the points $x_1, \ldots, x_N$ by lines given by factors $f_\beta(x_i, x_j)$.

Now note that, upon integrating over $x_1, \ldots, x_N$, we can perform the integrals separately over the subsets I_k and the corresponding factors

$$\sum_{\Gamma \in G_c(I_k)} \int_{\mathcal{R}} \cdots \int_{\mathcal{R}} \prod_{(i,j) \in \Gamma} f_\beta(x_i, x_j) \prod_{i \in I_k} d^\nu x_i$$

only depend on the number of points in I_k. The number of partitions of $\{1, \ldots, N\}$ with given sizes $|I_k| = n_k$ is given by $\dfrac{1}{p!} \dfrac{N!}{\prod_{k=1}^p n_k!}$, so

$$\int_{\mathcal{R}} d^\nu x_1 \cdots \int_{\mathcal{R}} d^\nu x_N \prod_{1 \leqslant i < j \leqslant N} e^{-\beta\phi(x_i - x_j)} = \sum_{p=1}^N \frac{1}{p!} \sum_{n_1, \ldots, n_p \geqslant 1: \sum n_k = N}$$

$$\times \frac{N!}{\prod_{k=1}^p n_k!} \prod_{k=1}^p \left(\sum_{\Gamma \in G_c(\{1, \ldots, n_k\})} \int_{\mathcal{R}} d^\nu x_1 \cdots \int_{\mathcal{R}} d^\nu x_{n_k} \prod_{(i,j) \in \Gamma} f_\beta(x_i, x_j) \right).$$

$$(36.8)$$

Summing over N, we have

$$Z_{\mathcal{R}}(\beta, \mu) = 1 + \sum_{N=1}^\infty \frac{z^N}{N!} \sum_{p=1}^N \frac{1}{p!} \sum_{n_1, \ldots, n_p \geqslant 1: \sum n_k = N} \frac{N!}{\prod_{k=1}^p n_k!}$$

$$\times \prod_{k=1}^p \left(\sum_{\Gamma \in G_c(\{1, \ldots, n_k\})} \int_{\mathcal{R}} d^\nu x_1 \cdots \int_{\mathcal{R}} d^\nu x_{n_k} \prod_{(i,j) \in \Gamma} f_\beta(x_i, x_j) \right).$$

Interchanging the sums over p and N, we can sum over each n_k individually up to infinity to obtain

$$Z_{\mathcal{R}}(\beta, \mu) =$$

$$= 1 + \sum_{p=1}^{\infty} \frac{1}{p!} \left(\sum_{n=1}^{\infty} \frac{z^n}{n!} \sum_{\Gamma \in G_c(\{1,\ldots,n\})} \int_{\mathcal{R}} d^\nu x_1 \cdots \int_{\mathcal{R}} d^\nu x_n \prod_{(i,j) \in \Gamma} f_\beta(x_i, x_j) \right)^p$$

$$= \exp \left(\sum_{n=1}^{\infty} \frac{z^n}{n!} \sum_{\Gamma \in G_c(\{1,\ldots,n\})} \int_{\mathcal{R}} d^\nu x_1 \cdots \int_{\mathcal{R}} d^\nu x_n \prod_{(i,j) \in \Gamma} f_\beta(x_i, x_j) \right).$$
(36.9)

The expressions

$$\psi(x_1, \ldots, x_n) = \sum_{\Gamma \in G_c(\{1,\ldots,n\})} \prod_{(i,j) \in \Gamma} f_\beta(x_i, x_j) \qquad (36.10)$$

are called **Ursell functions** (Ursell (1927)). In general, they are defined recursively by the relations

$$e^{-\beta E_N(x_1,\ldots,x_N)} = \sum_{p=1}^{N} \sum_{\{I_k\}_{k=1}^p \in \Pi_p(N)} \prod_{k=1}^{p} \psi(x_{I_k}), \qquad (36.11)$$

starting with $\psi(x) = 1$, and writing $x_I = \{x_i\}_{i \in I}$ for a subset $I \subset \{1, \ldots, N\}$. In terms of these Ursell functions, we therefore have

$$\ln Z_{\mathcal{R}}(\beta, \mu) = \sum_{n=1}^{\infty} \frac{z^n}{n!} \int_{\mathcal{R}} d^\nu x_1 \cdots \int_{\mathcal{R}} d^\nu x_n \, \psi(x_1, \ldots, x_n) \qquad (36.12)$$

and in the thermodynamic limit,

$$\omega(\beta, \mu) = -\frac{1}{\beta} \sum_{n=1}^{\infty} b_n(\beta) z^n,$$

$$b_n(\beta) = \lim_{l \to \infty} \frac{1}{V_l \, n!} \int_{\mathcal{R}_l} d^\nu x_1 \cdots \int_{\mathcal{R}_l} d^\nu x_n \, \psi(x_1, \ldots, x_n). \qquad (36.13)$$

This is the **Mayer expansion**.

Note that for the free (ideal) gas, i.e. if there is no interaction, $\phi(x) = 0$ and hence $f_\beta(x_i, x_j) = 0$, and the only contribution is from $n = 1$. Thus $\ln Z_{\mathcal{R}} = z V$ and therefore,

$$\omega_{IG}(\beta, \mu) = -\frac{z}{\beta} = -\frac{e^{\beta \mu}}{\beta}. \qquad (36.14)$$

This is just the ideal gas law: see equation (23.37). (The factor $1/\lambda_T^3$ is due to the kinetic energy, which we have ignored in the energy function E_N of equation (36.3). For a classical gas, it yields a factor $1/\lambda_T^3$ independent of the interaction: see remark 36.1.)

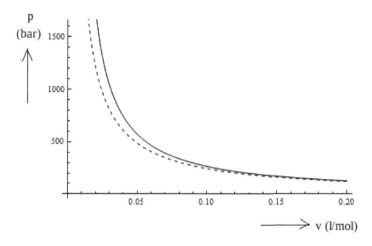

Figure 36.1. The corrected pressure of helium at $T = 293$ K compared to the ideal gas law (dashed).

The first correction to the ideal gas law is given by $-\frac{1}{\beta}b_2(\beta)z^2$ where

$$
\begin{aligned}
b_2(\beta) &= \lim_{l\to\infty} \frac{1}{2V_l} \int_{\mathcal{R}_l} d^\nu x_1 \int_{\mathcal{R}_l} d^\nu x_2 \, (e^{-\beta\phi(x_1-x_2)} - 1) \\
&= \frac{1}{2} \int_{\mathbb{R}^\nu} d^\nu x \, (e^{-\beta\phi(x)} - 1).
\end{aligned}
\tag{36.15}
$$

Hence

$$
w(\beta, \mu) \approx -\frac{1}{\beta}z(1 + b_2(\beta)z) + O(z^3).
\tag{36.16}
$$

In case the potential ϕ only depends on the distance $|x_1-x_2|$, we can integrate out the angle variables. In $\nu = 3$ dimensions, we get

$$
b_2(\beta) = 2\pi \int_0^\infty (e^{-\beta\phi(r)} - 1)\, r^2 \, dr.
\tag{36.17}
$$

For example in helium, the interaction potential is well approximated by the Lennard-Jones potential: see equation (13.9) and figure 13.1. With $n = 12$ and $m = 6$ one has for helium, $a = 2\epsilon_0 r_0^6$ and $b = \epsilon_0 r_0^{12}$ where $-\epsilon_0$ is the minimum of the potential, attained at $r = r_0$, with $\epsilon_0/k_B = 10.2$ K, and

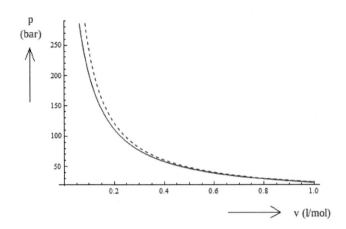

Figure 36.2. The corrected pressure of oxygen at $T = 293$ K
compared to the ideal gas law (dashed).

$r_0 = 2.65$ Å. Differentiating (36.16), we have,

$$\rho = -\frac{\partial \omega}{\partial \mu} = -\beta z \frac{\partial \omega}{\partial z} = z(1 + 2b_2(\beta)z). \tag{36.18}$$

For high temperatures $b_2 < 0$ and this identity only makes sense provided
$\rho < (8|b_2(\beta)|)^{-1}$, but in any case, the expansion only makes sense for small
values of ρ. At room temperature, $b_2 \approx -14$ Å^3 per atom, i.e. $b_2 = -8.4$ ml
mol^{-1}, so we need $v > 67$ ml mol^{-1}. For small ρ, $z \approx \rho(1 - 2b_2(\beta)\rho)$, and we
have

$$p(v, T) \approx \frac{k_B T}{v}\left(1 - \frac{b_2(\beta)}{v}\right). \tag{36.19}$$

Figure 36.1 shows a comparison with the ideal gas law at $T = 293$ K.

For oxygen, $r_0 = 3.88$ Å, and $\epsilon_0/k_B = 118$ K. The corresponding b_2 is
positive: at $T = 293$ K, $b_2 = 27.7$ Å^3. The resulting pressure is therefore
slightly lower than the ideal-gas pressure.

We now want to show that the Mayer expansion converges absolutely for
small values of the fugacity. We consider the case of positive potentials, and
follow Penrose (1967). Given a connected graph $\Gamma \in G_c(\{1, \dots, n\})$, we define
an associated tree graph as follows. We first assign a *weight* w_i to each vertex
in the graph given by the number of links in a shortest path from vertex i to
vertex 1. Then we delete all bonds from Γ between vertices of equal weight as
well as all bonds from a vertex $i \neq 1$ to vertices with weight $w_i - 1$ other than
the one with lowest index. This does not affect the weights of the vertices.
We claim that the resulting graph is a tree T on $\{1, \dots, n\}$. Indeed, there is

a unique path in T from any vertex i to 1, because there is a unique bond from i to a vertex of weight $w_i - 1$, and by induction a unique path to a vertex of weight $w_i - k$ for any $k \leqslant w_i$. The only vertex of weight 0 is 1. Conversely, given a tree T on $\{1, \dots, n\}$, we can construct the set $S(T)$ of connected graphs which reduce to this tree by the above procedure by joining some of the pairs of equal weight in T by additional links, and by joining some of the vertices $i \neq 1$ to vertices j of weight $w_j = w_i - 1$ (other than the one it was already connected to in T). The maximal such graph we call T^*. The total number of graphs in $S(T)$ is then 2^L, where L is the number of links in T^* not in T. This procedure is illustrated in figure 36.3. The dashed lines are the lines in $T^* - T$.

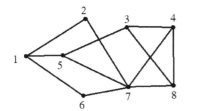 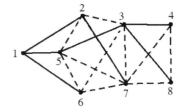

Figure 36.3. A connected graph on 8 vertices and the reduced tree graph.

We can now rewrite the Ursell functions (36.10) as follows

$$\psi(x_1, \dots, x_n) = \sum_{T \in \mathcal{T}(\{1,\dots,n\})} \prod_{(i,j) \in T} f_\beta(x_i, x_j) \sum_{\Gamma \in S(T)} \prod_{(i,j) \in \Gamma - T} f_\beta(x_i, x_j)$$

$$= \sum_{T \in \mathcal{T}(\{1,\dots,n\})} \prod_{(i,j) \in T} f_\beta(x_i, x_j) \prod_{(i,j) \in T^* - T} (1 + f_\beta(x_i, x_j)). \quad (36.20)$$

For positive potentials $\phi(x_i - x_j) \geqslant 0$, we have $1 + f_\beta(x_i, x_j) = e^{-\beta\phi(x_i, x_j)} \leqslant 1$ and hence

$$|\psi(x_1, \dots, x_n)| \leqslant \sum_{T \in \mathcal{T}(\{1,\dots,n\})} \prod_{(i,j) \in T} |f_\beta(x_i, x_j)| \qquad (\phi(x) \geqslant 0). \quad (36.21)$$

According to **Cayley's theorem**, the number of tree graphs on n vertices is given by n^{n-2}, so upon integration, we find that (writing $f_\beta(x) = f_\beta(0, x)$)

$$|b_n(\beta)| \leqslant \frac{1}{n!} \lim_{l \to \infty} \frac{1}{V_l} \int_{\mathcal{R}_l} d^\nu x_1 \cdots \int_{\mathcal{R}_l} d^\nu x_n \, |\psi(x_1, \dots, x_n)|$$

$$\leqslant \frac{1}{n!} n^{n-2} \left(\int_{\mathbb{R}^\nu} |f_\beta(x)| \, d^\nu x \right)^{n-1}. \quad (36.22)$$

Here we assume that the integral

$$C(\beta) = \int_{\mathbb{R}^\nu} |f_\beta(x)| \, d^\nu x \quad (36.23)$$

converges: this is the case if $\phi(x)$ decays at least as fast as $|x|^{-\nu-\epsilon}$ for some $\epsilon > 0$. Using the Stirling bound $n! \geqslant n^n e^{-n}$ (see theorem 19.2) we conclude that the Mayer expansion converges absolutely provided

$$|z| \leqslant (C(\beta)e)^{-1}. \tag{36.24}$$

More generally, one says that a potential is **stable** if there is a constant $B \geqslant 0$ such that for all $n \in \mathbb{N}$ and all points $x_1, \ldots, x_n$,

$$\sum_{1 \leqslant i < j \leqslant n} \phi(x_i - x_j) \geqslant -Bn. \tag{36.25}$$

One can prove that the Lennard-Jones potential is stable.

This implies that

$$\prod_{(i,j) \in T^* - T} (1 + f_\beta(x_i, x_j)) \leqslant e^{\beta B(n-1)}$$

and convergence of the Mayer expansion holds provided

$$|z| \leqslant (C(\beta)Be)^{-1}. \tag{36.26}$$

CHAPTER 37

Cluster Expansion and Polymer Models

We now return to the case of a classical spin system as in chapter 25. The partition function (25.3) is given by

$$Z_\Lambda(\beta) = \sum_{\{s_x\}_{x\in\Lambda}} e^{-\beta E_\Lambda(\{s_x\})}, \tag{37.1}$$

where $\Lambda \subset \mathbb{Z}^\nu$. We introduce the relation (36.11), i.e.

$$e^{-\beta E_\Lambda(s_\Lambda)} = \sum_{p=1}^{|\Lambda|} \sum_{\{X_k\}_{k=1}^p \in \Pi_p(\Lambda)} \prod_{k=1}^p \psi(s_{X_k}), \tag{37.2}$$

to write

$$Z_\Lambda(\beta) = \sum_{p=1}^{|\Lambda|} \sum_{\{X_k\}_{k=1}^p \in \Pi_p(\Lambda)} \prod_{k=1}^p \sum_{\{s_x\}_{x\in X_k}} \psi(s_{X_k}). \tag{37.3}$$

Here, as before, we denote $s_X = \{s_x\}_{x\in X}$ for any $X \subset \mathbb{Z}^\nu$. In this case, the sums $\sum_{\{s_x\}_{x\in X_k}} \psi(s_{X_k})$ do not just depend on the size of the subsets X_k. Moreover, we have to deal with the condition that the sets X_k must not overlap. The case of a lattice gas was first analysed by Gallavotti and Miracle-Sole (1968). The general case was expressed in terms of a so-called polymer system by Gruber and Kunz (1971). This is explained below.

Let us first single out the term $\Psi(s_x)$ of (25.1). It is obvious from equation (37.2) that we can write

$$\psi(s_{X_k}) = \exp\left[-\beta \sum_{x\in X_k} \Psi(s_x)\right] \overline{\psi}(s_{X_k}), \tag{37.4}$$

where $\overline{\psi}(s_{X_k})$ are the Ursell functions corresponding to the energies E_Λ with $\Psi = 0$. In particular, $\overline{\psi}(s_x) = 1$. For the sets X_k with $|X_k| = 1$, we therefore have only a factor $\sum_{s_x=1}^{q} \exp[-\beta\Psi(s_x)]$ in equation (37.3). Extracting this factor for each $x \in \Lambda$, we must compensate the factors with $|X_k| > 1$. Thus we can write

$$Z_\Lambda(\beta) = \left(\sum_{s=1}^{q} e^{-\beta\Psi(s)} \right)^{|\Lambda|} \overline{Z}_\Lambda(\beta), \qquad (37.5)$$

where

$$\overline{Z}_\Lambda(\beta) = 1 + \sum_{X \subset \Lambda} \sum_{p=1}^{|X|} \sum_{\{X_k\}_{k=1}^{p} \in \Pi_p(X)} \prod_{k=1}^{p} K(X_k), \qquad (37.6)$$

and

$$K(X) = \begin{cases} 0 & \text{if } |X| = 1, \\ \int \overline{\psi}(s_X) \prod_{x \in X} dF_\beta(s_x) & \text{if } |X| > 1, \end{cases} \qquad (37.7)$$

where we have defined the single-spin distribution function F_β by

$$F_\beta(u) = \frac{\sum_{s \leqslant u} e^{-\beta\Psi(s)}}{\sum_{s=1}^{q} e^{-\beta\Psi(s)}}. \qquad (37.8)$$

The right-hand side of equation (37.6) defines the partition function of a so-called **polymer system**. These were first studied by Kunz and Gruber (1971). One can imagine the *polymers* X_k floating in the 'sea' Λ. The condition that they do not intersect corresponds to a hard-core repulsion $v(X_k, X_l) = 0$ if $X_k \cap X_l = \emptyset$ and $v(X_k, X_l) = +\infty$ if $X_k \cap X_l \neq \emptyset$. We can then introduce *polymer Ursell functions* ψ_{pol} by

$$e^{-\sum_{1 \leqslant k < l \leqslant p} v(X_k, X_l)} = \sum_{k=1}^{p} \sum_{\{I_j\}_{j=1}^{k} \in \Pi_k(p)} \psi_{\text{pol}}(X_{I_1}) \dots \psi_{\text{pol}}(X_{I_k}),$$

$$\left(X_{I_j} = \bigcup_{i \in I_j} X_i \right). \qquad (37.9)$$

Inserting this into equation (37.6) the sets X_k can be summed over individually, and we obtain, analogous to equation (36.12),

$$\ln \overline{Z}_\Lambda(\beta) = \sum_{n=1}^{\infty} \frac{1}{n!} \sum_{X_1 \subset \Lambda} \cdots \sum_{X_n \subset \Lambda} \prod_{k=1}^{n} K(X_k) \, \psi_{\text{pol}}(X_1, \dots, X_n). \qquad (37.10)$$

This is called a **cluster expansion**. We would now like to show that the free energy density given by the limit

$$\begin{aligned} \overline{f}(\beta) &= -\lim_{\Lambda \to \mathbb{Z}^\nu} \frac{1}{\beta|\Lambda|} \ln \overline{Z}_\Lambda(\beta) \\ &= -\frac{1}{\beta} \lim_{\Lambda \to \mathbb{Z}^\nu} \frac{1}{|\Lambda|} \sum_{n=1}^{\infty} \frac{1}{n!} \sum_{X_1, \dots, X_n \subset \Lambda} K(X_1) \dots K(X_n) \, \psi_{\text{pol}}(X_1, \dots, X_n) \end{aligned}$$

$$(37.11)$$

exists if the **polymer activities** $K(X)$ are small. Of course, we already know from theorem 25.1, that this limit exists. However, having a convergent series expansion allows us to conclude that $\overline{f}(\beta)$ is an analytic function of β and the parameters of the interaction Φ and Ψ.

We shall prove the following theorem.

Theorem 37.1 *Suppose that*

$$\sum_{X \subset \mathbb{Z}^\nu : 0 \in X} |K(X)| \, e^{|X|} < 1 \tag{37.12}$$

and for all $n \in \mathbb{N}$,

$$\lim_{\rho \to \infty} \sum_{\substack{X \subset \mathbb{Z}^\nu : 0 \in X \\ X \cap S_\rho^c \neq \emptyset}} |X|^{n-1} |K(X)| = 0, \tag{37.13}$$

where S_ρ is the cube of side $\rho > 0$ centred at the origin. Then the limit of the free energy (37.11) exists and converges uniformly in any region of parameters in which the bound (37.12) holds and the limit (37.13) converges uniformly.

The proof is long and technical and will be postponed until the end of this chapter.

Note that for $|X| > 1$, $K(X)$ is given by

$$K(X) = \sum_{\Gamma \in G_c(X)} \int \prod_{(x,y) \in \Gamma} (e^{-\beta \Phi_{x-y}(s_x, s_y)} - 1) \prod_{x \in X} \mathrm{d}F_\beta(s_x). \tag{37.14}$$

For fixed X this is small if $\left| e^{-\beta \Phi_{x-y}(s_x,s_y)} - 1 \right|$ is small and decays as $|x - y| \to \infty$. Assuming a finite-range interaction as in chapter 25, we have that $e^{-\beta \Phi_{x-y}(s_x,s_y)} - 1 = 0$ unless $|x - y| \leqslant R$. In the sum over connected graphs, we therefore only need to consider those for which each link $(x, y) \in \Gamma$ satisfies $|x - y| \leqslant R$. For each $x \in X$ there are therefore at most $V(R) = (2R+1)^\nu - 1$ possible links. Defining

$$\|\Phi\| = \max_{|x-y| \leqslant R} \max_{s_x, s_y=1}^{q} |\Phi_{x-y}(s_x, s_y)|, \tag{37.15}$$

we have for each link $(x, y) \in \Gamma$,

$$\left| e^{-\beta \Phi_{x-y}(s_x,s_y)} - 1 \right| \leqslant \beta \|\Phi\| \, e^{\beta \|\Phi\|}.$$

With Penrose's decomposition (36.20), we now need to estimate the number of contributing trees. If we denote the coordination numbers of the tree, i.e. the number of links at each point $x \in X$, by p_x then $\sum_{x \in X} p_x = 2(|X| - 1)$ since the number of links in the tree is $|X| - 1$. Moreover, $p_x \leqslant V(R)$ and so the possible number of trees is bounded by

$$\sum_{\{p_x\}:\, p_x \geqslant 1;\, \sum p_x = 2(|X|-1)} V(R) \prod_{x \in X} \binom{V(R)}{p_x - 1}$$

$$\leqslant V(R)^{|X|-1} \sum_{\{q_x\}:\, q_x \geqslant 0,\, \sum q_x = |X|-2} \prod_{x \in X} \frac{1}{q_x!}$$

$$= \frac{|X|^{|X|-2}}{(|X|-2)!} V(R)^{|X|-1} \leqslant e^{|X|} V(R)^{|X|-1}.$$

Here we used Stirling's formula $n! \geqslant n^n \, e^{-n}$. There is a factor $\beta||\Phi||e^{\beta||\Phi||}$ for each link in the tree. The sum over the graphs $\gamma \in S(T)$ of the lines in $T^* \setminus T$ is bounded by $e^{\beta||\Phi||(V(R)-1)|X|}$ since the number of links in $T^* \setminus T$ is at most $(V(R)-1)|X|$. Thus we obtain

$$|K(X)| \leqslant (\beta||\Phi||V(R))^{|X|-1} e^{(1+\beta||\Phi||V(R))\,|X|}. \tag{37.16}$$

Moreover, $K(X) = 0$ unless $\operatorname{diam}(X) \leqslant |X|\, R$. Therefore $X \cap S_\rho^c \neq \emptyset \implies |X| > \rho/R$ so that (37.13) follows from equation (37.12). To show that the activity $K(X)$ satisfies equation (37.12) we need a bound on the number of subsets X containing 0 contributing to the sum.

Lemma 37.1 *There exists a constant c_ν such that the number of connected subsets $X \subset \mathbb{Z}^\nu$ containing 0 of size n is bounded by $e^{c_\nu n}$. If the connectivity is defined by the maximum-norm, i.e. x and y are neighbours if $\max_{i=1}^\nu |x_i - y_i| = 1$, then the same holds but with a larger constant d_ν.*

Proof. This is similar to counting the number of contours containing 0 in chapter 28 (cf. equation (28.31)). Given a set $X \subset \mathbb{Z}^\nu$ of size n containing 0, we number its sites as follows. First fix an ordering of the unit vectors of $\mathbb{Z}^\nu$ from 1 to 2ν. Then let 0 have number 1 and suppose that the first k points of X have already been numbered. Choose the already numbered site with the lowest number that has still got a neighbour in X which has not been numbered. Assign the number $k+1$ to its unnumbered neighbour with lowest order. This defines a unique map from the sets X containing 0 with n sites to numberings of sites in $\mathbb{Z}^\nu$. Conversely, suppose the neighbours of the first $k-1$ points of X have already been determined. Then k has at most $2\nu - 1$ unfilled neighbours left, for which there are $2^{2\nu-1} - 1$ possible fillings. Therefore the possible number of choices is certainly bounded by $2^{(2\nu-1)n}$, i.e. $c_\nu \leqslant (2\nu - 1)\ln 2$. If points are also considered neighbours along a diagonal, we get similarly, $d_\nu \leqslant (3^\nu - 2)\ln 2$.   ∎

For an interaction of range R, we can decompose the lattice into boxes of size R. Then the boxes containing points of X must be touching. If X covers p boxes, we can distribute its n points over these boxes. If the ith box contains n_i points of X, there are $\binom{R^\nu}{n_i} \leqslant R^{n_i \nu}$ ways of choosing these points. And there are $\binom{n-1}{p-1}$ ways of assigning numbers of sites $n_1, \ldots, n_p$ to the p boxes. (Place $p-1$ dividing lines in $n-1$ positions.) Therefore, the number of possible sets X is bounded by

$$\sum_{p=1}^{n} \binom{n-1}{p-1} e^{d_\nu p} R^{n\nu} \leqslant ((1+e^{d_\nu}) R^\nu)^n. \tag{37.17}$$

Setting

$$\lambda = (1+e^{d_\nu}) R^\nu e^{2+\beta||\Phi|| V(R)} \tag{37.18}$$

we conclude, inserting equations (37.17) and (37.16) into (37.12), that

$$\sum_{X \ni 0} e^{|X|} |K(X)| \leqslant \sum_{n=2}^{\infty} \lambda^n (\beta||\Phi|| V(R))^{n-1} = \frac{\beta||\Phi|| V(R) \lambda^2}{1 - \beta||\Phi|| V(R)\lambda}. \tag{37.19}$$

This is < 1 if

$$\beta||\Phi|| V(R) < \frac{1}{\lambda(1+\lambda)}. \tag{37.20}$$

This is a very rough bound but nevertheless of theoretical importance. In particular, the uniform convergence means that the free energy is analytic in any of the interaction parameters (and also β) which implies that *for high temperatures there is no phase transition.*

Let us consider the case of nearest-neighbour interaction in more detail. The number of connected graphs on n points containing 0 and consisting of nearest-neighbour lines only, can be estimated directly in the same way as in the above lemma 37.1. At each point $x \in X$ (except 0 and the last) there are at most $2\nu - 1$ empty neighbours where we can put the next point of X. Filling p of these we have for each the choice to insert a connecting line. Therefore the number of graphs is bounded by

$$\sum_{p_1=1}^{2\nu} \binom{2\nu}{p_1} (2^{p_1} - 1) \left(\sum_{p=1}^{2\nu-1} \binom{2\nu - 1}{p} 2^p \right)^{n-2} < 3^{(2\nu-1)n}.$$

Taking into account that each graph contains at least $n-1$ lines and at most $n\nu$, we get

$$\sum_{X \ni 0} e^{|X|} |K(X)| \leqslant \sum_{n=2}^{\infty} (\beta||\Phi||)^{n-1} 3^{(2\nu-1)n} e^{\beta||\Phi||n\nu}.$$

This yields the bound (37.20) with $\lambda = 3^{2^\nu - 1} e^{1+\beta||\Phi||\nu}$. For the Ising model, we get

$$\beta J e^{\beta J\nu} \lesssim e^{-2(1+(2\nu-1)\ln 3)}.$$

In 2 dimensions, $\beta J < 0.002$. This unimpressive result can be improved, however: see problem III-34. One reason for the poor estimate is the bound (37.16).

Here we used $|e^{-\beta\Phi_{x-y}(s_x,s_y)} - 1| \leqslant e^{\beta||\Phi||} - 1$. However, in fact, in the integral (37.14) $\Phi_{x-y}(s_x,s_y)$ takes both signs and the odd powers of β cancel. Thus, for example, the term $|X| = 2$ is

$$
\begin{aligned}
K(\{x,y\}) &= \int\int (e^{\beta J s_x s_y} - 1)\, dF_\beta(s_x)dF_\beta(s_y) \\
&= \frac{\sum_{s_x,s_y=\pm 1} e^{\beta J s_x s_y + \beta H(s_x + s_y)}}{\left(\sum_{s=\pm 1} e^{\beta H s}\right)^2} - 1 \\
&= \frac{e^{\beta J}\cosh(2\beta H) + e^{-\beta J}}{2\cosh^2(\beta H)} - 1. \qquad (37.21)
\end{aligned}
$$

In particular for $H = 0$, this is $\cosh(\beta J) - 1 \approx \frac{1}{2}(\beta J)^2$. Summing over nearest-neighbour pairs $\{x,y\}$ we have in case $H = 0$,

$$
\begin{aligned}
\overline{f}(\beta) &\sim -\frac{1}{\beta}\lim_{l\to\infty}\frac{1}{|\Lambda_l|}\sum_{\{x,y\}\subset\Lambda_l} K(\{x,y\}) \\
&= -\frac{1}{\beta}(\cosh(\beta J) - 1)\nu \approx -\frac{1}{2}\beta J^2\nu. \qquad (37.22)
\end{aligned}
$$

This is the first-order correction to the free energy density. (Note that the pre-factor in equation (37.5) yields the term $-(\ln 2)/\beta$. The approximation $-\frac{1}{2}\beta J^2\nu$ can in fact also be obtained *formally* by expanding the exponential in the partition function to second order in βJ. The above derivation is rigorous for small β.

Proof of Theorem 37.1

We first prove the convergence of individual terms:

Lemma 37.2 *Suppose that (37.13) holds. Then*

$$
\lim_{l\to\infty}\frac{1}{|\Lambda_l|}\sum_{X_1,\dots,X_n\subset\Lambda_l} K(X_1)\dots K(X_n)\,\psi_{\text{pol}}(X_1,\dots,X_n) \qquad (37.23)
$$

exists.

Proof. We go about the proof in the same way as in theorem 25.1 and first prove the convergence in case Λ_l is a sequence of increasing cubes. Define

$$
F_\Lambda^{(n)} = \sum_{X_1,\dots,X_n\subset\Lambda} K(X_1)\dots K(X_n)\,\psi_{\text{pol}}(X_1,\dots,X_n). \qquad (37.24)
$$

To show that $\lim_{l\to\infty} F_{\Lambda_l}^{(n)}/|\Lambda_l|$ exists, we consider a collection $\{\Lambda_i(a)\}_{i=1}^m$ of disjoint cubes of side a as in lemma 25.1. We then want to prove that

$$
\lim_{a\to\infty}\left|\frac{1}{ma^\nu}F_{\cup\Lambda_i(a)}^{(n)} - \frac{1}{a^\nu}F_{\Lambda_i(a)}^{(n)}\right| = 0.
$$

As in corollary 25.1, this implies that $F_{\Lambda_l}^{(n)}/|\Lambda_l|$ is a Cauchy sequence for any sequence of cubes Λ_l with side tending to infinity.

By Penrose's estimate (36.21), we have

$$
\begin{aligned}
|\psi_{\mathrm{pol}}(X_1, \ldots, X_n)| &\leqslant \sum_{T \in \mathcal{T}_n} \prod_{(k,k') \in T} |e^{-v(X_k, X_{k'})} - 1| \\
&= \sum_{T \in \mathcal{T}_n} \prod_{(k,k') \in T} (1 - \delta_{X_k \cap X_{k'}, \emptyset}).
\end{aligned}
\tag{37.25}
$$

Note that

$$
\frac{1}{a^\nu} F_{\Lambda_i(a)}^{(n)} = \frac{1}{m a^\nu} \sum_{i=1}^m F_{\Lambda_i(a)}^{(n)}
$$

and if $X_k \in \Lambda_i$ and $X_l \in \Lambda_j$ with $i \neq j$ then there must be a k' such that $X_{k'} \cap \Lambda_i \neq \emptyset$ and $X_{k'} \cap \Lambda_j \neq \emptyset$. We can therefore write

$$
\left| \frac{1}{m a^\nu} F_{\cup \Lambda_i(a)}^{(n)} - \frac{1}{a^\nu} F_{\Lambda_i(a)}^{(n)} \right| \leqslant \frac{1}{m a^\nu} \sum_{T \in \mathcal{T}_n} \sum_{\substack{X_1, \ldots, X_n \subset \Lambda: \\ \exists k, i \neq j: X_k \cap \Lambda_i(a) \neq \emptyset, X_k \cap \Lambda_j(a) \neq \emptyset}}
$$

$$
\times \prod_{(k_1, k_2) \in T} (1 - \delta_{X_{k_1} \cap X_{k_2}, \emptyset}) \prod_{k=1}^n |K(X_k)|.
\tag{37.26}
$$

Here we denoted $\Lambda = \bigcup_{i=1}^m \Lambda_i(a)$. Now let $\sum_{X_1, \ldots, X_n \subset \Lambda}^{(k)}$ denote the sum over subsets $X_1, \ldots, X_n$ such that there exist $i, j \in \{1, \ldots, m\}$ so that $i \neq j$ and $X_k \cap \Lambda_i(a) \neq \emptyset$ and $X_k \cap \Lambda_j(a) \neq \emptyset$. Then we have

$$
\left| \frac{1}{m a^\nu} F_{\cup \Lambda_i(a)}^{(n)} - \frac{1}{a^\nu} F_{\Lambda_i(a)}^{(n)} \right| \leqslant \frac{1}{m a^\nu} \sum_{T \in \mathcal{T}_n} \sum_{k=1}^n \sum_{X_1, \ldots, X_n \subset \Lambda}^{(k)}
$$

$$
\times \prod_{(k_1, k_2) \in T} (1 - \delta_{X_{k_1} \cap X_{k_2}, \emptyset}) \prod_{k=1}^n |K(X_k)|.
\tag{37.27}
$$

Let us now introduce weights in the same way as in the derivation of Penrose's estimate, but starting at the vertex k, i.e. $w_{k'}$ is the number of links in T connecting k' to k. Then, since $X_{k_1} \cap X_{k_2} \neq \emptyset$ if $(k_1, k_2) \in T$, we can bound the right-hand side by

$$
\frac{1}{m a^\nu} \sum_{k=1}^n \sum_{x_k \in \Lambda} \sum_{X_k \ni x_k}^{(k)} |K(X_k)| \sum_{T \in \mathcal{T}_n}
$$

$$
\times \prod_{(k_1, k_2) \in T: w_{k_1} < w_{k_2}} \left(\sum_{x_{k_1} \in X_{k_1}} \sum_{X_{k_2} \subset \Lambda: x_{k_1} \in X_{k_2}} |K(X_{k_2})| \right).
\tag{37.28}
$$

Note that the expression in brackets is the same for all k_1 such that $(k_1, k_2) \in T$. Therefore, if p_{k_1} is the *coordination number* of T at k_1 (the number of lines in T emanating from k_1) the factor in brackets occurs $p_{k_1} - 1$ times, and we can bound the sum over x_{k_2} by $|X_{k_1}|^{p_{k_1} - 1}$. By the condition (37.13),

$$C_{p-1} = \sum_{X \subset \mathbb{Z}^\nu : 0 \in X} |X|^{p-1} |K(X)| < +\infty$$

and we have

$$\prod_{(k_1, k_2) \in T : w_{k_1} < w_{k_2}} \left(\sum_{x_{k_1} \in X_{k_1}} \sum_{X_{k_2} \subset \Lambda : x_{k_1} \in X_{k_2}} |K(X_{k_2})| \right)$$

$$\leqslant |X_k|^{p_k - 1} \prod_{k_1 = 1; k_1 \neq k}^{n} \left(\sum_{X_{k_1} \subset \Lambda : 0 \in X_{k_1}} |X_{k_1}|^{p_{k_1} - 1} |K(X_{k_1})| \right)$$

$$\leqslant |X_k|^{p_1 - 1} \prod_{k_1 = 1; k_1 \neq k}^{n} C_{p_{k_1} - 1}.$$

Replacing the sum over trees now by n^{n-2} we have since $p_{k_1} - 1 \leqslant n - 2$,

$$\left| \frac{1}{ma^\nu} F_{\cup \Lambda_i}^{(n)}(a) - \frac{1}{a^\nu} F_{\Lambda_i}^{(n)}(a) \right|$$

$$\leqslant \frac{n^{n-1}}{ma^\nu} C_{n-2}^{n-1} \sum_{x \in \Lambda} \sum_{\substack{X \subset \Lambda : x \in X \\ \exists i \neq j : X \cap \Lambda_i(a) \neq \emptyset, \, X \cap \Lambda_j(a) \neq \emptyset}} |X|^{n-1} |K(X)|.$$

$$(37.29)$$

Now, by the condition (37.13), given $\epsilon > 0$, there exists $\rho < +\infty$ such that

$$\sum_{X \subset \mathbb{Z}^\nu : 0 \in X, \operatorname{diam}(X) > \rho} |X|^{n-1} |K(X)| < \epsilon.$$

We subdivide the sum over X into those sets with $\operatorname{diam}(X) \leqslant \rho$ and those with $\operatorname{diam}(X) > \rho$. Hence the right-hand side is bounded by

$$n^{n-1} C_{n-2}^{n-1} \sum_{X \ni 0; \operatorname{diam}(X) > \rho} |X|^{n-1} |K(X)|$$

$$+ \frac{n^{n-1}}{ma^\nu} C_{n-2}^{n-1} \sum_{x \in \Lambda} \sum_{\substack{X \ni x; \operatorname{diam}(X) \leqslant \rho \\ \exists i, j : X \cap \Lambda_i(a) \neq \emptyset, \, X_1 \cap \Lambda_j(a) \neq \emptyset}} |X|^{n-1} |K(X)|. \qquad (37.30)$$

The first term is bounded by $n^{n-1} C_{n-2}^{n-1} \epsilon$ and tends to 0 as $\epsilon \to 0$. The second term is bounded in the same way as in equation (25.10). Namely, the distance of x to the boundary of one of the $\Lambda_i(a)$ must be less than ρ, so

$$\sum_{x \in \Lambda} \sum_{\substack{X \ni x;\, \mathrm{diam}(X) \leqslant \rho \\ \exists i,j: X \cap \Lambda_i(a) \neq \emptyset,\, X_1 \cap \Lambda_j(a) \neq \emptyset}} |X|^{n-1}|K(X)| \leqslant m[a^\nu - (a - 2\rho)^\nu]C_{n-1}.$$

The second term is therefore bounded by $2\nu n^{n-1}C_{n-1}^n(\rho/a)$. We conclude that

$$\limsup_{a \to \infty} \left| \frac{1}{ma^\nu} F_{\cup \Lambda_i(a)}^{(n)} - \frac{1}{a^\nu} F_{\Lambda_i(a)}^{(n)} \right| \leqslant n^{n-1}C_{n-2}^{n-1} \epsilon \to 0.$$

This proves the analogue of lemma 25.1. As in corollary 25.1 this implies that $F_{\Lambda_l}^{(n)}/|\Lambda_l|$ is a Cauchy sequence if Λ_l is a sequence of cubes tending to infinity. For general Van Hove sequences an argument as in the proof of theorem 25.1 is needed, but we omit this here. ∎

Note that the convergence of the limit in lemma 37.2 is uniform in any region of the parameters of $K(X)$ in which the limit (37.13) holds uniformly. It follows that, if $K(X)$ is analytic in these parameters, then so is the limit. We now prove the same for the limit (37.11).

We use the following theorem from analysis:

> Suppose that $(f_{n,l})_{n,l \in \mathbb{N}}$ is a double sequence of analytic functions such that
>
> 1. $\lim_{l \to \infty} f_{n,l} = f_n$ exists and converges uniformly on a domain $\mathcal{D}$; and
>
> 2. there is constant $C < +\infty$ such that $\sum_{n=1}^\infty |f_{n,l}| \leqslant C$ for all $l \in \mathbb{N}$ uniformly on $\mathcal{D}$.
>
> Then $\lim_{l \to \infty} \sum_{n=1}^\infty f_{n,l} = \sum_{n=1}^\infty f_n$ uniformly on $\mathcal{D}$.

We take $f_{n,l} = F_{\Lambda_l}^{(n)}/|\Lambda_l|$. By the previous lemma, we have that $\lim_{l \to \infty} f_{n,l}$ exists uniformly on bounded regions where equation (37.13) holds uniformly.

To bound the sum $\sum_{n=1}^\infty |f_{n,l}|$ we use the same approach as in equation (37.28), tracing trees according to their coordination numbers. However, we need to be more accurate and use a more detailed version of Cayley's theorem:

Lemma 37.3 *The number of tree graphs on a set $\{1, \ldots, n\}$ of vertices with given coordination numbers $p_1, \ldots, p_n$ (where $p_1 + \cdots + p_n = 2(n-1)$) is given by*

$$N_n(p_1, \ldots, p_n) = \frac{(n-2)!}{\prod_{i=1}^n (p_i - 1)!}.$$

Tracing trees starting at X_1, we have,

$$
\begin{aligned}
|f_{n,l}| &\leqslant \frac{1}{|\Lambda_l|} \sum_{X_1,\ldots,X_n \subset \Lambda_l} \sum_{T \in \mathcal{T}_n} \prod_{(k,k') \in T} (1 - \delta_{X_k \cap X_{k'},\emptyset}) \prod_{k=1}^{n} |K(X_k)| \\
&\leqslant \frac{1}{|\Lambda_l|} \sum_{x_1 \in \Lambda_l} \sum_{X_1 \subset \Lambda_l : x_1 \in X_1} |K(X_1)| \\
&\quad \times \sum_{T \in \mathcal{T}_n} \prod_{(k,k') \in T : w_k < w_{k'}} \left(\sum_{x_k \in X_k} \sum_{X_{k'} \subset \Lambda_l : x_k \in X_{k'}} |K(X_{k'})| \right) \\
&\leqslant \frac{1}{|\Lambda_l|} \sum_{x_1 \in \Lambda_l} \sum_{X_1 \subset \Lambda_l : x_1 \in X_1} \sum_{\substack{p_1,\ldots,p_n \geqslant 1 \\ \sum_{i=1}^{n} p_i = 2(n-1)}} \frac{(n-2)!}{\prod_{i=1}^{n}(p_i-1)!} \\
&\quad \times |X_1|^{p_1} |K(X_1)| \prod_{k=2}^{n} \left(\sum_{X_k \subset \Lambda_l : 0 \in X_k} |X_k|^{p_k-1} |K(X_k)| \right).
\end{aligned}
\tag{37.31}
$$

Therefore,

$$
\begin{aligned}
\sum_{n=1}^{\infty} \frac{1}{n!} |f_{n,l}| &\leqslant \frac{1}{|\Lambda_l|} \sum_{n=1}^{\infty} \frac{1}{n(n-1)} \sum_{p_1=1}^{n-1} \sum_{x_1 \in \Lambda_l} \sum_{X_1 \subset \Lambda_l : x_1 \in X_1} \frac{|X_1|^{p_1}}{(p_1-1)!} |K(X_1)| \\
&\quad \times \sum_{\substack{p_2,\ldots,p_n \geqslant 1 \\ \sum_{i=2}^{n}(p_i-1)=n-p_1-1}} \prod_{k=2}^{n} \left(\sum_{X_k \in \Lambda : 0 \in X_k} \frac{1}{(p_k-1)!} |X_k|^{p_k-1} |K(X_k)| \right) \\
&\leqslant \frac{1}{|\Lambda_l|} \sum_{n=1}^{\infty} \frac{1}{n} \sum_{x_1 \in \Lambda_l} \sum_{X_1 \subset \Lambda_l : x_1 \in X_1} \sum_{p_1=1}^{\infty} \frac{|X_1|^{p_1}}{(p_1)!} |K(X_1)| \\
&\quad \times \prod_{k=2}^{n} \left(\sum_{X_k \in \Lambda : 0 \in X_k} \sum_{p_k=1}^{\infty} \frac{1}{(p_k-1)!} |X_k|^{p_k-1} |K(X_k)| \right) \\
&\leqslant \sum_{n=1}^{\infty} \frac{1}{n} \left(\sum_{X \in \mathbb{Z}^{\nu} : 0 \in X} e^{|X|} |K(X)| \right)^{n} < +\infty
\end{aligned}
\tag{37.32}
$$

if the condition (37.12) holds.   ∎

Problems to Part III

III-1. Show that equation (19.7) can also be derived as the limit of a Riemann sum. Now consider the solution of a polymer in a solvent which consists of small molecules. The **Flory-Huggins theory** represents the solvent as a lattice and the dissolved polymer as a collection of linear strains of equal length l. Each repeating unit of a strain occupies one position in the lattice. Let there be N lattice points, N_1 solvent molecules and N_2 dissolved polymer strains, so that $N = N_1 + N_2 l$. Argue that at low concentration ($N_2 l \ll N$) the number of different ways of arranging the $(k+1)$th polymer strain can be approximated by

$$\Omega_{k+1} \approx N\nu(\nu - 1)^{l-2}(1 - f_k),$$

where $f_k = lk/N$ is the fraction of occupied positions after the k-th strain has been placed, and ν is the coordination number of the lattice (number of neighbours). Hence show that the total number of different ways to arrange all strains can, in the case $N \gg N_2 l$, be approximated by

$$\Omega = \frac{\Omega_1 \dots \Omega_{N_2}}{N_2!} \approx \nu^{N_2}(\nu - 1)^{N_2(l-2)} \prod_{k=1}^{N_2} \left(\frac{N - (k - 1)l}{k} \right).$$

Now use the convergence of a Riemann sum to compute the entropy per lattice point (i.e. per unit volume) in terms of the volume fractions $v_1 = N_1/N$ and $v_2 = 1 - v_1$. Show that it is the sum of two terms: the **mixing entropy** and the entropy due to the folding of the polymer chains.

III-2. Derive equation (28.23) from (28.22) using (24.11).

III-3. Use the transfer matrix to derive directly the partition function of the one-dimensional Ising chain with free boundary conditions. Conclude that the free energy is again given by equation (28.12).

(Hint: You have to determine the eigenvectors of the transfer matrix.)

III-4. The q-state Potts model is a generalization of the nearest neighbour Ising model where the spin variables can take on q values: $s_x \in \{1, 2, \dots, q\}$ and where the energy function is given by

$$E(\{s_x\}) = -\frac{J}{2} \sum_x \sum_{y:\, |x-y|=1} \delta_{s_x, s_y}.$$

Write down the corresponding $q \times q$ transfer matrix A for the one-dimensional q-state Potts chain and hence compute the free energy density for this model.

III-5. A finite **Cayley tree** with $M+1$ levels and coordination number ν is a graph constructed as follows. A central vertex (level 0) is connected to $\nu \geqslant 3$ vertices at level 1. Each of these is in turn connected to $\nu - 1$ more vertices at level 2, and in general each vertex at level k ($1 \leqslant k \leqslant M - 1$) is connected to 1 vertex at level $k - 1$ and $\nu - 1$ vertices at level $k + 1$. The figure below shows a Cayley tree with 4 levels and $\nu = 3$.

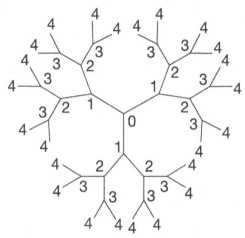

Figure III-5. Cayley tree with $\nu = 3$.

Show that the number of vertices of a Cayley tree is given by

$$N = \frac{\nu(\nu - 1)^M - 2}{\nu - 2}.$$

Now define an Ising model on the Cayley tree by the energy function

$$E(\{s_x\}) = -\frac{1}{2}J \sum_x \sum_{y: y \leftrightarrow x} s_x s_y,$$

where $x \leftrightarrow y$ means that x and y are vertices connected by a line in the Cayley tree. Compute the free energy density as follows. Define the partition functions for a branch of the tree starting at the central vertex:

$$\tilde{Z}_M(s_0) = \sum_{\{s_x\}} e^{-\beta \tilde{E}(\{s_x\})},$$

where the sum is over all spins in that branch other than the central spin s_0 and $\tilde{E}$ is the energy of a branch. Then show that

$$\tilde{Z}_M(s_0) = \sum_{s_1} e^{\beta J s_0 s_1} \left(\tilde{Z}_{M-1}(s_1)\right)^{\nu-1}.$$

Finally compute $Z_M = \sum_{s_0} \tilde{Z}_M(s_0)^\nu$ and hence the free energy density $f(\beta)$.

III-6. Show that the eigenvalues of the transfer matrix (28.7) are given by equation (28.13) and perform the derivatives of $f(\beta, J, H)$ with respect to H to compute m and χ as in equations (28.19) and (28.20).

III-7. Consider a spin system on a lattice where the spins can take on a symmetric set of values: $s_x = \pm 1$ or $s_x \in \{-q, -q+1, \ldots, q\}$. For any set of lattice sites X put $s^X = \prod_{x \in X} s_x$. Let the energy function be given by an expression of the type

$$E\left(\{s_x\}_{x \in \Lambda}\right) = -\sum_{X \subset \Lambda} J_X s^X,$$

where the constants $J_X \geq 0$. (Note that the Ising model is of this form.) The expectation of s^A for a finite subset $A \subset \Lambda$ with respect to the canonical distribution (24.3) (or (28.30)) is called a **correlation function** and denoted

$$\langle s^A \rangle_\Lambda = \frac{1}{Z_\Lambda} \sum_{\{s_x\}_{x \in \Lambda}} s^A \, e^{-\beta E(\{s_x\})}.$$

By expanding the exponential in the numerator, prove the correlation inequality $\langle s^A \rangle_\Lambda \geq 0$.

III-8. Given two *independent* sets of spin variables $\{s_x\}$ and $\{s'_x\}$ as in problem III-7, define new spin variables ξ_x and η_x by $\xi_x = \frac{1}{2}(s_x + s'_x)$ and $\eta_x = \frac{1}{2}(s_x - s'_x)$. Show that the energy function $E(\{s_x\}) + E(\{s'_x\})$ is still of the same form, when written in terms of the spin variables ξ_x and η_x and use this together with the correlation inequality derived in the previous problem to show that $\langle s^A s^B \rangle_\Lambda - \langle s^A \rangle_\Lambda \langle s^B \rangle_\Lambda \geq 0$. (Write the left-hand side as $\langle s^A(s^B - s'^B) \rangle_\Lambda = \langle (\xi + \eta)^A[(\xi + \eta)^B - (\xi - \eta)^B] \rangle_\Lambda$.)

III-9*. Use the correlation inequality derived in the previous problem to show that $\langle s^B \rangle_\Lambda$, when considered as a function of the couplings J_X, is monotonically increasing. Deduce that, if we impose positive boundary conditions at the boundary of Λ, then $\langle s^B \rangle_\Lambda$ is decreasing for increasing Λ and hence that the thermodynamic limit of correlation functions for the Ising model with positive boundary conditions exists and is unique.

III-10. Consider a one-dimensional spin chain with spins $s_i \in S$, where $S = \{1, \ldots, q\}$ and with energy function given by

$$E(\{s_i\}_{i=1}^N) = \sum_{i=1}^N \phi(s_i, s_{i+1})$$

where $\phi : S \times S \to \mathbb{R}$ is a given function (nearest-neighbour interaction).

(i) Write the transfer matrix A for this model and express the partition function Z_N in terms of A.

(ii) The Perron-Frobenius theorem says that the maximum eigenvalue $\lambda_{\max}$ of a matrix with strictly positive entries is strictly positive. Write the free energy density of the model in terms of the maximal eigenvalue of the transfer matrix A and conclude that $f(\beta)$ is analytic in $\beta > 0$ and in the parameters of the model, i.e. the values of the function ϕ.

III-11. Show that in d dimensions the number of closed contours of surface area $l \geqslant 2d$ in a region $\Lambda \subset \mathbb{Z}^d$ is bounded by $N(l) \leqslant 2d\,|\Lambda|\,3^{l-1}/l$. Hence argue that the Ising model on $\mathbb{Z}^d$ has spontaneous magnetization if βJ is large enough. In particular, obtain an upper bound for $\beta_c J$ in the cases $d = 2$ and $d = 3$. In the case $d = 2$, compare this with equation (28.25), and in the case $d = 3$ with the numerical estimate $\beta_c J \approx 0.4$.

III-12*. Modify the Peierls argument to show that at low temperatures the two-dimensional Potts model on a square lattice defined by the energy function in problem III-4 with $q = 3$ has (at least) 3 phases corresponding to the three possible boundary conditions. Taking the spin values to be $s_x = -1, 0, +1$, argue that one therefore needs two order parameters to distinguish these phases, which can be taken to be ρ_+ and ρ_-, the fractions of spins equal to ± 1 respectively. Introduce corresponding external fields H_+ and H_- and argue that the free energy is discontinuous if either H_+ or H_- is increased from negative to positive values while the other equals zero. What happens if the other field is not zero?

III-13. Derive equation (29.17) from (29.16) and prove that $\chi(\beta) > 0$ for all $\beta > 0$.

III-14. Compute the specific heat for the Weiss-Ising model and show that it approaches a finite value as $T \uparrow T_c$ and is zero for $T > T_c$. As it does not diverge at $T = T_c$ one writes $\alpha = 0$ for the critical exponent defined in equation (9.2). Compute also the other critical exponents β, γ and δ defined (for a magnetic system) by

$$m_0(T) \sim (T - T_c)^\beta \text{ at } H = 0$$
$$\chi(T) \sim (T - T_c)^{-\gamma} \text{ at } H = 0$$
$$m(H) \sim H^{1/\delta} \text{ at } T = T_c$$

and verify the scaling relations $\alpha + 2\beta + \gamma = 2$ and $\beta + \gamma = \beta\delta$. These relations are universally true. Determine the exponents α and β for the 2-dimensional Ising model and hence find also γ and δ. (There are ways of determining these also independently.)

III-15. Consider a spin system where the Ising spins s_i $(i = 1, 2, \ldots, N)$ have an indirect interaction with each other via a set of 'phantom neighbours' s_i^α $(\alpha = 1, \ldots, \nu)$ which themselves have a mean-field interaction:

$$E\left(\{s_i\}, \{s_i^\alpha\}\right) = -J\sum_{i=1}^{N}\sum_{\alpha=1}^{\nu} s_i s_i^\alpha - \frac{\lambda J}{N}\left(\sum_{i=1}^{N}\sum_{\alpha=1}^{\nu} s_i^\alpha\right)^2 - H\sum_{i=1}^{N} s_i.$$

Compute the free energy per spin

$$f(\beta) = -\frac{1}{\beta} \lim_{N \to \infty} \frac{1}{N} \ln Z_N(\beta).$$

Show that this model has a phase transition.

III-16*. The two-dimensional **ice model** is a model of water molecules H_2O on a square lattice where each O-atom occupies a lattice site and the H-O-H angle of each molecule is assumed to be $90°$ so that the hydrogen atoms lie along the bonds of the lattice. They have to form a hydrogen bond with the oxygen atom of a neighbouring molecule so that there must be only one H-atom on each bond. One can thus assign an arrow to each bond of the lattice pointing towards the O-atom with which the corresponding H-atom forms a molecule.

Figure III-16. Six-vertex configurations.

The model consists therefore essentially of an assignment of arrows to the bonds of a square lattice such that the ice condition holds: at every vertex there must be two inward-pointing and two outward-pointing arrows. The allowed configurations of arrows at each vertex are therefore as in figure III-16.

Assume that the lattice consists of M rows and N columns. There are then $2NM$ bonds. The problem is to compute the **residual entropy** of this model given by the usual equation (19.3):

$$s_0 = \lim_{M,N \to \infty} \frac{k_B}{NM} \ln \Omega$$

where Ω is the number of allowed configurations of arrows. This is the entropy at zero temperature; at higher temperatures there is additional entropy due to vibrational degrees of motion. (Note that the third law does not hold for ice!) Find an estimate (due to Pauling) for s_0 by assuming that the configurations at every vertex are independent but subject to the ice condition. Compare this value with the experimental value which is $s_0 = 3.4 \pm 0.2$ J mol^{-1} K^{-1}. (Ice is three-dimensional of course but the coordination number of its lattice structure is four as in the square lattice.)

The value of s_0 for square ice can in fact be computed exactly (see Lieb (1967)). The calculation uses the transfer matrix: consider a row of vertical bonds and denote the arrow configuration on this row by α. Let α' denote the configuration at the next row of vertical bonds and let $A(\alpha, \alpha')$ be the number of configurations of arrows on the intervening row of horizontal bonds. Argue that $\Omega = \text{Tr}\, A^M$ assuming periodic boundary conditions, and determine the matrix A (it takes the values 0, 1 or 2 depending on how α' differs from α). Denote the configuration of vertical arrows with down arrows at the positions

$x_1 < \cdots < x_n$ by $\alpha(x_1, \ldots, x_n)$. Show that A leaves the vector spaces spanned by these configurations with fixed n invariant. An eigenvector ψ of A can thus be written as

$$\psi = \sum_{1 \leqslant x_1 < \cdots < x_n \leqslant N} f(x_1, \ldots, x_n) \, \alpha(x_1, \ldots, x_n).$$

Show that the so-called **Bethe Ansatz**:

$$f(x_1, \ldots, x_n) = \sum_P a(P) \exp\left(i \sum_{j=1}^n k_{P(j)} x_j \right)$$

where the sum runs over all permutations P of $\{1, 2, \ldots, N\}$, satisfies the eigenvalue equation provided the wavenumbers $k_1, \ldots, k_n$ satisfy a certain transcendental set of equations, and the parameters $a(P)$ are chosen correctly (depending on $k_1, \ldots, k_N$). Determine the corresponding eigenvalues.

III-17. The expansion of the pressure in powers of $1/v$ is called the **virial expansion**. Rescale the volume in the lattice gas model of chapter 30 by a factor v_0, the molecular volume, to determine the equation of state (30.16) for cells of volume v_0. Then compare the first non-trivial coefficient in the virial expansion with that for the van der Waals model and relate λ and v_0 to a and b. Using his original values for a and b, $a = 0.004$ J m^3 mol^{-2} and $b = 2.0 \times 10^{-5}$ m^3 mol^{-1}, what value for the critical temperature of helium would Kamerlingh Onnes have found had he used the lattice gas model for helium instead of the van der Waals equation?

III-18. Construct a model for paramagnetism in metals (so called **Pauli paramagnetism**) as follows. Assume that the energy of each conduction electron in an external magnetic field H is given by $\epsilon(\vec{k}, s) = \epsilon(\vec{k}) - \mu_0 \mu_e H$, where $\mu_e = -s\mu_B$ is the magnetic moment depending on the direction $s = \pm$ of the spin. Compute the grand potential $w(\beta, \mu, H)$ for this model. Then deduce that the magnetization $m(\beta, \mu, H) = -\mu_0^{-1}(\partial w/\partial H)$ can be written in terms of the density of states $g(\epsilon_F)$ at the Fermi level assuming that the magnetic perturbation is much smaller than ϵ_F.

III-19. Show that the grand potential of a Fermi gas satisfies the relation $w = -\frac{2}{3}\tilde{u}$. Then, differentiating equation (31.11), and dividing by the density, derive an expression for the entropy per particle at low temperatures.

A dilute solution of ^{3}He in ^{4}He below the λ-transition can be modelled as a free Fermi gas because the superfluid ^{4}He acts as an inert background. The ^{3}He atoms in this background have an effective mass which is 2.5 times the normal value. The density of ^{4}He is 2.16×10^{22} atoms cm^{-3}. Compute the Fermi energy of a 6.8% solution and show that it is a degenerate Fermi gas below 0.1 K. Pure ^{3}He can also be modelled as a free Fermi gas but the effective mass is a factor of 3 higher than the normal value and the density is 1.6×10^{22} atoms cm^{-3}. Compute also the Fermi energy for this gas. Hence find an expression for the latent heat gained when a mole of ^{3}He atoms is moved from the concentrated phase to the dilute phase in a dilution refrigerator. Note that it is quadratic in T as opposed to exponential as in problem I-35. Compute the heat loss that can be sustained at a temperature of 10 mK when the ^{3}He is circulated at a rate of 10^{-4} mol s^{-1}.

III-20. In the **Einstein model** for phonons it is assumed that all normal modes have the same frequency: $\omega_i = \omega$. Compute the free energy density for this model and hence the specific heat c_V. Show that c_V tends to zero very rapidly as $T \to 0$ whereas it tends to the classical Dulong and Petit value for $T \to \infty$. Also verify equation (33.12) and show that it also tends to this value as $T \to \infty$.

III-21. The mean-field boson gas is a simplification of the Huang-Yang-Luttinger (HYL) model obtained by neglecting also U_2 so that the interaction energy is simply given by

$$U_1 = \frac{\lambda}{2V_l}N^2 \quad \text{(see equation (34.18))}.$$

Compute the grand canonical potential for this model using the fact that N/V_l satisfies the LDP with respect to the grand canonical distribution of the free gas with artificial parameter μ_0.

III-22. Compute the cumulant generating function for the two-dimensional random variables (Y_l, Z_l) with distribution function (34.24) and hence derive the rate function (34.25).

III-23*. Consider a Cayley tree as described in problem III-5. Denote the infinite tree by Γ and the finite tree up to level M by Γ_M. Define a lattice Laplacian on the sites of this tree as follows: If $\phi \in \ell^2(\Gamma)$ then

$$(\Delta_\Gamma \phi)(x) = \sum_{y:\, x \leftrightarrow y} \phi(y) - \nu\phi(x).$$

On the finite tree we can similarly define $\Delta_M = \Delta_{\Gamma_M}$ by restricting x and y to Γ_M (free boundary conditions). Prove by induction that the spectrum of $-\Delta_M$ is given by the eigenvalues $\lambda_{n,k,M}$ where $k = 1, \ldots, M+1$ and $n = 1, \ldots, k$ and where

$$\lambda_{n,k,M} = \nu - 2\sqrt{\nu - 1} \cos\left(\frac{n\pi}{k+1}\right)$$

for $k \leqslant M$ and

$$\lambda_{n,M+1,M} = \nu - 2\sqrt{\nu - 1} \cos(\tfrac{1}{2}\theta_n)$$

if $0 < \theta_1 < \cdots < \theta_{M+1} < 2\pi$ are the solutions of

$$\sin\left[\left(1 + \frac{1}{2}M\right)\theta_n\right] = \frac{1}{\nu - 1}\sin\left(\frac{1}{2}M\theta_n\right).$$

Show also that the corresponding multiplicities are

$$
m_k = \begin{cases} \nu(\nu-2)(\nu-1)^{M-k-1} & \text{if } 1 \leqslant k \leqslant M-1 \\ \nu-1 & \text{if } k = M \\ 1 & \text{if } k = M+1. \end{cases}
$$

Consider a system of free bosons on the tree with single-particle energy levels given by the eigenvalues $\lambda_{n,k,M}$ and with multiplicities m_k. Write down the grand partition function for this model and show that this model exhibits boson condensation.

III-24. Argue that the second Mayer coefficient $b_2(\beta)$ for a classical gas with Lennard-Jones interaction is negative for large temperatures.

III-25. Derive an expression analogous to (36.17) for $b_3(\beta)$.

III-26. Express the third coefficient in the **virial expansion** (36.19) of the pressure in powers of $1/v$ in terms of $b_2(\beta)$ and $b_3(\beta)$.

III-27*. Prove stability for a classical gas with hard-core interaction, i.e. for which there exists $r_0 > 0$ such that $\phi(x) = +\infty$ if $|x| \leqslant r_0$ (as well as $\phi(x) > -C|x|^{-\nu-\epsilon}$ for some constants $C > 0$ and $\epsilon > 0$).

III-28. The correlation functions $\rho_{\mathcal{R}}(x_1, \ldots, x_m)$ of a classical gas are defined by

$$
\begin{aligned}
\rho_{\mathcal{R}}(x_1, \ldots, x_m) =\ & \frac{1}{Z_{\mathcal{R}}(\beta,\mu)} \prod_{i=1}^{m} \chi_{\mathcal{R}}(x_i) \sum_{n=0}^{\infty} \frac{z^{m+n}}{n!} \\
& \times \int_{\mathcal{R}} \mathrm{d}^\nu y_1 \cdots \int_{\mathcal{R}} \mathrm{d}^\nu y_n e^{-\beta E_{m+n}(x_1,\ldots,x_m,y_1,\ldots,y_n)}.
\end{aligned}
$$

(Here $\chi_{\mathcal{R}}$ is the indicator function of $\mathcal{R}$.) Show that they satisfy the so-called **Kirkwood-Salsburg equations**

$$
\rho_{\mathcal{R}}(x_1) = \chi_{\mathcal{R}}(x_1) z \left[1 + \sum_{n=1}^{\infty} \frac{1}{n!} \int_{\mathcal{R}} \mathrm{d}^\nu y_1 \cdots \int_{\mathcal{R}} \mathrm{d}^\nu y_n K(x_1; \underline{y}) \rho_{\mathcal{R}}(\underline{y}) \right]
$$

and

$$
\begin{aligned}
\rho_{\mathcal{R}}(x_1, \ldots, x_m) =\ & \prod_{i=1}^{m} \chi_{\mathcal{R}}(x_i) z e^{-\beta W(x_1; x_2, \ldots, x_m)} \Bigg\{ \rho_{\mathcal{R}}(x_2, \ldots, x_m) + \\
& \sum_{n=1}^{\infty} \frac{1}{n!} \int_{\mathcal{R}} \mathrm{d}^\nu y_1 \cdots \int_{\mathcal{R}} \mathrm{d}^\nu y_n K(x_1; \underline{y}) \rho_{\mathcal{R}}(x_2, \ldots, x_m, \underline{y}) \Bigg\},
\end{aligned}
$$

where $\underline{y} = (y_1, \ldots, y_n)$, $W(x_1; x_2, \ldots, x_m) = \sum_{j=2}^{m} \phi(x_1 - x_j)$, and where the kernel K is given by

$$
K(x_1; \underline{y}) = \prod_{j=1}^{n} f_\beta(x_1, y_j).
$$

[Use the fact that, analogous to (36.7),

$$\prod_{j=1}^{n} e^{-\beta\phi(x_1-y_j)} = \sum_{I\subset\{1,\ldots,n\}} \prod_{i\in I} f_\beta(x_1,y_j).]$$

III-29. Derive equations (37.6) and (37.10). [For the latter, write

$$\sum_{\{X_k\}_{k=1}^P \in \Pi_p(X)} = \frac{1}{p!} \sum_{\substack{X_1,\ldots,X_p\subset X:\cup_{k=1}^P X_k=X \\ X_i\cap X_j\neq\emptyset}} .$$

Then notice that the sum over p can be extended to infinity setting $K(\emptyset) = 0$.]
Observe that for $|\Lambda| = 2$, (37.10) is the Taylor expansion of $\ln(1 + K(\Lambda))$.

III-30. Prove lemma 37.2 by induction on n and hence prove Cayley's theorem.

III-31. Show that the polymer activities for the one-dimensional Ising model given by the energy function

$$E_N(s_1,\ldots,s_N) = -J\sum_{i=1}^{N-1} s_i s_{i+1} - H\sum_{i=1}^{N} s_i$$

are zero unless the set $X \subset \{1,\ldots,N\}$ is an interval, in which case they are given by

$$K(\{1,\ldots,n\}) = A_+(\lambda_+ - 1)^{n-1} + A_-(\lambda_- - 1)^{n-1},$$

where $A_\pm$ are given in the answer to problem III-3.

III-32. Consider a classical spin model on the two-dimensional lattice $\mathbb{Z}^2$ with Ising spins $s_x = \pm 1$ but with 'interaction around a square (or plaquette)' given by

$$E(\{s_x\}_{x\in\Lambda}) = -J\sum_{x:\, B_x\subset\Lambda} s_{B_x},$$

where $B_x = \{x, x+e_1, x+e_2, x+e_1+e_2\}$ is an elementary square (e_1 and e_2 are the standard basis vectors) and where $s_{B_x} = s_x s_{x+e_1} s_{x+e_2} s_{x+e_1+e_2}$. Show that the corresponding Ursell functions $\psi(s_X) = 0$ unless X is a connected union of overlapping squares B_x, in which case

$$\psi(X) = \prod_{x:\, B_x\subset X} (e^{\beta J s_{B_x}} - 1).$$

III-33*. Expand the partition function for the ν-dimensional Ising model with zero magnetic field formally to 4-th order in powers of $\beta J |\Lambda|$. Hence obtain the power series expansion of $f(\beta)$ in powers of βJ up to 4-th order.

Determine also the limits of the first and second terms of equation (37.11) in the case of the zero-field Ising model, restricting the first term to the sets X_1 with $|X_1| \leqslant 4$ and the second term to sets X_1 and X_2 with $|X_1| = |X_2| = 2$. (Other terms are of higher order in β.) Note that for $|X_1| = 3$, there are two types of graphs with non-zero contribution, and for $|X_1| = 4$ there are 8 types of non-zero graphs (6 for $\nu = 2$). In the latter case the only graph which is not a tree is an elementary lattice square. Observe that it is only of order β^4. Expand the result up to order β^4 and compare with the expression obtained by formal expansion.

III-34*. Show that in the case of nearest-neighbour Ising model the bound (37.16) can be improved in two ways. First note that in bounding the number of possible graphs, we over-counted by allowing up to $2\nu - 1$ neighbours to be filled for every $x \in X$ independently, whereas the total number of neighbours cannot exceed $|X| - 1 = n - 1$. Introducing the condition $p_1 + \cdots + p_k = n - 1$ with $k \leqslant n - 1$, use Stirling's formula to obtain the bound $2\nu((2\nu - 1)e)^{n-1}$. (Cf. the argument below equation (37.15).) Secondly, each connected graph consists of a multiply-connected part with attached trees. The latter can be integrated out first, yielding factors $\cosh(\beta J) - 1$, whereas the number of links in the former is at least equal to the number of points. With these two improvements, obtain a new bound on $\sum_{X \ni 0} e^{|X|} |K(X)|$ and hence derive a larger value of βJ below which the cluster expansion converges.

III-35. Observe that $K(X)$ in equation (37.14) is also small if $dF_\beta(s_x)$ is concentrated at a single value of s_x and $\Phi_{x-y}(s, s) = 0$. For example, one can rewrite the interaction potential in the Ising model in the form $\Phi(s_x, s_y) = \frac{1}{2}J(s_x - s_y)^2$ upon multiplication of $Z_\Lambda(\beta)$ by a constant. Consider the more general q-state Potts model of spins s_x for $x \in \mathbb{Z}^\nu$ taking values in $\{1, \ldots, q\}$ with interaction energy

$$E(\{s_x\}) = -J \sum_x \sum_{y: |x-y|=1} (1 - \delta_{s_x, s_y}) - H \sum_x \delta_{s_x, q}.$$

Find a lower bound on βH analogous to equation (37.20) beyond which the cluster expansion (37.11) converges.

III-36. In Pirogov-Sinai theory, one uses an expansion analogous to the cluster expansion but converging for low temperatures, replacing the polymer model by a so-called **contour model**. Write the partition function of the zero-field 2-dimensional Ising model with positive boundary conditions in terms of contours as follows:

$$Z_\Lambda^+(\beta) = e^{\beta J \, N_2(\Lambda)} \sum_{\partial \subset \Lambda} e^{-\beta J \, |\partial|},$$

where $N_2(\Lambda) = |\{x \in \Lambda : x + e_1 \in \Lambda\}| + |\{x \in \Lambda : x + e_2 \in \Lambda\}|$ is the number of nearest-neighbour pairs in Λ ($N_2(\Lambda) \sim 2|\Lambda|$), and the sum is over collections ∂ of compatible contours (cf. the end of chapter 28), where $|\partial|$ denotes the total length of the contours in ∂. Introduce 'contour Ursell functions' ψ_{ctr} analogous to equation (37.9) to write an expression for $\ln Z_\Lambda^+(\beta)$ as in equation (37.10) and argue that the corresponding series for the free energy density converges uniformly for βJ large enough.

Probability Theory

This appendix is a collection of some results and concepts from probability theory. For a proper introduction to the subject see for example the books by Billingsley (1979) and Feller (1966).

Let us begin by studying experiments with a countable (finite or infinite) number of possible outcomes. For example, in throwing dice the possible outcomes are 1,2,3,4,5, and 6. The set of possible outcomes, $\{1, 2, 3, 4, 5, 6\}$ in this case, is called the **sample space** and often denoted Ω. Next we can assign probabilities to each of these outcomes. If the dice are fair it is reasonable to assume that each outcome occurs with equal probability, which is $\frac{1}{6}$ because the total probability must be 1 as it is certain that at least one of the possible outcomes actually happens. We can now ask for the probability that the outcome is at least 3. It is obvious that the answer is $\frac{4}{6}$ since there are 4 possible outcomes which are at least 3. In general, we can ask for the probability of a subset of the sample space. Such a subset is called an **event**. We thus arrive at the following *model* for a probabilistic experiment with countably many possible outcomes.

There is a set of possible outcomes Ω, called the **sample space**, which is countable. The subsets of Ω are called **events**. There is a function P which assigns to each event A a probability $P(A) \in [0, 1]$ such that

1. $P(\Omega) = 1$ and
2. The probability $P(A)$ of any event A is the sum of the probabilities of the elementary events contained in A.

An **elementary event** is a subset of Ω consisting of a single element.

Clearly, $P(\emptyset) = 0$. Notice also that it follows from point 2 above that if A and B are events such that $A \cap B = \emptyset$, i.e. if they are mutually exclusive, then $P(A \cup B) = P(A) + P(B)$. In particular, if $A^c = \Omega \setminus A$ is the complementary event then $P(A^c) = 1 - P(A)$. If A and B are general events there is also a formula for $P(A \cup B)$:

Theorem A.1 *If A and B are two events then the probability that A or B occurs, i.e. either A or B or both, is given by*

$$P(A \cup B) = P(A) + P(B) - P(A \cap B). \tag{A.1}$$

We also define the **conditional probability** of an event A **given** B:

$$P(A|B) = \frac{P(A \cap B)}{P(B)} \tag{A.2}$$

provided $P(B) \neq 0$. Two events are called **independent** if $P(A|B) = P(A)$, i.e. if $P(A \cap B) = P(A)P(B)$.

A **random variable** is a function $X : \Omega \to \mathbb{R}$. For example, the square of the number thrown with a die is a random variable. The **expectation (value)** of a random variable X is given by

$$\mathbb{E}(X) = \sum_{\omega \in \Omega} X(\omega) P(\{\omega\}). \tag{A.3}$$

For example, if $X(\omega) = \omega^2$ in the case of the dice then $\mathbb{E}(X) = \frac{1}{6}(1^2 + 2^2 + 3^2 + 4^2 + 5^2 + 6^2) = 15\frac{1}{6}$. The meaning of this number is that if the dice are thrown many times, say N times, then the average of the squares of the numbers thrown will be close to $15\frac{1}{6}$. This is the **law of large numbers**; we discuss it in more detail below. More generally one defines the **kth moment** of a random variable X by

$$\mathbb{E}(X^k) = \sum_{\omega \in \Omega} X(\omega)^k P(\{\omega\}). \tag{A.4}$$

The **distribution function** of a random variable X is defined by

$$F_X(x) = P\{\omega \in \Omega | X(\omega) \leqslant x\}. \tag{A.5}$$

The **standard deviation** $\sqrt{\mathbb{E}(X^2) - \mathbb{E}(X)^2}$ gives an indication of the distribution around the mean or expectation value. In the case of the dice with $X(\omega) = \omega^2$, $F_X(x) = \frac{1}{6}[\sqrt{x}]$ for $0 \leqslant x < 36$, $F_X(x) = 0$ for $x < 0$, and $F_X(x) = 1$ for $x \geqslant 36$. The standard deviation is 12.2, not much smaller than the expectation value, so the distribution is quite spread out.

Two random variables X and Y are called **independent** if for any two intervals $(x_1, x_2) \subset \mathbb{R}$ and $(y_1, y_2) \subset \mathbb{R}$, the events $A = \{\omega \in \Omega | X(\omega) \in (x_1, x_2)\}$ and $B = \{\omega \in \Omega | Y(\omega) \in (y_1, y_2)\}$ are independent. It is easy to prove the following theorem.

Theorem A.2 *Let X and Y be random variables with corresponding distribution functions F_X and F_Y. Then X and Y are independent if and only if*

$$F_{X,Y}(x, y) = F_X(x) F_Y(y) \tag{A.6}$$

*where the **joint distribution** $F_{X,Y}$ is defined by*

$$F_{X,Y}(x, y) = P\{\omega \in \Omega | X(\omega) \leqslant x \text{ and } Y(\omega) \leqslant y\}. \tag{A.7}$$

Now let us generalize the above concepts to uncountable sample spaces. For example, consider throwing darts at a dart board. Assuming that the dart player is sufficiently competent to be sure that he/she always hits the board, the probability of hitting the board is 1. But the probability of hitting any particular point on the board is zero! We have to consider *regions* of the board with non-zero area. If X is the random variable giving the distance from the centre of the board then the corresponding distribution function $F(r) = P\{X \leqslant r\}$ is the probability that the dart lands within a distance r from the centre. This will be a continuous function of r as opposed to the distribution function for the dice above. In general, one has the following characterization of distribution functions.

Theorem A.3 *A function $F : \mathbb{R} \to \mathbb{R}$ is a distribution function of a random variable if and only if it has the following properties:*

1. *F is non-decreasing,*
2. *F is continuous from the right: $\lim_{x \downarrow x_0} F(x) = F(x_0)$;*
3. *$\lim_{x \to -\infty} F(x) = 0$ and $\lim_{x \to \infty} F(x) = 1$.*

The definition of a probability function (called a **probability measure**) on an uncountable sample space is more complicated than that for a countable sample space. The reason is that it turns out not to be feasible to allow *every* subset of Ω to be assigned a probability. Instead, one defines a σ-**algebra** of subsets of Ω to be a collection $\mathcal{A}$ of subsets with the following closure properties.

1. $\Omega \in \mathcal{A}$;
2. If $A \in \mathcal{A}$ then $A^c \in \mathcal{A}$, and
3. If $A_n \in \mathcal{A}$ for $n = 1, 2, \ldots$ then $\cup_{n=1}^{\infty} A_n \in \mathcal{A}$.

Clearly, $\mathcal{A}$ then also contains countable intersections of its elements.
If Ω is a metric space (or more generally a topological space) then the smallest σ-algebra containing the open sets is called the **Borel σ-algebra**.

Given a sample space Ω with a σ-algebra $\mathcal{A}$ of subsets, a **probability measure** on Ω is a function $\mathbb{P} : \mathcal{A} \to [0, 1]$ such that

1. $\mathbb{P}(\Omega) = 1$, and
2. If $A_n \in \mathcal{A}$ $(n = 1, 2, \ldots)$ is a sequence of **disjoint** sets, that is $A_n \cap A_m = \emptyset$ if $n \neq m$, then

$$\mathbb{P}\left(\bigcup_{n=1}^{\infty} A_n\right) = \sum_{n=1}^{\infty} \mathbb{P}(A_n). \tag{A.8}$$

With this general definition of probability one can define a random variable as a function $X : \Omega \to \mathbb{R}$ such that $X^{-1}(a, b) \in \mathcal{A}$ for any interval $(a, b) \subset \mathbb{R}$. The distribution function of X is defined again by equation (A.5) and theorem A.3 holds. In fact, the distribution function is the more convenient way of describing a random variable. To generalize equations (A.3) and (A.4), we need the concept of integration with respect to a distribution function. This is done in analogy with the Riemann integral. Suppose first that X is a bounded random variable, that is there is a number $L > 0$ such that $F(x) = 0$ for $x < -L$ and $F(x) = 1$ for $x > L$. If $g : \mathbb{R} \to \mathbb{R}$ is a continuous function then we define the **Riemann-Stieltjes integral** of g with respect to the distribution function F as

$$\int g(x)\, \mathrm{d}F(x) = \lim_{n \to \infty} \sum_{i=1}^{n} g(x_i)(F(x_i) - F(x_{i-1})), \qquad (A.9)$$

where $x_i = -L + (2L/n)i \quad (i = 0, 1, \ldots, n)$ is a subdivision of $[-L, L]$. If X is not bounded then we can define $\int g(x)\mathrm{d}F(x) = \lim_{L \to \infty} \int_{-L}^{L} g(x)\mathrm{d}F(x)$. The **expectation (value)** of the random variable X is now defined by

$$\mathbb{E}(X) = \int x\, dF(x). \qquad (A.10)$$

More generally, the k**th moment** of X is defined as $\mathbb{E}(X^k)$. To see that equation (A.10) corresponds to (A.3) when X is in fact a discrete random variable consider the special case that $X = 0$ or $X = 1$, each with probability $\frac{1}{2}$. Then $F(x) = 0$ for $x < 0$; $F(x) = \frac{1}{2}$ for $0 \leqslant x < 1$ and $F(x) = 1$ for $x \geqslant 1$. We can take $L = 2$ and there are only two contributions to the sum in equation (A.9), namely from the intervals spanning $x = 0$ and $x = 1$: $i = [(n+1)/2]$ and $i = [(3n+1)/4]$. For both of these the jump in $F(x)$ is $\frac{1}{2}$ and in the limit $n \to \infty$ we obtain $\mathbb{E}(X) = \frac{1}{2}(0 + 1) = \frac{1}{2}$ in accordance with equation (A.3).

As above, we define two random variables X and Y to be **independent** when for any two open intervals (x_1, x_2) and (y_1, y_2) the events that $X \in (x_1, x_2)$ and $Y \in (y_1, y_2)$ are independent, that is

$$\mathbb{P}(\{\omega|\, X(\omega) \in (x_1, x_2) \text{ and } Y(\omega) \in (y_1, y_2)\})$$
$$= \mathbb{P}(\{\omega|\, X(\omega) \in (x_1, x_2)\})\, \mathbb{P}(\{\omega|\, Y(\omega) \in (y_1, y_2)\}). \qquad (A.11)$$

Theorem A.2 still holds. Moreover, we can define the double integral with respect to $F_{X,Y}(x, y)$ and hence the expectation of XY. For the latter the following holds:

Theorem A.4 *Let X and Y be independent random variables. Then $\mathbb{E}(XY)$ exists if both $\mathbb{E}(X)$ and $\mathbb{E}(Y)$ exist and in that case $\mathbb{E}(XY) = \mathbb{E}(X)\,\mathbb{E}(Y)$.*

Let us mention two special random variables of particular importance: If $X_1, \ldots, X_n$ are independent random variables which each take only the values 1 and 0 with probability p and $q = 1 - p$ respectively, then the random variable $X = X_1 + \cdots + X_n$ has the **binomial distribution** defined by

$$\mathbb{P}(X = k) = \binom{n}{k} p^k q^{n-k} \qquad (k = 0, 1, \ldots, n). \qquad (A.12)$$

This is a discrete random variable. A random variable X is said to have **normal** distribution with standard deviation σ if the corresponding distribution function is given by

$$G_\sigma(x) = \int_{-\infty}^{x} e^{-u^2/2\sigma^2} \frac{du}{\sqrt{2\pi}\,\sigma}. \tag{A.13}$$

If $\sigma = 1$ one speaks of the **standard normal distribution**.

One of the essential results in probability theory is the weak law of large numbers. It is proved using the **Chebyshev inequality**:

Theorem A.5 *Let X be a random variable taking values in a subset $D \subset \mathbb{R}$ and suppose that f is an increasing non-negative function on D. Then*

$$\mathbb{P}\{\omega \mid X(\omega) \geq a\} \leq \frac{\mathbb{E}(f(X))}{f(a)} \tag{A.14}$$

for any $a \in D$ such that $f(a) > 0$.

Proof.

$$\mathbb{E}(f(X)) = \int_{-\infty}^{\infty} f(x)\, dF_X(x) \geq \int_{a}^{\infty} f(x)\, dF_X(x)$$

$$\geq f(a) \int_{a}^{\infty} dF_X(x) = f(a)\, \mathbb{P}(X \geq a).$$

∎

The **weak law of large numbers** says the following.

Theorem A.6 *Let $(X_k)_{k=1}^{\infty}$ be a sequence of independent random variables with the same distribution function and with finite second moment. Then for any $\epsilon > 0$,*

$$\lim_{n\to\infty} \mathbb{P}\left(\left|\frac{X_1 + \cdots + X_n}{n} - \mathbb{E}(X_1)\right| \geq \epsilon\right) = 0. \tag{A.15}$$

Proof. It is an easy computation to show that if $\mathbb{E}(X_i) = \mu$ and $\mathbb{E}((X_i - \mu)^2) = \sigma^2$ then $S_n = X_1 + \cdots + X_n$ satisfies $\mathbb{E}(S_n) = n\mu$ and $\mathbb{E}((S_n - n\mu)^2) = n\sigma^2$. Applying equation (A.14) to $X = |S_n - n\mu|$ with $a = n\epsilon$ and $f(x) = x^2$ gives

$$\mathbb{P}(|S_n - n\mu| \geq n\epsilon) \leq \frac{n\sigma^2}{(n\epsilon)^2}$$

from which the theorem follows. ∎

This result shows that when the same experiment is repeated many times it becomes exceedingly unlikely that the average value strays more than any $\epsilon > 0$ from the expected value. In fact a stronger result holds: the **strong law of large numbers**.

Theorem A.7 *Let $(X_k)_{k=1}^{\infty}$ be a sequence of independent random variables with the same distribution, which has a mean μ. Then*

$$\mathbb{P}\left[\lim_{n\to\infty} \frac{X_1 + \cdots + X_n}{n} = \mu\right] = 1. \tag{A.16}$$

We prove here only a special case, namely when the fourth moment of the distribution exists. Let F be the distribution function of X_k and suppose that $\int x^4 \, dF(x) < +\infty$. In that case equation (A.16) follows from the following lemma.

Lemma A.1 (Borel-Cantelli lemma) *Let $A_1, A_2, \ldots$ be events in a probability space and define*

$$B = \cap_{n=1}^{\infty} \cup_{k=n}^{\infty} A_k.$$

Then the following hold.
(i)   *If $\sum_{n=1}^{\infty} \mathbb{P}(A_n) < +\infty$ then $\mathbb{P}(B) = 0$.*
(ii)   *If the events A_n are independent and $\sum_{n=1}^{\infty} \mathbb{P}(A_n) = +\infty$ then $\mathbb{P}(B) = 1$.*

Proof. It is easy to prove using equation (A.8) that if the sets are not disjoint the left-hand side of (A.8) is less than or equal to the right-hand side. Therefore

$$\mathbb{P}(B) \leqslant \mathbb{P}\left(\bigcup_{k=n}^{\infty} A_k\right) \leqslant \sum_{k=n}^{\infty} \mathbb{P}(A_k)$$

and this tends to zero in case (i). To prove (ii) it suffices to show that $\mathbb{P}(\cup_{k=n}^{\infty} A_k) = 1$ for all n. But

$$1 - \mathbb{P}\left(\bigcup_{k=n}^{\infty} A_k\right) \leqslant 1 - \mathbb{P}\left(\bigcup_{k=n}^{N} A_k\right) = \mathbb{P}\left(\bigcap_{k=n}^{N} A_k^c\right) = \prod_{k=n}^{N} \mathbb{P}(A_k^c)$$

for any finite N if the A_k are independent. If $\sum_{n=1}^{\infty} \mathbb{P}(A_n) = +\infty$ then the product in the right-hand side tends to zero as $N \to \infty$. ∎

Proof of Theorem A.7. Expanding the fourth power it is easy to compute that

$$\mathbb{E}\left[\left(\sum_{i=1}^{n}(X_i - \mu)\right)^4\right] = n\mathbb{E}[(X_i - \mu)^4] + 3n(n-1)\sigma^4 \leqslant Cn^2$$

for some constant C, where $\sigma^2 = \mathbb{E}[(X_i-\mu)^2]$. By theorem A.5 with $f(x) = x^4$ it follows that

$$\mathbb{P}\left[\left|\sum_{i=1}^{n}(X_i - \mu)\right| > \epsilon n\right] \leqslant \frac{Cn^2}{(\epsilon n)^4}$$

for any $\epsilon > 0$. By the above lemma, it now follows that with probability one, only finitely many of the events $A_n(\epsilon) = \{\omega : \left|\frac{1}{n}\sum_{i=1}^{n} X_i - \mu\right| > \epsilon\}$ occur.

This means that with probability one, ω belongs to $A_n(\epsilon)^c$ for n large enough. As this holds for all $\epsilon > 0$, $\frac{1}{n}\sum X_i$ tends to μ with probability one. ∎

To formulate the final important theorem we need to introduce the concept of weak convergence. We say that a sequence $(F_n)_{n=1}^{\infty}$ of distribution functions of random variables X_n ($n = 1, 2, \ldots$) **converges weakly** to a distribution function F if for every bounded continuous function $g : \mathbb{R} \to \mathbb{R}$,

$$\lim_{n\to\infty} \int g(x)\,\mathrm{d}F_n(x) = \int g(x)\,\mathrm{d}F(x). \tag{A.17}$$

Theorem A.8 *A sequence of distribution functions $(F_n)_{n=1}^{\infty}$ converges weakly to F if and only if*

$$\lim_{n\to\infty} F_n(x) = F(x) \tag{A.18}$$

for every x at which F is continuous.

It is easy to see with the help of this theorem that theorem A.6 expresses the fact that the distribution of the random variable $X = (X_1 + \cdots + X_N)/N$ converges weakly to that of the random variable equal to μ with probability 1.

With this definition one now has the following surprising result, called the **central limit theorem**.

Theorem A.9 *Let $X_1, X_2, \ldots$ be a sequence of independent random variables with identical distributions having mean μ and variance σ^2. Denote by F_n the distribution function of the random variable $n^{-1/2}(S_n - n\mu)$, where $S_n = X_1 + \cdots + X_n$. Then F_n converges weakly to the distribution function G_σ of the normal distribution given by equation (A.13).*

The proof of this theorem makes use of the following.

Theorem A.10 *A sequence of random variables $(X_n)_{n=1}^{\infty}$ with distribution functions $(F_n)_{n=1}^{\infty}$ converges weakly to X with distribution function F if and only if the corresponding **characteristic functions** defined by*

$$c_n(t) = \mathbb{E}\left(e^{itX_n}\right) \tag{A.19}$$

converge to the characteristic function $c(t) = \mathbb{E}(e^{itX})$ for all $t \in \mathbb{R}$.

This is proved essentially by expanding $g(x)$ in a Fourier series. Now, the characteristic function of G_σ is

$$c_\sigma(t) = \int_{-\infty}^{\infty} e^{itx} e^{-x^2/2\sigma^2} \frac{\mathrm{d}x}{\sqrt{2\pi}\sigma} = e^{-\sigma^2 t^2/2}. \tag{A.20}$$

On the other hand, if the distribution function of each X_i is denoted F then the characteristic function of $(S_n - n\mu)/\sqrt{n}$ is

$$c_n(t) = \int \cdots \int \exp\left(i\frac{t}{\sqrt{n}}\sum_{i=1}^{n}(x_i - \mu)\right) dF(x_1)\ldots dF(x_n)$$

$$= \prod_{i=1}^{n}\int \exp\left(i\frac{t}{\sqrt{n}}(x_i - \mu)\right) dF(x_i)$$

$$= \left(\mathbb{E}\left[1 + i\frac{t}{\sqrt{n}}(X_1 - \mu) - \frac{t^2}{2n}(X_1 - \mu)^2 + \ldots\right]\right)^n$$

$$\approx \left(1 - \frac{t^2}{2n}\mathbb{E}[(X_1 - \mu)^2]\right)^n$$

$$= \left(1 - \frac{t^2\sigma^2}{2n}\right)^n$$

which clearly tends to (A.20).   ∎

There is an analogue of theorem A.10 for the Laplace transform.

Theorem A.11 *Let $(X_n)_{n=1}^{\infty}$ be a sequence of random variables such that their Laplace transforms $L_n(t) = \mathbb{E}(e^{tX_n})$ exist for all $t \in \mathbb{R}$. Let X be another random variable for which the Laplace transform $L(t) = \mathbb{E}(e^{tX})$ exists for all $t \in \mathbb{R}$. Then the sequence $(X_n)_{n=1}^{\infty}$ converges weakly to X if and only if $\lim_{n\to\infty} L_n(t) = L(t)$ for all $t \in \mathbb{R}$.*

Quantum Mechanics

In this appendix, we collect some results from quantum mechanics supplementing the introduction to quantum theory in chapter 20. For a proper introduction to the subject see for example the books by Schiff (1969) or Messiah (1961).

The motion of a single particle in a potential $V(\vec{x}, t)$ depending on the position $\vec{x}$ and the time t is described by a **wave function** $\Psi(\vec{x}, t)$ which satisfies the **Schrödinger equation**

$$i\hbar \frac{\partial \Psi}{\partial t} = -\frac{\hbar^2}{2m} \Delta \Psi + V(\vec{x}, t) \Psi. \qquad (B.1)$$

Here m is the mass of the particle and Δ is the Laplace operator:

$$\Delta = \frac{\partial^2}{\partial x^2} + \frac{\partial^2}{\partial y^2} + \frac{\partial^2}{\partial z^2}.$$

This equation was postulated by Schrödinger in 1926 because if $V = 0$ then the plane wave

$$\Psi(\vec{x}, t) = \exp\left(\frac{i}{\hbar}(\vec{p} \cdot \vec{x} - Et)\right) \qquad (B.2)$$

satisfies this equation. Such a wave has a frequency $\nu = E/(2\pi\hbar) = E/h$ which is just Einstein's relation (18.1) and it has a wavelength $\lambda = h/p$. The latter relation had already been postulated by De Broglie in 1923 and verified experimentally by Compton in 1925.

A special case of Schrödinger's equation obtains if the potential V is independent of the time. In that case one can write the solution in the form

$$\Psi(\vec{x}, t) = \psi(\vec{x})e^{-iEt/\hbar} \qquad (B.3)$$

where $\psi(\vec{x})$ is a solution of the **time-independent Schrödinger equation**:

$$\left(-\frac{\hbar^2}{2m}\Delta + V(\vec{x})\right)\psi(\vec{x}) = E\psi(\vec{x}). \qquad (B.4)$$

This is an **eigenvalue equation**; in general it does not have solutions for every value of the energy E. The operator in brackets acting on ψ is called the **Hamiltonian** $\mathcal{H}$ so equation (B.4) is just (18.2). The wavefunction $\Psi(\vec{x}, t)$ has the following interpretation: $|\Psi(\vec{x}, t)|^2 \mathrm{d}^3 x$ *represents the probability of finding the particle in a volume element* $\mathrm{d}^3 x$ *around the position $\vec{x}$ at time t.* This interpretation is justified because

$$\frac{\mathrm{d}}{\mathrm{d}t}\left(\int |\Psi(\vec{x}, t)|^2 \mathrm{d}^3 x \right) = 0,$$

as follows immediately from equation (B.3). (In fact, it also holds in general.) Notice that if Ψ is a solution of the Schrödinger equation then $\lambda \Psi$ is also a solution for any constant λ so that we can normalize Ψ (and similarly ψ) by

$$\int |\Psi(\vec{x}, t)|^2 \mathrm{d}^3 x = 1. \tag{B.5}$$

To solve the (time-independent) Schrödinger equation we need **boundary conditions**. First of all, in order to be able to normalize Ψ (or equivalently, $\psi(\vec{x})$) as in equation (B.5) the integral has to exist, that is, ψ must be **square-integrable**. The space of square-integrable functions is usually denoted $L^2(\mathbb{R}^3)$. It is a **Hilbert space**: a space in which a scalar product is defined. The scalar product of two square-integrable functions $u, v \in L^2(\mathbb{R}^3)$ is defined by

$$\langle u|v \rangle = \int \overline{u(\vec{x})} v(\vec{x}) \mathrm{d}^3 x. \tag{B.6}$$

(The bar indicates complex conjugation.) Secondly, the wavefunction ψ must be continuously differentiable in order that the second derivatives are well-defined. (One might think that it must be twice differentiable, but in fact the second derivative can have jumps as the only requirement is that it is square-integrable.)

Let us now consider the examples mentioned in the text in turn. The first example is the **harmonic oscillator**: a particle moving on a line attracted to the origin by a force $F = -kx$ proportional to the excursion x. The corresponding potential is $V(x) = \frac{1}{2}kx^2$ (see equation (18.5)) and the Schrödinger equation is therefore

$$-\frac{\hbar^2}{2m}\psi''(x) + \frac{1}{2}kx^2\psi(x) = E\psi(x). \tag{B.7}$$

To solve this equation, one first introduces a change of variable

$$z = \alpha x \quad \text{with} \quad \alpha = \sqrt{\frac{2m\omega}{\hbar}} \quad \text{and} \quad \omega = \sqrt{\frac{k}{m}}. \tag{B.8}$$

This leads to the equation $-\psi''(z) + \frac{1}{4}z^2\psi(z) = \frac{E}{\hbar\omega}\psi(z)$. Next one substitutes $\psi(z) = v(z)\Omega_0(z)$ where

$$\Omega_0(z) = \frac{1}{(2\pi)^{1/4}}e^{-z^2/4}. \tag{B.9}$$

The resulting equation is $-v''(z) + zv'(z) = \left(\frac{E}{\hbar\omega} - \frac{1}{2}\right)v(z)$. It has two kinds of solution: polynomials and very rapidly increasing functions. The latter can be excluded because $\psi \in L^2(\mathbb{R})$. The polynomial solutions are given by the **Hermite polynomials**:

$$H_0(z) = 1, \qquad H_1(z) = z, \qquad H_{n+1}(z) = zH_n(z) - nH_{n-1}(z). \qquad \text{(B.10)}$$

(Note that another convention is in common use.) It is easy to check that these polynomials satisfy the equation for v using the relation

$$H_n'(z) = nH_{n-1}(z). \qquad \text{(B.11)}$$

The corresponding values of E are the energy eigenvalues (18.3) and the wavefunctions are therefore given by

$$\psi_n(x) = H_n(\alpha x)\Omega_0(\alpha x). \qquad \text{(B.12)}$$

The next example is a free particle on a line confined to an interval $[0, L]$. In that case we impose **Dirichlet boundary conditions**: $\psi(0) = \psi(L) = 0$. The Schrödinger equation in this case is simply the free one:

$$-\frac{\hbar^2}{2m}\psi''(x) = E\psi(x). \qquad \text{(B.13)}$$

Given the boundary conditions the solutions are standing waves:

$$\psi(x) = \sin\left(\frac{n\pi x}{L}\right) \qquad \text{(B.14)}$$

and the eigenvalues (18.6) follow immediately. In the three-dimensional case one has analogously

$$\psi(\vec{x}) = \sin\left(\frac{n_1\pi x}{L}\right)\sin\left(\frac{n_2\pi y}{L}\right)\sin\left(\frac{n_3\pi z}{L}\right). \qquad \text{(B.15)}$$

We have seen that the corresponding spectrum (18.7) is degenerate. This is due to the **symmetry** of the system; if the dimensions in the x-, y-, and z-direction are not equal the degeneracy disappears.

The example of a Coulomb potential is much more complicated. Before we analyse it let us discuss in some more detail the general structure of quantum mechanics. Notice first that the operator

$$\vec{p} = \frac{\hbar}{i}\vec{\nabla} \qquad \text{(B.16)}$$

when applied to the wavefunction (B.2) yields exactly the classical momentum of the particle. This operator is therefore identified as the **momentum operator** in quantum mechanics. Notice also that this is consistent with the first term in the Schrödinger equation which then becomes the usual expression for the kinetic energy

$$E_{\text{kin}} = \frac{\vec{p}^2}{2m}. \qquad \text{(B.17)}$$

In general, every observable quantity in quantum mechanics corresponds to a **self-adjoint operator** on the Hilbert space $L^2(\mathbb{R}^3)$. A self-adjoint operator is a linear map $A : L^2(\mathbb{R}^3) \to L^2(\mathbb{R}^3)$ which satisfies the following relation:

$$\int \overline{u(\vec{x})}(Av)(\vec{x})\mathrm{d}^3 x = \int \overline{(Au)(\vec{x})}v(\vec{x})\mathrm{d}^3 x \qquad (B.18)$$

for all $u, v \in L^2(\mathbb{R}^3)$. In terms of the scalar product (B.6) this reads simply $\langle u \,|\, Av \rangle = \langle Au \,|\, v \rangle$. The reason for this relation is as follows. If one measures the observable corresponding to the operator A then the only possible values this observable can have are the **eigenvalues** of the operator A, that is, values λ for which there exists a non-zero wavefunction ψ such that $A\psi = \lambda\psi$. After the measurement the particle then resides in the state given by the so-called **eigenfunction** ψ. The condition (B.18) guarantees that the eigenvalues λ are all real valued.

REMARK B.1
In fact, we have been inaccurate in the definition of operators and self-adjoint operators in particular. Unbounded operators, that is operators that can have arbitrarily large values cannot be defined on the whole Hilbert space. For example, the momentum operator $\vec{p}$ can only be defined on functions u which have square-integrable partial derivatives. In general, these operators are only defined on a dense subset of $L^2(\mathbb{R}^3)$. This subset is called the **domain** of the operator. An operator is defined properly if both its operation and its domain are given. Given an operator A with domain $D(A)$ one defines the **adjoint** A^* of A as the operator with domain

$$D(A^*) = \{v \in L^2(\mathbb{R}^3)|\, \exists w \in L^2(\mathbb{R}^3) : \langle w \,|\, u \rangle = \langle v \,|\, Au \rangle \text{ for all } u \in L^2(\mathbb{R}^3)\}$$

and one puts $A^*v = w$ in that case. Next one says that A is **self-adjoint** if $A^* = A$ and $D(A^*) = D(A)$. We shall ignore this technical point in the following.

Label the eigenvalues of the operator A by λ_n and the corresponding eigenfunctions ψ_n. Suppose that the particle is originally in a state ψ. Assuming that all these states are normalized, the probability of finding the value λ_n upon measuring A is

$$\mathbb{P}(\lambda_n) = |\langle \psi_n \,|\, \psi \rangle|^2. \qquad (B.19)$$

In analogy with equations (A.3) and (A.10) the **expectation value** of the operator A in the state ψ is given by

$$\mathbb{E}_\psi(A) = \sum_n \lambda_n \mathbb{P}(\lambda_n) = \langle \psi \,|\, A\psi \rangle. \qquad (B.20)$$

In general, if two observables A and B are measured then the probability of obtaining the eigenvalues λ_n for A and μ_m for B, respectively depends on which of these observables is measured first. This can be easily deduced from the fact that *the state of the particle changes upon measuring A or B*. There is an exception, however: if *A and B commute then they can be diagonalized simultaneously* so that there exists a common set of eigenfunctions $\psi_{n,m}$ for both A and B. Let us be more clear about this. If there is an operator B commuting with A then the eigenvalues of A are in general **degenerate**; for each eigenvalue λ_n of A there is a linear space V_n of eigenfunctions with the same eigenvalue λ_n:

$$A\psi = \lambda_n \psi \quad \text{for all } \psi \in V_n. \tag{B.21}$$

The eigenspaces V_n and $V_{n'}$ for different eigenvalues are **mutually orthogonal**; i.e. for every $u \in V_n$ and $v \in V_{n'}$, $\langle u \,|\, v \rangle = 0$. Moreover, *each V_n is left invariant by B*, for if $\psi \in V_n$ then $AB\psi = BA\psi = \lambda_n B\psi$ so $B\psi \in V_n$. This means that B can be diagonalized separately in each V_n.

Examples of two operators which do not commute are the momentum operator $\vec{p}$ and the position operator $\vec{x}$. The latter is simply given by $(\vec{x}\psi)(\vec{x}) = \vec{x}\psi(\vec{x})$, that is, it multiplies the wave function by the components $x_1 = x$, $x_2 = y$ and $x_3 = z$ of its argument. It is easy to see that

$$[x_j, p_k] = i\hbar \delta_{j,k}, \tag{B.22}$$

where the square brackets indicate the **commutator**:

$$[A, B] = AB - BA \tag{B.23}$$

which measures to what extent A and B do not commute. Corresponding components of the position and the momentum therefore do not commute and this means according to the discussion above that it matters which is measured first. In fact, it also means that they cannot both be known to arbitrary accuracy. To understand this we first define the **variance** of an operator A in a state given by the normalized wavefunction ψ in analogy with the expectation:

$$(\Delta A)^2 = \langle \psi \,|\, A^2 \psi \rangle - \langle \psi \,|\, A\psi \rangle^2. \tag{B.24}$$

Using the fact that $\langle (\lambda x_j + ip_j)\psi \,|\, (\lambda x_j + ip_j)\psi \rangle \geqslant 0$ for all real λ it is an easy calculation to find that

$$(\Delta x_j)^2 (\Delta p_j)^2 \geqslant \frac{1}{4}\hbar^2. \tag{B.25}$$

Another important quantity is the **orbital angular momentum**. It is given by

$$\vec{L} = \vec{x} \times \vec{p} = \frac{\hbar}{i}\vec{x} \times \vec{\nabla}. \tag{B.26}$$

Its components satisfy the commutation relations

$$[L_1, L_2] = i\hbar L_3 \qquad [L_2, L_3] = i\hbar L_1 \qquad [L_3, L_1] = i\hbar L_2. \tag{B.27}$$

It follows that

$$[\vec{L}^2, L_j] = 0 \text{ for } j = 1, 2, 3. \tag{B.28}$$

This means that $\vec{L}^2$ and L_j can be diagonalized simultaneously. One usually chooses $j = 3$. In terms of polar coordinates, the operators $\vec{L}^2$ and L_3 are given by

$$\vec{L}^2 = -\hbar^2 \left[\frac{1}{\sin\theta} \frac{\partial}{\partial\theta} \left(\sin\theta \frac{\partial}{\partial\theta} \right) + \frac{1}{\sin^2\theta} \frac{\partial^2}{\partial\phi^2} \right] \tag{B.29}$$

and

$$L_3 = \frac{\hbar}{i} \frac{\partial}{\partial\phi}. \tag{B.30}$$

The corresponding eigenfunctions are called **spherical harmonics** and denoted $Y_l^m(\theta, \phi)$, and the corresponding eigenvalues are $l(l+1)\hbar^2$ and $m\hbar$:

$$\vec{L}^2 Y_l^m = l(l+1)\hbar^2 Y_l^m \qquad \text{and} \qquad L_3 Y_l^m = m\hbar Y_l^m. \tag{B.31}$$

Here l and m are integers subject to the constraints: $l \geqslant 0$ and $|m| \leqslant l$. The angular momentum is therefore quantized.

Let us now come back to the example of the hydrogen atom. In terms of polar coordinates the Hamiltonian can be written as

$$H = -\frac{\hbar^2}{2m}\Delta + V(\vec{x}) = -\frac{\hbar^2}{2mr^2}\frac{\partial}{\partial r}r^2\frac{\partial}{\partial r} + \frac{\vec{L}^2}{2mr^2} - \frac{e^2}{2\pi\epsilon_0 r}. \tag{B.32}$$

It commutes with $\vec{L}^2$ as well as L_3 due to rotation invariance. The eigenfunctions can therefore be written as $R(r)Y_l^m(\theta, \phi)$. The resulting equation for the radial function $R(r)$ then yields the eigenvalues (18.11) where $n \geqslant l + 1$. A given energy level thus has a degeneracy determined by the allowed values of l and m which are: $l = 0, 1, \ldots, n-1$ and $m = -l, -l+1, \ldots, l$, that is a total of $\sum_{l=0}^{n-1}(2l+1) = n^2$. (In fact there is an additional factor of 2 owing to the electron spin as we have seen.)

The operators corresponding to electron spin were postulated by Pauli to satisfy the same commutation relations (B.27) but allowing only the value $\frac{1}{2}$ for l. (Notice that this value is not allowed in the case of the orbital angular momentum $\vec{L}^2$ above!) It is easily seen that the following 2×2 matrices satisfy these conditions: $\vec{S} = \frac{1}{2}\hbar\vec{\sigma}$, where

$$\sigma^{(x)} = \begin{pmatrix} 0 & 1 \\ 1 & 0 \end{pmatrix} \qquad \sigma^{(y)} = \begin{pmatrix} 0 & -i \\ i & 0 \end{pmatrix} \qquad \sigma^{(z)} = \begin{pmatrix} 1 & 0 \\ 0 & -1 \end{pmatrix}. \tag{B.33}$$

The latter are known as the **Pauli matrices**. The eigenstates corresponding to the eigenvalues $\pm\frac{1}{2}\hbar$ of $S^{(z)}$ are simply $\chi_+ = \begin{pmatrix} 1 \\ 0 \end{pmatrix}$ and $\chi_- = \begin{pmatrix} 0 \\ 1 \end{pmatrix}$. They are also eigenstates of $\vec{S}^2$: $\vec{S}^2\chi_\pm = \frac{3}{4}\hbar^2\chi_\pm$. The complete eigenstates of the hydrogen atom are thus given by $R_n(r)Y_l^m(\theta, \phi)\chi_\pm$.

Note that in adding the spin degree of freedom to the Hamiltonian of the hydrogen atom, we have in fact extended the Hilbert space to two-component vector-valued functions: $\mathcal{H} = L^2\left(\mathbb{R}^3; \mathcal{C}^2\right)$. If we want to describe two or more particles we need functions depending on the coordinates of all these particles. For example, for two particles $\mathcal{H} = L^2(\mathbb{R}^3 \times \mathbb{R}^3)$ if they are spinless particles, and $L^2(\mathbb{R}^3 \times \mathbb{R}^3; \mathbb{C}^2 \times \mathbb{C}^2)$ if they are spin-$\frac{1}{2}$ particles. The Hamiltonian for two interacting particles has the form

$$H = -\frac{\hbar^2}{2m_1}\Delta_1 - \frac{\hbar^2}{2m_2}\Delta_2 + V_1(\vec{x}_1) + V_2(\vec{x}_2) + V_{12}(\vec{x}_1 - \vec{x}_2). \qquad \text{(B.34)}$$

If the particles are non-interacting the last term vanishes and the eigenfunctions can be written as products $\psi(\vec{x}_1, \vec{x}_2) = \psi_1(\vec{x}_1)\psi_2(\vec{x}_2)$, where ψ_i $(i = 1, 2)$ is an eigenfunction of $H_i = -(\hbar^2/2m_i)\Delta_i + V_i(\vec{x}_i)$. The corresponding energy levels are sums of energy levels of the two particles as stated in general in postulate QM 2 of chapter 18. If the particles are *indistinguishable* the wavefunction must be invariant under exchange of the coordinates of the particles: $\mathcal{S}\psi = \psi$, where $\mathcal{S}$ is the **particles exchange operator**:

$$\mathcal{S}\psi(\vec{x}_1, \vec{x}_2) = \psi(\vec{x}_2, \vec{x}_1). \qquad \text{(B.35)}$$

The Hamiltonian must commute with $\mathcal{S}$ in that case, i.e. $m_1 = m_2$ and $V_1 = V_2$. The exchange-invariant eigenfunctions for non-interacting particles are:

$$\psi_S(\vec{x}_1, \vec{x}_2) = \frac{1}{\sqrt{2}}(\psi_1(\vec{x}_1)\psi_2(\vec{x}_2) + \psi_1(\vec{x}_2)\psi_2(\vec{x}_1)) \qquad \text{(B.36)}$$

and

$$\psi_A(\vec{x}_1, \vec{x}_2) = \frac{1}{\sqrt{2}}(\psi_1(\vec{x}_1)\psi_2(\vec{x}_2) - \psi_1(\vec{x}_2)\psi_2(\vec{x}_1)). \qquad \text{(B.37)}$$

The latter is of course not strictly invariant but an overall sign-change has no physical effect. If the particles have spin the situation becomes more complicated. In that case we must also exchange their spin coordinates. For spin-$\frac{1}{2}$ particles there are four product spin states: $\chi_+(1)\chi_+(2)$, $\chi_+(1)\chi_-(2)$, $\chi_-(1)\chi_+(2)$, and $\chi_-(1)\chi_-(2)$ (the number in brackets indicates the particles). The first and last are already $\mathcal{S}$-invariant but the other two must be combined to form symmetric and anti-symmetric combinations:

$$\chi_{S,A} = \frac{1}{\sqrt{2}}(\chi_+(1)\chi_-(2) \pm \chi_-(1)\chi_+(2)).$$

These spin states have to be combined with the spatial wavefunctions (B.36) and (B.37) to obtain a total of 8 possible wavefunctions of definite symmetry. Remarkably, not all these combinations can occur! It turns out that for systems of identical spin-$\frac{1}{2}$ particles only wavefunctions which are *anti-symmetric* under exchange of space and spin coordinates are allowed. This fundamental law of nature is the **connection between spin and statistics**. For two non-interacting spin-$\frac{1}{2}$ particles the possible wavefunctions are therefore

$$\psi_{1,1} = \psi_A(\vec{x}_1, \vec{x}_2)\chi_+(1)\chi_+(2), \qquad \text{(B.38a)}$$

$$\psi_{1,0} = \psi_A(\vec{x}_1, \vec{x}_2)\chi_S, \tag{B.38b}$$

$$\psi_{1,-1} = \psi_A(\vec{x}_1, \vec{x}_2)\chi_-(1)\chi_-(2), \tag{B.38c}$$

and

$$\psi_{0,0} = \psi_S(\vec{x}_1, \vec{x}_2)\chi_A. \tag{B.38d}$$

If we introduce the total spin operators

$$\vec{S} = \vec{S}_1 + \vec{S}_2 \tag{B.39}$$

then a simple calculation shows that

$$S^{(z)}\chi_+(1)\chi_+(2) = \hbar\chi_+(1)\chi_+(2)$$
$$S^{(z)}\chi_-(1)\chi_-(2) = -\hbar\chi_-(1)\chi_-(2),$$

and $S^{(z)}\chi_{A,S} = 0$, and also that

$$\vec{S}^2\chi_+(1)\chi_+(2) = 2\hbar^2\chi_+(1)\chi_+(2) \qquad \vec{S}^2\chi_S = 2\hbar^2\chi_S$$
$$\vec{S}^2\chi_-(1)\chi_-(2) = 2\hbar^2\chi_-(1)\chi_-(2) \qquad \vec{S}^2\chi_A = 0.$$

This shows that the first three states (B.38a), (B.38b), and (B.38c) are the three possible spin-1 states, whereas (B.38d) is a spin-0 state. One speaks of a **triplet** and a **singlet** of states. In general, the addition of an odd number of spin-$\frac{1}{2}$ operators leads to half-odd integer spins and addition of an even number leads to integer spins.

For systems of a large number of particles it is convenient to introduce the so-called **creation and annihilation operators**. Let us consider a system of spin-$\frac{1}{2}$ particles which according to the above are fermions so that the wavefunction for the total system must be anti-symmetric in the coordinates of the particles:

$$\Psi(\vec{x}_1, \sigma_1; \ldots; \vec{x}_k, \sigma_k; \ldots; \vec{x}_l, \sigma_l; \ldots; \vec{x}_N, \sigma_N)$$
$$= -\Psi(\vec{x}_1, \sigma_1; \ldots; \vec{x}_l, \sigma_l; \ldots; \vec{x}_k, \sigma_k; \ldots; \vec{x}_N, \sigma_N).$$

Note that both the position and the spin coordinates of the k-th and l-th particles must be interchanged. There is a straightforward generalization of the states (B.37) and (B.38). Let a complete set of (eigen)states for a single particle be labelled by $\vec{k}$ and σ. Then the following is a typical N-particle state:

$$\Psi = \frac{1}{\sqrt{N!}}\sum_P (-1)^{|P|}\psi_{\vec{k}_{P(1)},\sigma_{P(1)}}(\vec{x}_1)\ldots\psi_{\vec{k}_{P(N)},\sigma_{P(N)}}(\vec{x}_N), \tag{B.40}$$

where the sum runs over all $N!$ permutations of the labels and where all pairs $(\vec{k}_i, \sigma_i)$ are necessarily different. It is customary to introduce the following notation for this state: $\Psi = |\{n_{\vec{k},\sigma}\}\rangle$, where $\{n_{\vec{k},\sigma}\}$ denotes a set of **occupation**

numbers for each possible single-particle state such that $n_{\vec{k},\sigma} = 1$ if $(\vec{k}, \sigma)$ is one of the states $(\vec{k}_i, \sigma_i)$ $(i = 1, \ldots, N)$ and $n_{\vec{k},\sigma} = 0$ otherwise. The **creation operator** $b^*_{\vec{k},\sigma}$ is then defined as follows:

$$b^*_{\vec{k}_1,\sigma_1}|\{n_{\vec{k},\sigma}\}\rangle = \delta_{n_{\vec{k}_1,\sigma_1},0}|\{n'_{\vec{k},\sigma}\}\rangle, \tag{B.41}$$

where $n'_{\vec{k},\sigma} = 1$ if $(\vec{k}, \sigma) = (\vec{k}_1, \sigma_1)$ and $n'_{\vec{k},\sigma} = n_{\vec{k},\sigma}$ otherwise. The **annihilation operator** $b_{\vec{k}_1,\sigma_1}$ is defined by

$$b_{\vec{k}_1,\sigma_1}|\{n_{\vec{k},\sigma}\}\rangle = \delta_{n_{\vec{k}_1,\sigma_1},1}|\{n''_{\vec{k},\sigma}\}\rangle, \tag{B.42}$$

where $n''_{\vec{k},\sigma} = 0$ if $(\vec{k}, \sigma) = (\vec{k}_1, \sigma_1)$ and $n''_{\vec{k},\sigma} = n_{\vec{k},\sigma}$ otherwise.

These operators act on a very large Hilbert space, called the **Fock space**, which is the direct sum of all N-particle spaces for arbitrary N. Remember that the N-particle Hilbert space is a space of square-integrable functions of N position variables (and spin variables). Note that the creation operator increases the number of particles by one and the annihilation operator decreases it by one. As the notation suggests, the creation operator is the adjoint of the corresponding annihilation operator. Moreover, it is easily checked that they satisfy the **anti-commutation relations**

$$\left(b^*_{\vec{k},\sigma}\right)^2 = 0, \left(b_{\vec{k},\sigma}\right)^2 = 0 \text{ and } \left\{b_{\vec{k},\sigma}, b^*_{\vec{k}',\sigma'}\right\} = \delta_{\vec{k},\vec{k}'}\delta_{\sigma,\sigma'}. \tag{B.43}$$

Here the curly brackets denote the **anti-commutator**:

$$\{A, B\} = AB + BA. \tag{B.44}$$

The product operator $b^*_{\vec{k},\sigma} b_{\vec{k},\sigma}$ measures the number of particles in state $\vec{k}, \sigma$ (0 or 1). The merit of introducing the creation and annihilation operators is that an operator for a system of many particles can be written in terms of these, irrespective of the actual number of particles present. For example, consider a system of identical fermions moving in a cubic box of sides L. The single particle energy levels are then given by equation (18.7) and the corresponding eigenfunctions $\psi_{n_1,n_2,n_3,\sigma}$ are given by equation (B.15) (for spin-$\frac{1}{2}$ particles we have to add a spin index σ). The Hamiltonian for the total system can now be written in the form

$$H_{\text{many particle}} = \sum_{n_1,n_2,n_3,\sigma} E(n_1, n_2, n_3)\, b^*_{n_1,n_2,n_3,\sigma} b_{n_1,n_2,n_3,\sigma}. \tag{B.45}$$

Next suppose there is an additional interaction between the particles. If each particle exerts a force on each of the other particles given by a potential $V(\vec{x}' - \vec{x})$ depending on the distance then we can define

$$\hat{v}(\underline{n}''', \underline{n}'', \underline{n}', \underline{n}) = \int\int \overline{\psi_{\underline{n}'''}(\vec{x}')\, \psi_{\underline{n}''}(\vec{x})} V(\vec{x}' - \vec{x})\psi_{\underline{n}'}(\vec{x}')\psi_{\underline{n}}(\vec{x})\, \mathrm{d}^3 x\, \mathrm{d}^3 x'$$

$$\tag{B.46}$$

where $\underline{n}$ stands for (n_1, n_2, n_3, σ), and we have

$$H_{\text{interaction}} = \sum_{\underline{n}''', \underline{n}'', \underline{n}', \underline{n}} \hat{v}(\underline{n}''', \underline{n}'', \underline{n}', \underline{n}) \, b_{\underline{n}'''}^* b_{\underline{n}''}^* b_{\underline{n}'} b_{\underline{n}}. \tag{B.47}$$

For bosons, one can introduce similar creation and annihilation operators. The difference is that there can be more than one bosonic particle in the same state. This means that the numbers $n_{\vec{k}}$ can now be arbitrary non-negative integers. The **bosonic creation and annihilation operators** act as follows:

$$a_{\vec{k}}^* |n_{\vec{k}}\rangle = \sqrt{n_{\vec{k}} + 1} \, |n_{\vec{k}} + 1\rangle \qquad \text{and} \qquad a_{\vec{k}} |n_{\vec{k}}\rangle = \sqrt{n_{\vec{k}}} \, |n_{\vec{k}} - 1\rangle. \tag{B.48}$$

They satisfy the **commutation relations**

$$[a_{\vec{k}}, a_{\vec{k}'}] = 0 \qquad [a_{\vec{k}}^*, a_{\vec{k}'}^*] = 0 \qquad [a_{\vec{k}}, a_{\vec{k}'}^*] = \delta_{\vec{k}, \vec{k}'}. \tag{B.49}$$

Convexity

In this appendix, we prove some more general results about convex functions. Vectors will be denoted by roman letters without arrows.

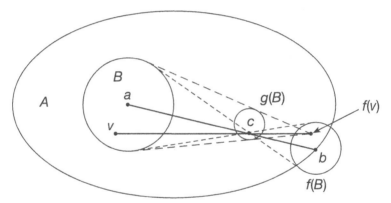

Figure C.1 The proof of lemma C.1.

We start with the following technical result:

Lemma C.1 *Let A be a convex subset of $\mathbb{R}^k$ with non-empty interior. Suppose $a \in \mathrm{int}(A)$ and $b \in \overline{A}$. Then every internal point of the segment from a to b, that is every point $c = \lambda a + (1 - \lambda)b$ with $\lambda \in (0, 1)$, is an interior point of A.*

Proof. Set $t = \lambda/(1 - \lambda)$ and define the mapping $f(x) = c - t(x - c)$. Then $f(a) = b$, $f(c) = c$, and $\lambda x + (1 - \lambda)f(x) = c$ for all x. Since $a \in \mathrm{int}(A)$ there is $r > 0$ such that the ball B with radius r and centre a is contained in A. Then $f(B)$ is a ball with centre b and because $b \in \overline{A}$, there exists $v \in B$ such that $f(v) \in A$. Now define another map g by $g(x) = \lambda x + (1 - \lambda)f(v)$. The somewhat complicated situation is illustrated in figure C.1. Then $g(B)$ is an open neighbourhood of $g(v) = c$. But $g(B) \subset A$. ∎

Theorem C.1 *If $A \subset \mathbb{R}^k$ is convex then its closure $\overline{A}$ is also convex. If, moreover, A has non-empty interior then $\text{int}(A)$ is convex, $\overline{A} = \overline{\text{int}(A)}$, and $\text{int}(\overline{A}) = \text{int}(A)$.*

Proof. Suppose $x, y \in \overline{A}$ and $\lambda \in [0, 1]$. Then there exist sequences $x_n, y_n \in A$ converging to x and y, respectively. It follows that $\lambda x_n + (1 - \lambda)y_n \to \lambda x + (1 - \lambda)y$ so that the latter is also in $\overline{A}$. This proves the first statement. If $a, b \in \text{int}(A)$ then every interior point of the segment from a to b is also in $\text{int}(A)$ so this set is convex. Obviously, $\overline{\text{int}(A)} \subset \overline{A}$. Now suppose $b \in \overline{A}$. Choose $a \in \text{int}(A)$ arbitrarily. Then $c = \lambda a + (1 - \lambda)b \in \text{int}(A)$ if $\lambda > 0$ and $c \to b$ as $\lambda \to 0$, so $b \in \overline{\text{int}(A)}$. Obviously also $\text{int}(A) \subset \text{int}(\overline{A})$. Suppose $c \in \text{int}(\overline{A})$. Then there exists a ball B with centre c and radius $r > 0$ contained in $\overline{A}$. Because $\overline{A} = \overline{\text{int}(A)}$, there exists $a \in \text{int}(A) \cap B$. If $a = c$ then we are done. Otherwise, let $b = 2c - a$. Then $b \in B \subset \overline{A}$. By the lemma, therefore, $c = \frac{1}{2}(a + b) \in \text{int}(A)$. ∎

Nex we give some more definitions.

> *A closed convex set with non-empty interior is called a **convex body**.*
> *A **linear variety** in $\mathbb{R}^k$ is a subset of the form $V = b + N$, where N is a linear subspace of $\mathbb{R}^k$, that is $x, y \in N \Rightarrow \lambda x + \mu y \in N$ for all real λ, μ.*
> *A **hyperplane** in $\mathbb{R}^k$ is a linear variety with dimension $k - 1$.*

EXAMPLE C.1: *Hyperplanes in $\mathbb{R}^3$.*
A hyperplane in $\mathbb{R}^3$ is a plane. It is given by an equation of the form $a_1 x_1 + a_2 x_2 + a_3 x_3 = \lambda$. The vector $a = (a_1, a_2, a_3)$ is perpendicular to the plane.

Lemma C.2 *$H \subset \mathbb{R}^k$ is a hyperplane if and only if there exist $a \in \mathbb{R}^k, a \neq 0$ and $\lambda \in \mathbb{R}$ such that*

$$H = \{x \in \mathbb{R}^k \mid \langle a, x \rangle = \lambda\}. \tag{C.1}$$

Here $\langle a, x \rangle$ is the scalar product given by $\langle a, x \rangle = \sum_{i=1}^{k} a_i x_i$.

Proof. Let $H = b + N$, where N is $k - 1$-dimensional. There exist $k - 1$ independent vectors $v_1, \ldots, v_{k-1} \in N$. Choose $v_k \notin N$. Then every $x \in \mathbb{R}^k$ can be written uniquely in the form $x = \sum_{i=1}^{k} \lambda_k v_k$. Define a linear map $f : \mathbb{R}^k \to \mathbb{R}$ by $f(x) = \lambda_k$. Then $f(x) = 0$ for all $x \in N$ and hence $f(x) = f(b)$ for $x \in H$. Define $\lambda = f(b)$. Finally, let $a_i = f(e_i)$, where $\{e_i \mid i = 1, \ldots, k\}$ is the canonical basis of $\mathbb{R}^k$. It is now straightforward to show that $H = \{x \in \mathbb{R}^k \mid f(x) = \lambda\} = \{x \in \mathbb{R}^k \mid \langle a, x \rangle = \lambda\}$.

Conversely, suppose $H = \{x \in \mathbb{R}^k \mid \langle a, x \rangle = \lambda\}$ with $a \neq 0$. Then at least one component of a is non-zero. We may assume that it is a_1. Define vectors v_i $(i = 1, \ldots, k - 1)$ by $(v_i)_1 = -a_{i+1}/a_1$, $(v_i)_{i+1} = 1$, and $(v_i)_j = 0$ for $j \neq 1, i + 1$. It is easy to see that these vectors are independent and span a $(k - 1)$-dimensional space N. Now $H = b + N$ where $b = \lambda a / \|a\|^2$. ($\|a\| = \sqrt{\langle a, a \rangle}$ is the norm of the vector a.) ∎

The following theorem is intuitively obvious but it is in fact a very important result.

Theorem C.2 *Let A be a non-empty, open, convex subset of $\mathbb{R}^k$. Let M be a linear variety in $\mathbb{R}^k$ such that $M \cap A = \emptyset$. (M could be a single point.) Then there exists a hyperplane $H \supset M$ such that $H \cap A = \emptyset$.*

We need one more simple lemma.

Lemma C.3 *If $X \subset \mathbb{R}^k$ is connected and S is a non-empty proper subset of X (i.e. $S \subset X$, $S \neq X$) then S has at least one boundary point in X.*

Proof. We must show that $\overline{S} \cap \overline{X \setminus S} \cap X \neq \emptyset$. But, if this intersection were empty then $X = (X \cap \overline{S}) \cup (X \cap \overline{X \setminus S})$ is a disconnection of X. ∎

Proof of Theorem C.2 We may assume that $0 \in M$. Let $H \supset M$ be a linear variety of maximal dimension such that $H \cap A = \emptyset$. We must show that H has dimension $k - 1$. Suppose not. Then there are two independent vectors v_1 and v_2 such that $F \cap H = \{0\}$ where F is the 2-dimensional space spanned by v_1 and v_2: $F = \{\lambda v_1 + \mu v_2 \mid \lambda, \mu \in \mathbb{R}\}$. Let B be the projection of A onto F along H, i.e. $B = \{x \in F \mid \exists v \in H : x + v \in A\}$ and let $C = \bigcup_{r>0} rB \subset F$.

C is convex as is easily checked. Moreover $0 \notin C$ so, if $x \in C$ then $-x \notin C$. C is therefore a non-empty proper subset of $G = F \setminus \{0\}$.

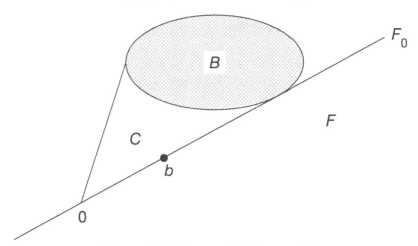

Figure C.2 The proof of theorem C.2.

By lemma C.3, the boundary of C in G contains at least one point, b say. As C is open, $b \notin C$. Moreover, $b \in \overline{C} \cap G$. Let $F_0 = \mathbb{R}b$ be the 1-dimensional space spanned by b. Suppose $F_0 \cap B = \emptyset$. Then $(H + F_0) \cap A = \emptyset$, which would mean that H was not a linear variety with maximal dimension having this property. We conclude that $F_0 \cap B \neq \emptyset$. Let $\lambda b \in F_0 \cap B$. If $\lambda > 0$ then, by the definition of C, $\lambda b \in C$, and hence $b \in C$, a contradiction. Also, $\lambda \neq 0$ because $0 \notin B$.

Hence $\lambda < 0$. It follows that $-b \in C$. But $b \in \overline{C}$ and C is convex and open, so then lemma C.1 implies that $0 = \frac{1}{2}b + \frac{1}{2}(-b) \in C$. This too is a contradiction. We must therefore conclude that the original assumption is false, that is, the dimension of H is not less than $k - 1$. ∎

REMARK C.1: *Infinite-dimensional spaces.*
The above results are also valid in infinite-dimensional spaces. Infinite dimensional spaces are studied in the subject called **functional analysis.** In particular the last theorem is of great importance in that field. It is the celebrated **Hahn-Banach theorem.**

We shall need one corollary to theorem C.2. First let us give another definition.

*If the hyperplane H in $\mathbb{R}^k$ is given by the equation $\langle a, x \rangle = \lambda$ as in (C.1) then the **closed half-spaces** determined by H are given by*

$$H^+ = \{x \in \mathbb{R}^k \,|\, \langle a, x \rangle \geqslant \lambda\} \ and \ H^- = \{x \in \mathbb{R}^k \,|\, \langle a, x \rangle \leqslant \lambda\} \qquad (C.2)$$

*and the **open half-spaces** are given by the interiors of these sets. A set $A \subset \mathbb{R}^k$ is said to **lie to one side** of H if $A \subset H^+$ or $A \subset H^-$ and to **lie strictly to one side** of H if $A \subset \text{int}(H^+)$ or $A \subset \text{int}(H^-)$.*
*H is said to be **tangent** to A if A lies to one side of H and $\overline{A} \cap H \neq \emptyset$.*

Lemma C.4 *A convex subset A of $\mathbb{R}^k$ lies strictly to one side of H if and only if $A \cap H = \emptyset$.*

Proof. This is left as an exercise. ∎

Theorem C.3 *Let A be a convex body in $\mathbb{R}^k$ and b a boundary point of A. Then there exists a hyperplane H tangent to A and passing through b. Moreover, A is the intersection of all closed half-spaces that contain A and are determined by hyperplanes tangent to A.*

Proof. By definition of a convex body, $\text{int}(A)$ is a non-empty, open convex set and $b \notin \text{int}(A)$. Therefore $\{b\}$ is a 0-dimensional variety disjoint from $\text{int}(A)$ and by theorem C.2 there exists a hyperplane H through b such that $\text{int}(A) \cap H = \emptyset$. By the preceding lemma $\text{int}(A)$ lies strictly to one side of H and by continuity, A lies to one side of H as $A = \overline{\text{int}(A)}$. But $b \in A \cap H$ so H is tangent to A.

Let $z \notin A$. We want to prove that there is a hyperplane H tangent to A such that z lies in the half-plane opposite to A. We obtain a suitable point of contact for this hyperplane by choosing $a \in \text{int}(A)$ and connecting it to z. There is a unique point b at which the line segment from a to z intersects the boundary of A. If H is a hyperplane tangent to A at b, we may assume $A \subset S = \{x \in \mathbb{R}^k \,|\, \langle u, x \rangle \leqslant \lambda\}$.

Now suppose $z \in S$. Then, by lemma C.1, $\text{int}(S)$ contains all interior points of the line segment from a to z since S is convex. In particular, $b \in \text{int}(S)$. This contradicts $b \in H$. ∎

Let us now turn to convex functions. We have the following.

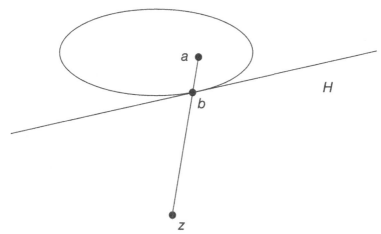

Figure C.3 Tangent hyperplane to a convex body.

Lemma C.5 *Let D be an open region in $\mathbb{R}^k$. A function $g : D \to \mathbb{R}$ is continuous if and only if the sets $D^+ = \{(x,h) \in D \times \mathbb{R} \mid h > g(x)\}$ and $D^- = \{(x,h) \in D \times \mathbb{R} \mid h < g(x)\}$ are open in $\mathbb{R}^k \times \mathbb{R}$.*

Proof. Suppose g is continuous and let $(x_0, h_0) \in D^+$. Set $\epsilon = h_0 - g(x_0)$. Then $\epsilon > 0$ and there exists a ball B with centre x_0 and radius $r > 0$ in D such that $|g(x) - g(x_0)| < \epsilon/2$ for $x \in B$. Hence $B \times (h_0 - \epsilon/2, h_0 + \epsilon/2) \subset D^+$ because if $x \in B$ and $|h - h_0| < \epsilon/2$ then $h > h_0 - \epsilon/2 = g(x_0) + \epsilon/2 > g(x)$. Thus D^+ is open. Similarly, D^- is open.

Conversely, suppose that D^+ is open. We show that g is **upper semicontinuous**, i.e. $\limsup_{x \to x_0} g(x) \leqslant g(x_0)$ for $x_0 \in D$. Indeed, suppose $x_n \to x_0$ in D and let $\lambda = \limsup_{n \to \infty} g(x_n)$. If $h > g(x_0)$ then $(x_0, h) \in D^+$ so there exists a ball B around x_0 with radius $r > 0$ and a positive ϵ such that $B \times (h - \epsilon, h + \epsilon) \subset D^+$. Now, $x_n \in B$ for n large enough and hence $(x_n, h) \in D^+$, that is $h > g(x_n)$. It follows that $h \geqslant \limsup g(x_n)$ and since $h > g(x_0)$ was arbitrary, $g(x_0) \geqslant \limsup_{n \to \infty} g(x_n)$. In the same way, one proves that, if D^- is open then g is **lower semi-continuous**, i.e. $g(x_0) \leqslant \liminf_{x \to x_n} g(x_0)$. ∎

Theorem C.4 *A convex function $g : \mathbb{R}^k \to \mathbb{R} \cup \{+\infty\}$ is continuous at every point of the interior of its essential domain.*

Proof. Redefining $g(x) = \infty$ for x on the boundary, we may assume that the essential domain D of g is open. We prove that D^+ and D^- are open. Let E be the epigraph of g. By lemma 7.2, E is convex. Let $(x_0, h_0) \in D^-$. It is easy to see that $(x_0, g(x_0))$ is a boundary point of E and by theorem C.3 there exists a tangent hyperplane H to E at $(x_0, g(x_0))$. Let $H = \{y \in \mathbb{R}^{k+1} \mid \langle u, y \rangle = \lambda\}$ for some $u \in \mathbb{R}^{k+1}$ and $\lambda \in \mathbb{R}$. Obviously, $u_{k+1} \neq 0$ because D is open and E lies to one side of H. We may assume $u_{k+1} > 0$. As E lies to one side of H, we must have $E \subset H^+ = \{y \in \mathbb{R}^{k+1} | \langle u, y \rangle \geq \lambda\}$. It follows that $(x_0, h_0) \in \{(x, h) | x \in D, \langle u, (x, h) \rangle < \lambda\} \subset D_-$ and hence D^- is open.

To show that D^+ is open, assume the contrary. Then there exists $(x_0, h_0) \in D^+ \setminus \text{int}(D^+)$. As D^+ is convex, $\text{int}(D^+)$ is also convex by theorem C.1. By theorem C.2 there exists a hyperplane H through (x_0, h_0) such that $\text{int}(D^+)$ lies strictly to one side of H. Again, if $H = \{y \in \mathbb{R}^{k+1} | \langle u, y \rangle = \lambda\}$ we may assume $u_{k+1} > 0$. As $(x_0, h_0) \in H$ and $h_0 > g(x_0)$, $\langle u, (x_0, g(x_0)) \rangle < \lambda$. This implies that $(x_0, g(x_0)) \notin D^+$, a contradiction. ∎

Theorem C.5 *Let $g : \mathbb{R}^k \to \mathbb{R} \cup \{+\infty\}$ be a function with convex essential domain D. Assume that $g^*(t_0) < +\infty$ for at least one point $t_0 \in \mathbb{R}^k$. Then the epigraph of g^{**} is given by the closed convex hull of the epigraph of g. In particular, if g is convex then $g^{**}(x) = g(x)$ for $x \in \text{int}(D)$.*

Proof. The last statement follows from the first in combination with theorem C.1 and lemma 7.2. To prove the first statement, notice first that, by the definition of $g^*(t)$, $\langle t, x_0 \rangle - g(x_0) \leq g^*(t)$ for all $t \in \mathbb{R}^k$, so $g(x_0) \geq \langle t, x_0 \rangle - g^*(t)$ and hence $g(x_0) \geq g^{**}(x_0)$. This means that if $(x_0, h_0) \in \text{epi}(g)$ then $(x_0, h_0) \in \text{epi}(g^{**})$. To complete the proof it suffices, by theorem C.3 and lemma 7.1, to show that if a point (x_0, h_0) is separated from the epigraph of g by a hyperplane H then it is also separated from the epigraph of g^{**} by that hyperplane. Suppose, therefore, that a hyperplane $H = \{(x, h) \in \mathbb{R}^{k+1} | \langle z, (x, h) \rangle = \lambda\}$ is given such that $\langle z, (x, h) \rangle \geq \lambda$ for all (x, h) with $h \geq g(x)$ and $\langle z, (x_0, h_0) \rangle < \lambda$. We distinguish two cases.

First assume that $z_{k+1} = 0$. Then $x_0 \notin \overline{D}$ and we want to prove that $x_0 \notin D^{**} = D(g^{**})$, i.e. $g^{**}(x_0) = +\infty$. We can write $z = (a, 0)$ with $a \in \mathbb{R}^k$ and we have: $\langle a, x \rangle \geq \lambda$ for $x \in D$ and $\langle a, x_0 \rangle < \lambda$. Since $g^*(t_0) < +\infty$, $g(x) \geq \langle t_0, x \rangle - g^*(t_0)$. Then, by the definition of g^*, $g^*(-pa + t_0) \leq -p\lambda + g^*(t_0)$ for all $p > 0$. Again, by the definition of g^{**}, $g^{**}(x_0) \geq \langle t, x_0 \rangle - g^*(t)$ for all t. Taking $t = -pa + t_0$ we obtain $g^{**}(x_0) \geq p(\lambda - \langle x_0, a \rangle) + \langle t_0, x_0 \rangle - g^*(t_0)$. Taking $p \to \infty$ we find that $g^{**}(x_0) = +\infty$.

Now suppose $z_{k+1} \neq 0$. Then $z_{k+1} > 0$ because the epigraph must lie above the hyperplane. Let $z = (a, r)$. Then, for any x and any $h \geq g(x)$, $\langle a, x \rangle + rh \geq \lambda$. Taking $h = g(x)$ we can write this as $-\frac{1}{r}\langle a, x \rangle - g(x) \leq -\lambda/r$. By the definition of g^* we then have $g^*(-a/r) \leq -\lambda/r$ and inserting this into the definition of g^{**} we find $g^{**}(x) \geq -\langle a, x \rangle/r + \lambda/r$. Hence, if $h \geq g^{**}(x)$ then $\langle a, x \rangle + rh \geq \langle a, x \rangle + rg^{**}(x) \geq \lambda$. The epigraph of g^{**} therefore lies above the hyperplane. ∎

REMARK C.2: *Further properties.*
From the first part of the proof it also follows that the essential domain of g^{**} is given by the points $x_0 \in \mathbb{R}^k$ such that $\liminf_{x \to x_0, x \in D} g(x) < +\infty$. In fact one can show that in case g is convex, the limit exists so that

$$D(g^{**}) = \{x \in \mathbb{R}^k \mid \lim_{\substack{x \to x_0 \\ x \in D}} g(x) < +\infty\} \text{ and } g^{**}(x_0) = \lim_{\substack{x \to x_0 \\ x \in D}} g(x). \quad \text{(C.3)}$$

Like theorem C.2, theorem C.5 also holds in infinite-dimensional spaces.

References

Anderson, M. H., Ensher, J. R., Matthews, M. R., Wieman, C. E. & Cornell, E. A. 1995, *Science* **269**, 198–202 (1995).

Asmussen, S. 1987 *Applied Probability and Queues.* (New York: Wiley)

Bardeen, J., Cooper, L. N. & Schrieffer, J. R. 1957, *Phys. Rev.* **108**, 1175–1204.

Baxter, R. J. 1982, *Exactly Solved Models in Statistical Mechanics.* (New York: Academic Press)

Billingsley, P. 1979, *Probability and Measure.* (New York: Wiley)

Billmeyer, F. W. Jr. 1984, *Textbook of Polymer Science.* 3rd edn. (London: Chapman and Hall)

Binney, J. J., Dowrick, N. J., Fisher, A. J. & Newman, M. E. J. 1992, *The Theory of Critical Phenomena. An Introduction to the Renormalization Group.* (Oxford: Clarendon Press)

Blakemore, J. S. 1966, *Solid State Physics.* (Cambridge: Cambridge University Press)

Brailsford, F. 1966, *Physical Principles of Magnetism.* (New York: Van Nostrand)

Cegła, W., Lewis, J. T. & Raggio, G. A. 1988, *Commun. Math. Phys.* **118**, 337–354.

Chandrasekhar, S. 1935, *Mon. Not. R. Astron. Soc.* **95** 207.

Chandrasekhar, S. 1994, *Liquid Crystals.* 2nd edn. (Cambridge: Cambridge University Press)

Conolly, B. 1975, *Lecture Notes on Queueing Systems.* (Chichester: Ellis Horwood; New York: Wiley)

Dobrushin, R. L. 1965, *Theory Probab. Appl.* **10**, 209–230.

Dorlas, T. C., Lewis, J. T. & Pulé, J. V. 1993, *Commun. Math. Phys.* **156**, 37–65.

Duffield, N. G. & O'Connell, N. 1995, *Math. Proc. Cambridge Phil. Soc.* **118**, 363–374.

Duffield, N. & Pulé, J. V. 1988, *Commun. Math. Phys.* **118**, 475–494.

Feller, W. 1966, *An Introduction to Probability Theory and Its Applications.* Vols. 1 and 2. (New York: Wiley)

Gallavotti, G. & Miracle-Sole, S. 1968, *Commun. Math. Phys.* **7**, 274–288.

Gedde, U. W. 1995, *Polymer Physics.* (London: Chapman & Hall)

Glynn, P. W. & Whitt, W. 1993, *J. Appl. Probab.* **31A**, 131–156.

Griffiths, R. B. 1964, *Phys. Rev.* **136**, A437–A438.

Groeneveld, J. 1962 *Phys. Rev. Lett.* **3**, 50–51.

Gruber C. & Kunz, H. 1971, *Commun. Math. Phys.* **22**, 133–161.

Guinier, A. 1984, *The Structure of Matter. From the Blue Sky to Liquid Crystals.* (London: Edward Arnold)

Hardy, G. H. & Littlewood, J. E. 1914, *Acta Math.* **37**, 155–190.

Heisenberg, W. 1928, *Z. Phys.* **49**, 619–636.

Huang, K., Yang C. N. & Luttinger, J. M. 1957, *Phys. Rev.* **105**, 776–784.

Hugenholtz, N. M. 1982, *C*-Algebras and Statistical Mechanics*, in: *Proc. Symp. Pure Math.*, Vol. **38** Part 2, Ed. R. V. Kadison (Providence, RI: American Mathematical Society) pp. 407–465.

Khintchine, A. 1924, *Fundam. Math.* **6**, 9–20.

Kittel, C. 1976, *Introduction to Solid State Physics.* 5th edn. (New York: Wiley)

Landau, L. D. & Lifshitz, E. M. 1972, *Mechanics and Electrodynamics; A Shorter Course of Theoretical Physics.* Vol. 1 (Oxford: Pergamon Press)

Lee, T. D. & Yang, C. N. 1952, *Phys. Rev.* **87**, 404–419.

Lieb, E. H. 1967, *Phys. Rev.* **162**, 162—172.

Ma, S. K. 1971, *Modern Theory of Critical Phenomena.* (Reading, MA: W. A. Benjamin)

Mattis, D. C. 1965, *The Theory of Magnetism. An Introduction to the Theory of Cooperative Phenomena.* (New York: Harper & Row)

Mayer, J. E. 1937 *J. Chem. Phys.* **5**, 67.

Messiah, A. 1961, *Quantum Mechanics.* Vols. 1 and 2 (Amsterdam: North Holland)

McEliece, R. J. 1977, *The Theory of Information and Coding.* (Encyclopedia of Mathematics and Its Applications.) (Reading, MA: Addison-Wesley)

Montroll, E. W. & Mayer, J. E. 1941, *J. Chem. Phys.* **9**, 262.

Morrish, A. H. 1965, *The Physical Principles of Magnetism.* (New York: Wiley)

Onsager, L. 1944, *Phys. Rev.* **65**, 117–149.

Osheroff, D. D., Richardson, R. C. & Lee, D. M. 1972, *Phys. Rev. Lett.* **28**, 885–888.

Pauli, W. 1927, *Z. Phys.* **41**, 81–102.

Peierls, R. 1936, *Proc. Cambridge Phil. Soc.* **32**, 477–481.

Penrose, O. 1963, *J. Math. Phys.* **4** 1322–1320.

Penrose, O. 1967, in: *Statistical Mechanics: Foundations and Applications.* Proc. of the IUPAP meeting, Copenhagen 1966. Ed. T. A. Bak. (New York: Benjamin)

Pfeuty, P. & Toulouse, G. 1977, *Introduction to the Renormalization Group and Critical Phenomena.* (New York: Wiley)

Phillips, W. D. & Cohen-Tannoudji, C. 1990, *Phys. Today* **43**, 33–40.

Priestley, E. B., Wojtowicz, P. J. & Sheng, P. (eds) 1975, *Introduction to Liquid Crystals.* (New York: Plenum)

Ruelle, D. 1963, *Ann. Phys.* **25** 109–120.

Ruelle, D. 1969, *Statistical Mechanics. Rigorous Results.* (London: W. A. Benjamin)

Roman, S. 1992, *Coding and Information Theory*. (Berlin: Springer)

Saxena, S. K., Shen, G. & Lazor, P. 1993, *Science* **260**, 1312–1314.

Schiff, L. I. 1969, *Quantum Mechanics*. (New York: McGraw-Hill)

Schultz, T. D., Mattis, D. C. & Lieb, E. H. 1964, *Rev. Mod. Phys.* **36**, 856–871.

Sinai, Ya. G. 1982, *Theory of Phase Transitions: Rigorous Results*. (Oxford: Pergamon Press)

Söderlind, P., Moriarty, J. A. & Wills, J. 1996, *Phys. Rev.* B **53** 14063.

Tanaka, T. 1981, *Sci. Am.* **244**, 110–123.

Thouless, D. J. 1960, *Phys. Rev.* **117**, 1256–1260.

Ursell, H. D. 1927, *Proc. Camb. Philos. Soc.* **23** 685.

van den Berg, M., Dorlas, T. C., Lewis, J. T. & Pulé, J. V. 1990, *Commun. Math. Phys.* **128**, 231–245.

van den Berg, M., Lewis, J. T. & Pulé, J. V. 1988, *Commun. Math. Phys.* **118**, 61–85.

Walas, S. M. 1985, *Phase Equilibria in Chemical Engineering*. (London: Butterworth)

Wilks, J. 1970, *an Introduction to Liquid Helium*. (Oxford: Clarendon Press)

Williams, D. E. G. 1966, *The Magnetic Properties of Matter*. (London: Longmans Green)

Wilson, K. G. & Kogut, J. 1974, *Phys. Rep.* C **12**, 75–200.

Yang, C. N. 1952, *Phys. Rev.* **85**, 808–816.

Young, R. J. & Lovell, P. A. 1991, *Introduction to Polymers*. 2nd edn. (London: Chapman & Hall)

Further Reading

GENERAL INTRODUCTIONS TO STATISTICAL MECHANICS

Chandler, D. 1987, *Introduction to Modern Statistical Mechanics*. (Oxford Univ. Press)

Guénault, A. M. 1995, *Statistical Physics* 2nd edn. (London: Chapman & Hall)

Huang, K. 1987, *Statistical Mechanics* 2nd edn. (New York: Wiley)

Mandl, F. 1988, *Statistical Physics* 2nd edn. (New York: Wiley)

ADVANCED MATHEMATICAL TREATMENTS

Ellis, R. S. 1985, *Entropy, Large Deviations and Statistical Mechanics*. (Berlin: Springer Verlag)

Israel, R. B. 1979, *Convexity in the Theory of Lattice Gases*. (Princeton, NJ: Princeton Univ. Press)

Ruelle, D. 1969, *Statistical Mechanics. Rigorous Results*. (London: W. A. Benjamin)

Sewell, G. L. 1986, *Quantum Theory of Collective Phenomena*. (Oxford : Clarendon Press)

Thompson, C. J. 1972, *Mathematical Statistical Mechanics*. (Princeton, NJ: Princeton Univ. Press)

CLUSTER EXPANSIONS

Hill, T. J. 1956, *Statistical Mechanics*. (New York: McGraw-Hill)

Mayer, J. E. & Goeppert-Mayer, M. 1977, *Statistical Mechanics*. (New York: Wiley)

Bridges, D. C. 1986, *A Short Course on Cluster Expansions*. Les Houches Summer School 1984, Session XLIII, Part I. *Critical Phenomena, Random Systems, Gauge Theories*. Eds. K. Osterwalder & R. Stora. (Amsterdam: North-Holland)

Ruelle, D. 1964, *Cluster Properties of the Correlation Functions of Classical Gases*. *Rev. Mod. Phys.* **4**, 580–584.

Glimm, J. Jaffe, A. & Spencer, T. 1973, *The particle structure of the weakly coupled $P(\phi)_2$ model and other applications of high-temperature expansions*. In: *Constructive Quantum Field Theory*, Eds. G. Velo & A. S. Wightman. (New York: Springer)

PIROGOV–SINAI THEORY

Sinai, Ya. G. 1982, *Theory of Phase Transitions: Rigorous Results*. (Oxford: Pergamon)

BACKGROUND MATERIAL

Mathematical analysis

Rudin, W. 1964, *Principles of Mathematical Analysis*. (New York: McGraw-Hill)

Probability theory

Billingsley, P. 1979, *Probability and Measure*. (New York: Wiley)

Feller, W. 1966, *An Introduction to Probability Theory and Its Applications* Vols 1 and 2. (New York: Wiley)

Mechanics and electrodynamics

Landau, L. D. and Lifshitz, E. M. 1972, *Mechanics and Electrodynamics; A Shorter Course of Theoretical Physics* Vol. 1. (Oxford: Pergamon)

Quantum mechanics

Liboff, R. L. 1992, *Introductory Quantum Mechanics*. (Reading, MA: Addison-Wesley)

Messiah, A. 1961, *Quantum Mechanics* Vols. 1 and 2. (Amsterdam: North-Holland)

Schiff, L. I. 1969, *Quantum Mechanics*. (New York: McGraw-Hill)

THERMODYNAMICS

Applications in engineering

Eastop. T. D. and McConkey, A. 1986, *Applied Thermodynamics for Engineering Technologists* 4th edn. (London: Longman Scientific and Technical)

Chemical thermodynamics

Atkins, P. W. 1992, *The Elements of Physical Chemistry*. (Oxford: Oxford University Press)

PHASE DIAGRAMS

Moffatt, W. G. 1979, 1986. *The Handbook of Binary Phase Diagrams* Vols. 1, 2, 3, 4, 5. (Schenectady, NY: Genium)

Walas, S. M. 1985 *Phase Equilibria in Chemical Engineering*. (London: Butterworth)

LIQUID HELIUM

Wilks, J. 1970, *An Introduction to Liquid Helium.* (Oxford: Clarendon)

Dilution refrigerator

Betts, D. S. 1968, *Contemp. Phys.* **9**, 97–114

Lounasmaa, O. V. 1974, *Experimental Principles and Methods Below 1 K.* (London: Academic)

Radebaugh, R. and Siegwarth, J. D. 1971, *Cryogenics* **11**, 368–384

THE RENORMALIZATION GROUP AND CRITICAL PHENOMENA

Binney, J. J., Dowrick, N. J., Fisher, A. J. and Newman, M. E. J. 1992, *The Theory of Critical Phenomena. An Introduction to the Renormalization Group.* (Oxford: Clarendon)

Ma, S. K. 1971, *Modern Theory of Critical Phenomena.* (London: W A Benjamin)

Pfeuty, P. and Toulouse, G. 1977, *Introduction to the Renormalization Group and Critical Phenomena.* (New York: Wiley)

Wilson, K. G. and Kogut, J. 1974, *Phys. Rep.* C **12**, 75–200

SOLID STATE THEORY

Ashcroft, N. W. and Mermin, N. D. 1976, *Solid State Physics.* (New York: Holt, Rinehart & Winston)

Blakemore, J. S. 1985, *Solid State Physics*, (Cambridge: Cambridge University Press)

Kittel, C. 1976, *Introduction to Solid State Physics* 5th edn. (New York: Wiley)

THE BIG BANG

Peebles, P. J. E. 1993, *Principles of Physical Cosmology.* (Princeton, NJ: Princeton University Press) (Thorough exposition)

Silk, J. 1980, *The Big Bang. The Creation and Evolution of the Universe.* (San Francisco, CA: W H Freeman) (Layman's introduction)

THE CONSTITUTION OF STARS

Reddish, V. C. 1974, *The Physics of Stellar Interiors. An Introduction.* (Edinburgh: Edinburgh University Press)

MAGNETISM

Mattis, D. C. 1965, *The Theory of Magnetism. An Introduction to the Theory of Cooperative Phenomena.* (New York: Harper & Row)

Morrish, A. H. 1965, *The Physical Principles of Magnetism.* (New York: Wiley)

Williams, D. E. G. 1966, *The Magnetic Properties of Matter*. (London: Longmans Green)

INFORMATION THEORY

McEliece, R. J. 1977, *The Theory of Information and Coding (Encyclopedia of Mathematics and Its Applications)*. (Reading, MA: Addison-Wesley)

Roman, S. 1992, *Coding and Information Theory*. (Berlin: Springer)

QUEUING THEORY

Asmussen, S. 1987, *Applied Probability and Queues*. (New York: Wiley)

Conolly, B. 1975, *Lecture Notes on Queueing Systems*. (Chichester: Ellis Horwood; New York: Wiley)

Answers to the Problems

PART I

I-1. 1.013×10^5 Pa.

I-4. 25 kJ.

I-5. 1.3 kg.   78.6 °C.

I-6. $\Delta m = 0.66$ kg $= 656$ g.

I-7. $T' = 86\,\mathrm{K} = -187\,°\mathrm{C}$. $(\Delta T = -207\,°\mathrm{C}.)$

I-8. $\Delta T = 0.03$ K.

I-9. $\Delta \ell = 19.5$ cm. $T' = 80\,°\mathrm{C}$.

I-12. (ii) $\int_C \omega = -2\pi$.

I-14. $\Delta s = 73$ J kg^{-1} K^{-1}.

I-15. (i) $q = 727$ kJ kg^{-1}.   (ii) $w = 500$ kJ kg^{-1}.

I-16. $\eta = 67.6\%$. The work ratio is 17.5%.

I-22. $f^*(t) = 0$ if $|t| \leqslant 1$ and $+\infty$ otherwise.

I-23. $f^*(t) = |t|$ if $|t| \leqslant 2$ and $f^*(t) = \frac{1}{4}t^2 + 1$ if $|t| > 2$. $f^{**}(x) = x^2 - 1$ if $|x| \geqslant 1$ and $f^{**}(x) = 0$ for $|x| < 1$.

I-25. $f(v, T) = -c_V\, T \ln T - k_\mathrm{B}\, T\, \ln v + A\,T$, where $A = c_V(1 - \ln c_V) - s_0$.

I-28. During melting, $\delta q > 0 \implies ds > 0$. For $^3\mathrm{He}$, $dp/dT < 0$ at α; so $\Delta s < 0$ by equation (9.8).

I-29.

$$p_c = \frac{a}{27b^2} \qquad v_c = 3b \qquad \tilde{p} = \frac{8\tilde{T}}{3\tilde{v} - 1} - 3\tilde{v}^2.$$

I-30. $s(u, v) = c_V \ln(u + av^{-1}) + k_\mathrm{B} \ln(v - b) + s_0$.

I-31. $v_{\mathrm{initial}} = 0.0665$ l mol^{-1}. $v_{\mathrm{final}} = 0.08$ l mol^{-1}. $T' = 255$ K.

I-33. 66 °C.

I-34. The coefficient of performance is 5.3.

I-35. 100 μW. 10^{-8} Pa.

I-36. $\eta = 39.6\%$.

I-38.

$$\left(p + \frac{a}{b^2} + \frac{3k_{\mathrm{B}}T}{2b}\right)^2 = \frac{8ak_{\mathrm{B}}T}{b^3}; \qquad T_{\mathrm{inv}} = 913 \text{ K}.$$

I-40. $T = 6043$ K $= 5770$ °C.

PART II

II-3. $W = -\mu B_0 \cos\theta = -\vec{\mu}\cdot\vec{B}_0$.

II-7. $\mu_B = 9.27 \times 10^{-24}$ A m^2. $B_0 \ll 0.015$ T.

II-8. $T_1 = T_2\, H_1/H_2$.

II-9.

$$f(T) = -k_{\mathrm{B}}T\, \ln(e^{\epsilon/k_{\mathrm{B}}T} + e^{-\epsilon/k_{\mathrm{B}}T}).$$
$$u(T) = -\epsilon\tanh(\epsilon/k_{\mathrm{B}}T).$$
$$c_V(T) = k_{\mathrm{B}}\,(\epsilon/k_{\mathrm{B}}T)^2\mathrm{sech}^2(\epsilon/k_{\mathrm{B}}T).$$

II-10. Take $r = 2$, $a_n = (-1)^n$ and $b_n = -a_n$. Then $\liminf(a_n \vee b_n) = 1$ but $\liminf a_n = \liminf b_n = -1$.

II-11. $I(x) = 1 - \sqrt{1-x^2}$.

II-15. $\beta \vee 0$.

II-16. $-\frac{1}{2}$.

II-17. 3.

II-18. $f(\beta,\rho) = -\beta^{-1}\big[\ln(\rho^{-1} - b) + 1\big]$. $p(v,T) = k_{\mathrm{B}}T/(v - b)$.

II-19. $I(\vec{x}) = \displaystyle\sum_{k=1}^{r} x_k \ln\left(\frac{x_k}{p_k}\right)$ if $x_k \geqslant 0$ for all k and $\sum_{k=1}^{r} x_k = 1$; and $I(\vec{x}) = +\infty$ otherwise.

II-20.

$$H(p_1,\ldots,p_r) = -\sum_{k=1}^{r} p_k \log_2 p_k.$$

$$H(p_1,\ldots,p_r) = \lim_{\Delta_1,\ldots,\Delta_r \to 0} \lim_{N\to\infty} \left(\frac{1}{N}\log_2 \Omega_N(\Delta_1,\ldots,\Delta_r)\right),$$

where $\Omega_N(\Delta_1, \ldots, \Delta_r) = \#\{(n_1, \ldots, n_r) : \sum_{k=1}^{r} n_k = N$
and $N^{-1} n_k \in (p_k - \Delta_k, p_k + \Delta_k)$ for all $k\}$.

II-24. $\Omega = \displaystyle\sum_{n_+=m}^{N \wedge [(N+m)/2]} \dfrac{N!}{n_+! \, (n_+ - m)! \, (N - 2n_+ + m)!}$;

$$s(u) = - k_B \frac{u}{\epsilon} \left\{ \ln \left(\sqrt{4 - 3\left(\frac{u}{\epsilon}\right)^2} + \frac{u}{\epsilon} \right) - \ln \left[2 \left(1 - \frac{u}{\epsilon} \right) \right] \right\}$$

$$+ k_B \ln \left(1 + \sqrt{4 - 3\left(\frac{u}{\epsilon}\right)^2} \right) - k_B \ln \left(1 - \frac{u^2}{\epsilon^2} \right).$$

$$f(T) = - k_B T \ln \left(e^{\epsilon/k_B T} + 1 + e^{-\epsilon/k_B T} \right).$$

II-25. $f(T, H) = -k_B T \ln \left(\dfrac{\sinh \left(\gamma H (J + \frac{1}{2})/k_B T \right)}{\sinh \left(\frac{1}{2} \gamma H / k_B T \right)} \right)$.

$m(T, H) = \dfrac{\mu}{v} J B_J(z)$, where $\gamma = \mu_0 \mu$, $\mu = g \dfrac{q\hbar}{2m}$ and $z = \dfrac{\gamma J H}{k_B T}$.

$\chi(T, H) = \dfrac{\gamma^2 J^2}{\mu_0 v k_B T} \dfrac{dB_J}{dz}$, where

$$\frac{dB_J}{dz} = \frac{1}{4J^2} \operatorname{cosech}^2 \left(\frac{z}{2J} \right) - \left(\frac{2J+1}{2J} \right)^2 \operatorname{cosech}^2 \left(\frac{2J+1}{2J} z \right).$$

For $z \ll 1$, $\dfrac{dB_J}{dz} \approx \dfrac{1}{3} \dfrac{J+1}{J}$. Hence $\chi \approx \dfrac{1}{3} \dfrac{\gamma^2}{\mu_0 v k_B T} J(J+1)$.

II-26. $f(T) = k_B T \ln 2 \sinh \left(\dfrac{\hbar\omega}{2k_B T} \right)$.

$c_V = k_B \left(\dfrac{\hbar\omega}{2k_B T \sinh(\hbar\omega/2k_B T)} \right)^2$.

As $T \to \infty$, $c_V \to k_B$; as $T \to 0$, $c_V \sim k_B \left(\dfrac{\hbar\omega}{k_B T} \right)^2 e^{-\hbar\omega/k_B T}$.

$c_V(N_2) = 3.3$ J K^{-1} mol^{-1}.

II-28.
$$\rho = \frac{1}{4\pi^2} \left(\frac{2m}{\hbar^2} \right)^{3/2} \int_0^\infty \frac{\epsilon^{1/2} \, d\epsilon}{e^{\beta(\epsilon - \mu)} \pm 1}.$$

$\tilde{s}(\beta, \mu) = -k_B \beta \left(\frac{5}{2} \omega(\beta, \mu) + \mu\rho \right)$.

II-29.
$$\tilde{\phi}_{MB}(v) dv = 4\pi \left(\frac{m}{2\pi k_B T} \right)^{3/2} e^{-mv^2/2k_B T} v^2 \, dv.$$

II-30.

$$\omega(\beta,\mu) = (2\pi)^{-3}\beta^{-1}\sum_{r=0}^{\infty}\int d^3k \, \ln(1 - e^{\beta(\mu-\epsilon(k)-\hbar\omega(r+\frac{1}{2}))}).$$

In the classical limit, $\omega(\beta,\mu) \sim -\dfrac{e^{\beta\mu}}{\beta\lambda_T^3}\dfrac{1}{2}\mathrm{cosech}\,(\frac{1}{2}\beta\hbar\omega)$.

PART III

III-1.

$$\lim_{N\to\infty}\left(\frac{1}{N}\ln N! - \ln N\right) = \int_0^1 \ln(1-x)\,dx = -1.$$

$$s = k_B\left[\frac{v_2}{l}\left[\ln\nu + (l-2)\ln(\nu-1)\right] - \frac{v_2}{l}\ln\left(\frac{v_2}{l}\right) - \frac{v_1}{l}\ln v_1\right].$$

III-3. $Z_N = A_+\lambda_+^{N-1} + A_-\lambda_-^{N-1}$, where

$$A_\pm = 1 \pm \frac{e^{-2\beta J}(e^{\beta H} - e^{-\beta H}) + 2e^{-\beta J}(\lambda_+ - e^{\beta(J+H)})}{e^{-2\beta J} + (\lambda_+ - e^{\beta(J+H)})^2}.$$

III-4. $A = (e^{\beta J} - 1)I + qP$, where I is the $q \times q$ unit matrix and P is the projection onto the vector $(1,\ldots,1)$. Hence $f(\beta) = -\frac{1}{\beta}\ln(e^{\beta J} + q - 1)$.

III-5. $Z_M(\beta) = 2(2\cosh\beta J)^{N-1}$. $f(\beta) = -\frac{1}{\beta}\ln(e^{\beta J} + e^{-\beta J})$.

III-10.   (i) $A = \left(e^{-\beta\phi(s,s')}\right)_{s,s'\in S}$.   (ii) $f(\beta) = -\frac{1}{\beta}\ln\lambda_{\max}$ where the maximum eigenvalue $\lambda_{\max}$ of A is strictly larger than the other eigenvalues and therefore an analytic branch of the eigenvalue spectrum.

III-11.

$$\frac{1}{N}\mathbb{E}(N_-) \leqslant \frac{1}{3d^{d-1}}\sum_{k=d}^{\infty}k^{d-1}\kappa^{2d}$$

where κ is given by equation (28.37). For $d = 2$ this equals 0.5 for $\kappa = 0.77$, i.e. $\beta J = 0.68$ compared with the exact value $\beta_c J = 0.44$. For $d = 3$, it equals 0.5 for $\kappa = 0.777$, i.e. $\beta J = 0.66$ compared with the numerical value $\beta_c J \approx 0.4$.

III-14. $c_V = k_B\dfrac{(2\beta Jm + \beta H)^2}{(1-m^2)^{-1} - 2\beta J}$.

At $H = 0$, as $\beta \to \beta_c$, $c_V \to \frac{3}{2}k_B$.

$\beta = \frac{1}{2};$ $\gamma = 1;$ $\delta = 2.$

$\alpha_{\text{Ising2}} = 0;$ $\beta_{\text{Ising2}} = \frac{1}{8};$ $\gamma_{\text{Ising2}} = \frac{7}{4};$ $\delta_{\text{Ising2}} = 15.$

III-15. $f(\beta) = -\lambda Jx^2 - \frac{1}{\beta}\ln\left(e^{\beta H}c_+^\nu + e^{-\beta H}c_-^\nu\right)$ where

$$c_\pm = \cosh[\beta J(2\lambda x \pm 1)]\qquad s_\pm = \sinh[\beta J(2\lambda x \pm 1)]$$

and where x is a solution of

$$x = \nu \frac{c_+^{\nu-1} s_+ e^{\beta H} + c_-^{\nu-1} s_- e^{-\beta H}}{c_+^\nu e^{\beta H} + c_-^\nu e^{-\beta H}}.$$

The critical temperature is given by the solution of

$$2\lambda\nu\beta_c J \left(1 + (\nu - 1)\tanh^2(\beta_c J)\right) = 1.$$

III-16. $s_{\text{Pauling}} = R_0 \ln \frac{3}{2} = 3.37$ J K^{-1} mol^{-1}.
Equations for k_j $(j = 1, \ldots, n)$:

$$e^{ik_j N} = -\prod_{i \neq j} \left(-\frac{1 + e^{i(k_i + k_j)} - e^{ik_j}}{1 + e^{i(k_i + k_j)} - e^{ik_i}} \right).$$

Eigenvalues:

$$\prod_{j=1}^n (1 - e^{ik_j})^{-1} \left[1 + \exp\left(i \sum_{j=1}^n k_j \right) \right].$$

This eventually leads to the following value for the residual entropy (Lieb (1967)): $s_0 = \frac{3}{2} R_0 \ln \frac{4}{3} = 3.58$ J K^{-1} mol^{-1}.

III-17.

$$p = -\frac{\lambda v_0}{v^2} - \frac{1}{\beta v_0} \ln\left(1 - \frac{v_0}{v} \right).$$

$$p \sim \frac{1}{\beta v} \left[1 + (\frac{1}{2} - \beta\lambda)\left(\frac{v_0}{v}\right) + \frac{1}{3}\left(\frac{v_0}{v}\right)^2 + \ldots \right].$$

For the van der Waals equation:

$$p \sim \frac{1}{\beta v} \left[1 + \left(1 - \frac{\beta a}{b}\right)\left(\frac{b}{v}\right) + \left(\frac{b}{v}\right)^2 + \ldots \right].$$

$b = \frac{1}{2} v_0$; $\quad a/b = 2\lambda$; $\quad \beta_c = 2/\lambda = 4b/a$; $\quad T_c \approx 6$ K.

III-18.

$$\omega = -(2\pi)^{-3}\beta^{-1} \sum_{s=\pm} \int d^3 k \, \ln\{1 + \exp\left[\beta(\mu - \epsilon(k) - \mu_0 \mu_B s H)\right]\}.$$

$$m \approx \mu_0 \mu_B H \frac{\partial\rho}{\partial\mu} \approx \mu_0 \mu_B^2 H g(\epsilon_F).$$

III-19. $s = \frac{1}{3}\pi^2 k_B^2 \epsilon_F^{-1} T$.
$\epsilon_F^{(d)} \approx 5.35 \times 10^{-24}$ J; $\qquad \epsilon_F^{(c)} \approx 2.23 \times 10^{-23}$ J.

$\Delta q = T \Delta s \approx \frac{1}{3}\pi^2 k_{\rm B} R_0 ((\epsilon_F^{(d)})^{-1} - (\epsilon_F^{(c)})^{-1}) T^2.$
$\Delta q \approx 0.53\,\mu{\rm W}.$

III-20. $f(\beta) = \frac{3}{2}\hbar\omega + 3\beta^{-1}\ln(1 - e^{-\beta\hbar\omega}).$ $c_V = 3k_{\rm B}\dfrac{(\beta\hbar\omega)^2 e^{\beta\hbar\omega}}{(e^{\beta\hbar\omega} - 1)^2}.$

As $\beta \to 0$, $c_V \to 3k_{\rm B}$.   As $\beta \to \infty$, $c_V \sim 3k_{\rm B}(\beta\hbar\omega)^2 e^{-\beta\hbar\omega}$.

III-21. $\omega(\beta, \mu) = \inf_{x \geqslant 0}\{\frac{1}{2}\lambda x^2 - \mu x + f_{\rm FG}(x)\}.$
There is condensation for $\mu > \lambda\rho_c$.

III-22.

$$C(t_1, t_2) = \lim_{l \to \infty}\frac{1}{\beta V_l}\ln\int\int e^{\beta V_l(t_1 y + t_2 z)}\,dF_l(y, z)$$

$$= \begin{cases} \omega(\beta, \mu_0) - \omega(\beta, t_1 + \mu_0) & \text{if } t_1 + \mu_0 < 0 \text{ and } t_1 + t_2 + \mu_0 < 0, \\ +\infty & \text{otherwise.} \end{cases}$$

III-23. See van den Berg *et al.* (1992).
III-25.

$$b_3(\beta) = 2b_2(\beta)^2 + \frac{4}{3}\pi\int_0^\infty r_1^2 dr_1 \int_0^\infty r_2^2 dr_2 \int_0^{2\pi} d\varphi$$
$$(e^{-\beta\phi(r_1)} - 1)(e^{-\beta\phi(r_2)} - 1)(e^{-\beta\phi(r_{12})} - 1),$$

where $r_{12} = \sqrt{r_1^2 + r_2^2 - 2r_1 r_2 \cos(\varphi)}$.
III-26.
$$p(v, T) = \frac{k_{\rm B}T}{v}\left[1 - \frac{b_2(\beta)}{v} + \frac{4b_2(\beta)^2 - 2b_3(\beta)}{v^2}\right].$$

III-27. See Penrose (1963).
III-28. See Ruelle (1968).
III-33. $\overline{f}(\beta) \sim -\frac{1}{\beta}(\frac{1}{2}\nu\beta^2 - \frac{7}{12}\nu\beta^4 + \frac{1}{2}\nu^2\beta^4).$

$$\lim_{l \to \infty}\frac{1}{|\Lambda_l|}\sum_{X \subset \Lambda_l} K(X) \approx \nu(\cosh(\beta J) - 1) + \nu(\cosh(\beta J) - 1)^2$$
$$+ 2\nu(\nu - 1)(\cosh(\beta J) - 1)^2 + \nu(\nu - 1)(\cosh(\beta J) - 1)^2(\cosh^2(\beta J) + 1)$$
$$+ (\nu + 10\nu(\nu - 1))(\cosh(\beta J) - 1)^3 + 32\binom{\nu}{3}(\cosh(\beta J) - 1)^3.$$

(The last line is of order β^6.)

$$\lim_{l \to \infty}\frac{1}{2|\Lambda_l|}\sum_{X_1, X_2 \subset \Lambda_l} K(X_1)K(X_2)\psi_{\rm pol}(X_1, X_2) \approx -\frac{1}{2}\nu(4\nu - 1)(\cosh(\beta J) - 1)^2.$$

III-34. $|K(X)| \leqslant (\beta J)^{|X|}e^{\beta J\nu|X|}$ and hence $\beta J e^{\beta J\nu} < \frac{1}{(4\nu^2 - 1)e}$.

For $\nu = 2$ this gives $\beta J < 0.02$. This can be improved further by computing the terms $n = 2$ and $n = 3$ separately.

III-35.

$$\frac{q-1}{q-1+e^{\beta H}} < \frac{1}{\lambda(\lambda+1)} \qquad \lambda = 3^{2\nu-1}.$$

III-36.

$$\ln Z_\Lambda(\beta) = \beta J N_2(\Lambda) + \sum_{n=1}^{\infty} \frac{1}{n!} \sum_{\gamma_1 \subset \Lambda} \cdots \sum_{\gamma_n \subset \Lambda} e^{-2\beta J \sum_{i=1}^{n} |\gamma_i|} \psi_{\mathrm{ctr}}(\gamma_1, \ldots, \gamma_n).$$

Convergence for $2\beta J > 1 + \ln 3$.

Index

Printed in the United States
By Bookmasters